A *New York Times* Edito

People's Best Books of F

Chicago Tribune's 28 Books You N

Publishers Weekly's Most Anticipate

Booklist's Top Ten Sci-Tech Books of 2019

Praise for WILDHOOD

"Natterson-Horowitz and Bowers go deep and wide in addressing the raft of species-spanning equivalents. The authors make clear that, in a fundamental sense, adolescent animals and teen humans encounter the same sorts of challenges—and that what may strike elders of any species as nutty, exasperating behavior is not only inevitable for most creatures in that stage of development but truly valuable."

—Duncan Strauss, *The Washington Post*

"Adolescence isn't just for humans. Here an evolutionary biologist offers up rollicking tales of young animals navigating risk, social hierarchy, and sex with all the bravura (and dopiness) of our own teenage beasts."

—*People*

"The vivid storytelling and fascinating scientific digressions in *Wildhood* make it a pleasurable read. It's also a book parents may find reassuring: The authors show that the often painful struggles human and animal adolescents go through are a way of developing the skills and experience that will make it possible for them to function as adults. But there's also another implicit message in *Wildhood* about the interconnection among the planet's species. The awareness that we're all in this together ought to motivate humans to stop ravaging the planet so it can continue to be a place where adolescents of many different species can find their ways into adulthood."

—Patrick J. Kiger, *Los Angeles Times*

"Harvard evolutionary biologist Barbara Natterson-Horowitz and science journalist Kathryn Bowers draw fascinating connections between human and animal young adulthood."

—Meredith Wolf Schizer, *Newsweek*

"Bestselling authors of *Zoobiquity*, Barbara Natterson-Horowitz and Kathryn Bowers, paired up again to research what we can learn about adolescent behavior from their counterparts in the animal kingdom . . . in their newest book, *Wildhood*. Their five-year study found many similarities between the thrill-seeking and sometimes inexplicable-seeming choices of teens and those of adolescent animals developing in the wild."

—Laura Pearson, *Chicago Tribune*

"Do you want to know how an evolutionary lens can influence positive change in human culture and society? Read *Wildhood*. Do you want a master class in making other species' behavior relevant to our daily lives? Read *Wildhood*. Are you a teenager, have you been a teenager, do you know any teenagers? Read *Wildhood*. Read *Wildhood* for the same reason you engage in literature, art, and music: to become something other than what you were before."

—Holly Dunsworth, *Evolution*

"An enduring story plot finds a youth suddenly alone in the world, struggling to find shelter from the elements, safety from predators, food, and new friends. These struggles usually involve some tough lessons but ultimately lead to knowledge, a new identity, self-reliance, and maybe even love. In *Wildhood*, this theme comes to exhilarating life as evolutionary biologist Barbara Natterson-Horowitz and science writer Kathryn Bowers describe the challenges faced by adolescent animals. There is much here for the nature lover, the parent seeking advice, and the college freshman tackling 'adulting.' By laying out the adolescent experience of so many species in rich detail, the authors normalize and celebrate the beauty and complexity of our own species' journey into the big wide world."

—Linda Wilbrecht, *Science*

"Take the authors up on their invitation to observe animals in the wild and in your own household, and you'll never look at other beings the same again. *Wildhood* is for parents, nature lovers, and the curious alike. You'll be wild for it."

—Terri Schlichenmeyer, *Times Record*

"Reading [*Wildhood*], I was surprised to see that many of the adolescent behaviors humans exhibit are wired in for adolescents of most species. This may not provide much consolation for you as you try to guide your teen through the dangers of risk-taking, but it provides insights into how much your teen is exhibiting normal adolescent behavior shared with birds and monkeys. Most importantly, it's a reminder that this is usually not about you."

—Mark Phillips, *Marin Independent Journal*

"An incredibly fascinating read, *Wildhood* illuminates what humans can learn from the animal world and how all species are more connected to one another than they may appear."

—*Booklist* (starred review)

"Human teens have much in common with their counterparts throughout the animal kingdom—and those commonalities are eye-opening as described in the latest from biologist Natterson-Horowitz and science journalist Bowers. Reassuring . . . should appeal to anyone who's ever raised an adolescent, human or otherwise."

—*Publishers Weekly* (starred review)

"A lucid, entertaining account of how creatures of many kinds learn to navigate the complex world that adulthood opens."

—*Kirkus Reviews*

"This compelling account of how strongly human adolescent behaviors are rooted in our wild animal past should intrigue general science readers and fans of *Zoobiquity*."

—*Library Journal*

"The wild adventure of adolescence has never been analyzed in such depth. In lively personalized accounts that keep our attention, the authors explain how the transition to independence works in each species, and why it looks so similar across the board."

—Frans de Waal, PhD, author of *Mama's Last Hug* and *Our Inner Ape*

"Our teenage years can be many things, from fraught and frustrating to exhilarating and joyful. In *Wildhood*, Natterson-Horowitz and Bowers show that these years are something else altogether—essential for humans and animals in general. Read their enlightening journey and you will never see the transition to adulthood the same way again."

—Neil Shubin, PhD, author of *Your Inner Fish* and *The Universe Within*

"One of the most insightful books ever written about this critically important stage of life. Unfailingly fascinating—and sometimes downright mind-blowing—this is a remarkably original account of the nature, meaning, and purpose of adolescence in today's world."

—Laurence Steinberg, PhD, author of *Age of Opportunity: Lessons from the New Science of Adolescence*

"A masterpiece. This is a spellbinding lens on the ways creatures with big bodies yet little life experience figure out how to survive and thrive. Read *Wildhood*!"

—Wendy Mogel, PhD, author of *Voice Lessons for Parents* and *The Blessing of a Skinned Knee*

"A deeply researched and beautifully written description of the fundamental tasks of adolescence. The authors' account of the trials faced by teenagers across the animal kingdom inspires compassion for young people and a deep appreciation for what they must accomplish on the journey into adulthood."

—Lisa Damour, PhD, author of *Untangled: Guiding Teenage Girls Through the Seven Transitions Into Adulthood*

"The authors offer a life-changing perspective on adolescents venturing out into the world. A treasure trove of scientific exploration and practical implications for how we understand and support youth."

—Daniel J. Siegel, MD, author of *Brainstorm: The Power and Purpose of the Teenage Brain*

"This fascinating book tells the compelling story of adolescence across species, framed in the convincing context of evolutionary and adaptive explanations."

—Sarah-Jayne Blakemore, PhD, author of *Inventing Ourselves: The Secret Life of the Teenage Brain*

"*Wildhood*'s tour of the natural history of adolescence is original, entertaining, and constructive. The transition from youth to adulthood might never be easy, but this comparative biology is full of ideas for understanding it better."

—Richard Wrangham, PhD, author of *The Goodness Paradox* and *Catching Fire*

"Those travails of adolescence? It isn't just you. Or your culture. Or even your species. *Wildhood* uses riveting stories about the challenges overcome by specific whales, wolves, and more to put the challenges of adolescence in a universal evolutionary context for the first time. Groundbreaking and fascinating."

—Randolph M. Nesse, MD, author of *Good Reasons for Bad Feelings*

"*Wildhood* links coming-of-age neurobiology with ecology and evolutionary biology to create a powerful new lens for understanding the science (and art) of growing up. At times counterintuitive, at times paradigm-shattering, this illuminating new book generates dozens of hypotheses for raising, educating, counseling and treating, and living life as an adolescent human."

—Gene Beresin, MD, professor of psychiatry, Harvard Medical School

"Wise, entrancing, and astounding."

—Daniel E. Lieberman, PhD, author of *The Story of the Human Body: Evolution, Health, and Disease*

ALSO BY BARBARA NATTERSON-HOROWITZ
AND KATHRYN BOWERS

Zoobiquity: The Astonishing Connection
Between Human and Animal Health

WILDHOOD

The Astounding Connections Between Human and Animal Adolescents

Barbara Natterson-Horowitz, MD,
and Kathryn Bowers

SCRIBNER
New York London Toronto Sydney New Delhi

Scribner
An Imprint of Simon & Schuster, Inc.
1230 Avenue of the Americas
New York, NY 10020

Wildhood was previously published with the subtitle
The Epic Journey from Adolescence to Adulthood in Humans and Other Animals.

First Scribner trade paperback edition July 2020

Interior design by Kyle Kabel
Graphics by Oliver Uberti

Manufactured in the United States of America

1 3 5 7 9 10 8 6 4 2

Library of Congress Control Number: 2018278391

ISBN 978-1-5011-6469-9
ISBN 978-1-5011-6470-5 (pbk)
ISBN 978-1-5011-6471-2 (ebook)

To our parents,
Idell and Joseph Natterson
Diane and Arthur Sylvester

What is Wildhood?

Wildhood, the shared experience of adolescence across species, begins with the physical changes of puberty and ends when an individual has acquired four essential life skills. To become successful adults, all Earth's animals must learn how to: stay safe; negotiate social status; navigate sexuality; and live as adults.

WHEN IS WILDHOOD?

Because of vastly different lifespans, wildhood can last from a few days for a fruit fly to fifty years for a Greenland shark. (These fish can live an astounding 400 years and don't enter puberty until around 150.) Below, wildhoods of twenty-three species are depicted within their lifespans. Ranges are extrapolated from life history data; individual onset and duration will vary.

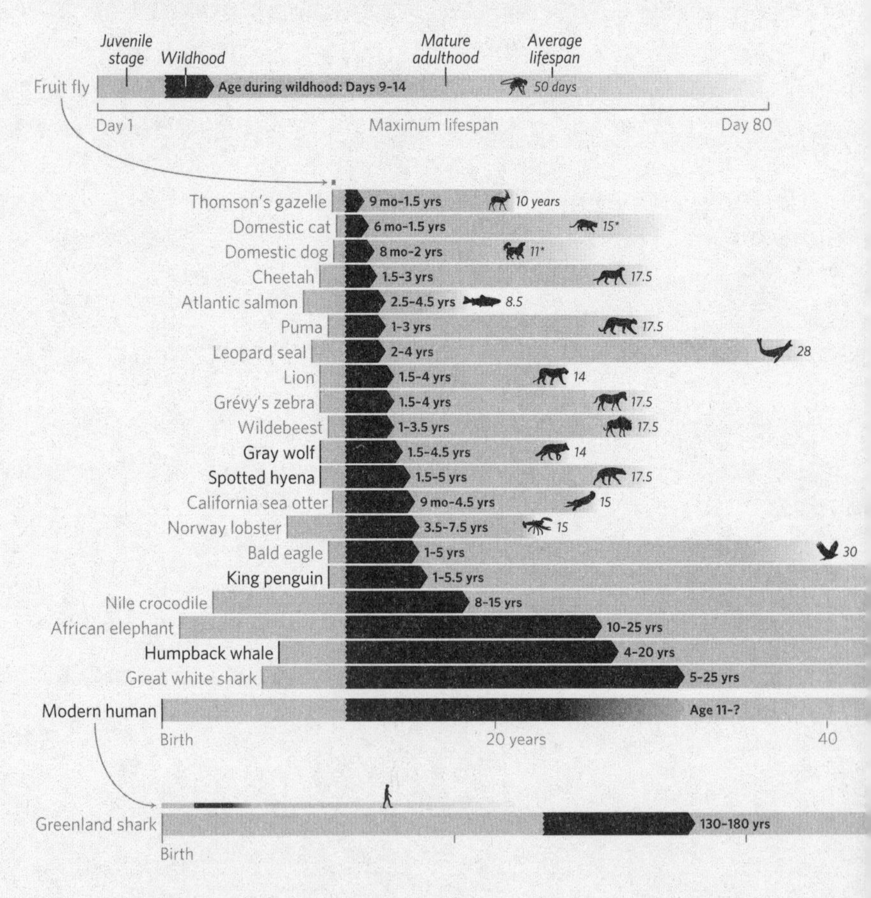

*LIFE EXPECTANCIES OF HUMANS AND PETS ARE BASED ON DOMESTIC LIVING.

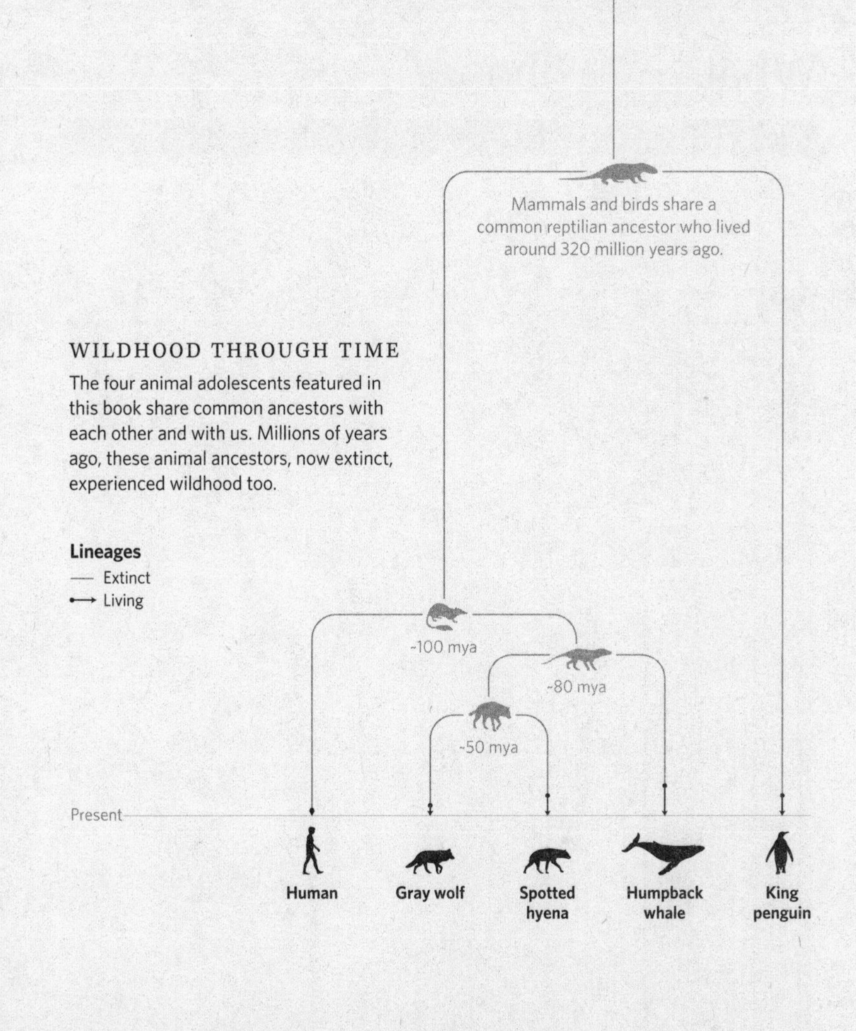

WILDHOOD THROUGH TIME

The four animal adolescents featured in this book share common ancestors with each other and with us. Millions of years ago, these animal ancestors, now extinct, experienced wildhood too.

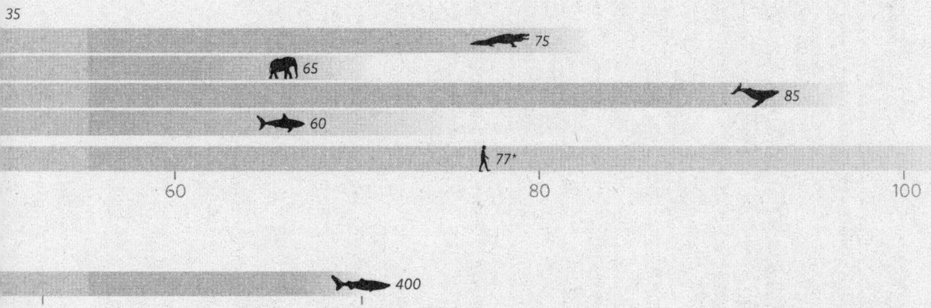

Contents

Prologue 1

PART I SAFETY

1 Dangerous Days 21

2 The Nature of Fear 29

3 Knowing Your Predators 39

4 The Self-Confident Fish 67

5 School for Survival 79

PART II STATUS

6 The Age of Assessment 93

7 The Rules of Groups 105

8 Privileged Creatures 119

9 The Pain of Social Descent 129

10 The Power of an Ally 147

PART III SEX

11 Animal Romance 161

12 Desire & Restraint 171

13 The First Time 187

14 Coercion & Consent 195

PART IV SELF-RELIANCE

15 Learning to Launch 211

16 Making a Living 231

17 The Great Alone 247

18 Finding a Self 259

Epilogue 263

Acknowledgments 269

Glossary of Terms 271

Notes 279

Note About the Illustrations 331

Index 333

Prologue

Our quest to understand the nature of adolescence began on a cold California beach in 2010. We stood on a sand dune, contemplating a stretch of the Pacific, an expanse with an intriguing nickname: the Triangle of Death.

We'd been drawn there by a marine biologist's unusual story. The Triangle of Death, he'd told us, owed its reputation to a horde of particularly lethal inhabitants: great white sharks. Hundreds of these colossal predators live in this region, and they're so notoriously ravenous that even the local sea life has learned to stay out of their way. Lush kelp forests grow up and down the California coast, but not in the Triangle, so any animal foolish or unlucky enough to venture in there has nowhere to hide. So treacherous are those waters that even the scientists who work in them don't get out of their boats.

But, the biologist said, that wasn't the most interesting part. Counterintuitively and at great danger to itself, one animal does regularly enter the Triangle of Death: the California sea otter. But not all of them. Only one specific kind joyrides into the death zone, and it's not the mature adults. It's certainly not the baby pups. No, the magnificent knuckleheads that swim into the cold, barren, shark-filled Triangle of Death are adolescents. Sometimes they die in a flash of teeth and a swirl of blood. But more often than not, these thrill-seeking animal "teens" emerge with hard-won experience, newfound confidence, and more sea smarts than they had as parent-protected, dependent juveniles.

At the time, we were researching our first book, *Zoobiquity*, which explores the ancient and essential connection between human and animal health. (We work as a team: Barbara is a visiting professor in the Department of Human Evolutionary Biology at Harvard and professor of medicine in the Division of Cardiology at UCLA. Kathryn, a science writer, is a certified animal behaviorist. Together we've designed and taught courses at Harvard and UCLA.) Gazing over the Triangle of Death, we were struck by how these adolescent otters sounded a lot like teenagers we knew: taking risks, seeking out danger, doing scary things their parents had grown out of. After regarding the ocean for a few more moments, we walked back up the beach, over a dune, and onto a spit of land overlooking a different scene.

In a cove protected from the whitecaps, kayakers slowly paddled through the calm water. This inlet, called Moss Landing, is a prime location for observing wildlife, including sea otters. The extended families of the adolescents drawn to the Triangle of Death come here to feed, relax, and socialize.

On that day there were dozens of the sleek creatures floating on their backs, twisting and twirling through the water. The panorama resembled open-swim hour at a public pool—with young and old otters cavorting together. Older animals swimming leisurely made way for splashing groups of youth. We saw otters diving for sea urchins and learning to break them open, play-fighting in pairs and groups, and testing out the nose-grabbing behavior these animals use when they court. Although it looked like carefree recreation, we'd later learn that this was actually a cove full of teachable moments for the younger members of the group.

As we watched, pandemonium suddenly broke out. The water exploded in a churn of white as a group of otters powered at top speed from one end of the inlet to the other. What just happened? we asked our biologist guide. Was it a shark? Had a predator entered the shallow bay?

No, the biologist responded, pointing. That kayak got too close. And look—they didn't all get spooked. There, still floating comfortably, undisturbed, was one cluster of otters. The gray fur on their heads showed they were mature adults, experienced and discerning. The skittish ones who'd taken flight were the adolescents who couldn't yet tell the difference between a great white and a Sea Ghost 130.

Swimming up to sharks one moment, then fleeing from a plastic boat the next: these inexperienced adolescents were both overly bold and overly cautious. But we observed that these animal adolescents were also exuberantly socializing with their peers, experimenting with sexual behaviors, and fumbling with how to feed themselves. The parallels with our own species, and even with our own younger selves, were remarkable.

It also crossed our minds, as it has often since we started researching animal-human overlaps, that we might be anthropomorphizing this otter behavior—reading too much into the antics of these wild mammals. From the beginning of our research together, we'd made it a point to avoid projecting human qualities onto other species, thinking it a central scientific danger. But as we learned more about work from fields including neurobiology, genomics, and molecular phylogeny, we realized the bigger danger might be denying humans' real and demonstrable connections with other animals, in body and behavior. The real threat, we recognized, wasn't perhaps anthropomorphism, but its opposite, what primatologist and ethologist Frans de Waal calls "anthropodenial."

Over and over in our work we've refuted claims of human uniqueness: Wild animals can and do get so-called human diseases like heart failure, lung cancer, eating disorders, and addiction. They can develop insomnia and anxiety. Some overeat when they're stressed. They're not all heterosexual. Some are timid; some are bold. Nearly every time we've encountered claims of human exceptionalism, we've found them to be incorrect.

And there, in the water right in front of us, was another striking parallel. Although it happens at any point from a few days to many years after they're born, all animals have a "teenage" period. Boys and girls don't become men and women overnight. And the transition from foal to stallion, joey to kangaroo, or sea otter cub to sea otter elder is just as distinct, just as necessary, and just as extraordinary. All animals need time, experiences, practice, and failure to become mature adults.

That day at the Triangle of Death we caught a glimpse of animal adolescence. And once we'd seen it there, we began to see it everywhere.

A NEW WAY OF SEEING

It was, almost quite literally, as if we'd taken off a blindfold. Our physical vision hadn't changed, but our perception had, and suddenly a whole new way of understanding what it means to grow up revealed itself. Flocks of birds, pods of whales, groups of young people, our own children—even memories of our own adolescence and young adulthood—would never look the same.

Over the next several years we focused our research on understanding animals in this in-between phase, the ones too physically grown to be juveniles but not yet possessing the experience to be considered mature.

Watching a herd of wildebeest crossing a crocodile-infested river, we noticed that the first ones in the water were big but gangly adolescents. Oblivious to danger, exuberant with inexperience, they leapt right in while their more prudent elders held back, swimming across safely once the crocodiles were occupied chasing the adolescents.

In Manhattan, Kansas, of all places, we came face-to-face with two young adult hyenas and observed how one bullied the other although they were the same age and size. All it took were two young individuals to form a clear social hierarchy.

Approached by a group of wide-eyed lemurs in a forest preserve in North Carolina, we were charmed by the one that came right up to us. He was an adolescent named Nacho, and his fearlessness both endeared him to us and—had we been poachers instead of scientists—endangered his own safety.

We listened to orphaned wild wolves learning to howl, their changing adolescent voices warbly and cracking. We watched panda adolescents learning to peel bamboo, a first step in eventually feeding themselves. One extraordinary afternoon, we observed herds of wild horses, white rhinos, and zebras. We zeroed in on the adolescents within those herds and saw how they postured and shoved each other as they jockeyed for a place in their groups.

Some of our searches were more successful than others. Adolescent Canadian bison in Prince Albert National Park near the Arctic Circle stayed invisible in spite of the twenty miles we hiked through mud and mosquito-dense wetland hoping to spot them. The young adult bear whose warm scat we found on the trail during that same trek also didn't materialize. We came close while tracking an adolescent mountain lion in Los Angeles. As we paused to rest, our guide opened a trail camera and showed us that we were standing in the very spot the lion had slinked through just a few hours earlier.

A PLANETWIDE TRIBE

Biologists have long been aware that animals—human and not—go through physical and behavioral changes between infancy and adulthood. But the risk-taking, the socializing and sexual experimenting, the leaving home to seek one's fortune or to find oneself, not to mention the angst and mood swings, the romantic and turbulent emotions—even the raging hormones and rapidly changing "teenage" brain—surely all that was uniquely human? No, we would learn, it is categorically not.

While every individual's adolescent experience will differ in its details—some will be triumphant; some will be tragic; most will be somewhere in between—when we started looking at adolescence across species, a universality presented itself. Regardless of the animal, its position on Earth, or the historical era it lives in, all individuals on this journey face the same core challenges. And successfully surmounting those challenges, we argue, is the definition of maturity.

While on this journey, adolescents from bottlenose dolphins to red-tailed hawks, clownfish to humans, have, in many ways, more in common with one another than with their mature parents or immature younger siblings. They share what author Andrew Solomon has called a "horizontal identity." In his book *Far from the Tree*, Solomon contrasts vertical identities, those between you and your ancestors, with horizontal identities, those among peers with whom you share similar attributes but no family ties. Expanding Solomon's concept to include other species, we suggest that adolescents share a horizontal identity: temporary membership in a planet-wide tribe of adolescents.

This global journey, and the ways successful adolescents navigate it, is the subject of this book. Its premise: human adolescence is rooted in our wild animal past, and the joys, the tragedies, the passions, and the purpose aren't inexplicable; they make exquisite evolutionary sense.

COMING OF AGE ON PLANET EARTH

At Harvard in the spring of 2018 we first offered "Coming of Age on Planet Earth," a course for undergraduates based on the research in this book. On the first day of class we had our students grab their backpacks and follow us through the Peabody archaeology museum, past the cases of kachina dolls and towering Mayan steles to the Tozzer Library of anthropology. Waiting for us, mounted and elevated on a long wooden table, was a first edition of Margaret Mead's *Coming of Age in Samoa*. In 1925, at the age of twenty-three (by today's measure an adolescent herself), Mead traveled to the South Pacific nation to study adolescence in another culture as a way of better understanding it in modern Americans. Mead's comparative approach completely transformed the field of anthropology, particularly her focus on culture, rather than biology, as the primary shaper of human individuals and societies. While her work was later criticized (many say unfairly) for relying on methods that were at times more impressionistic than data-driven, she remains a leading intellectual force of twentieth-century understanding of human development, especially adolescence.

In the late nineteenth century, scholarly interest in adolescence had been sparked by an American psychologist, G. Stanley Hall, who borrowed the German literary term *sturm und drang* (storm and stress) to describe the age. Throughout the twentieth century, psychoanalysts, including Sigmund and Anna Freud, Erik Erikson, and John Bowlby, advanced nurture-based explanations for the challenges of childhood and adolescence, while cognitive psychologist Jean Piaget saw more of a role for biology along with environment in shaping adolescent minds. The Nobel laureate Nikolaas Tinbergen, a founder of the field of animal behavior and an ornithologist by training, saw animal roots in human development. In this era, adolescence was often viewed as a malady: those afflicted with it were studied as if some disease had caused their restlessness, rebellion, risk-taking, and unhappiness.

Advances in neuroscience changed that, starting in the 1960s. Marian Diamond's work on brain plasticity and Robert Sapolsky's on the coevolution of social and emotional brain development shifted the view of human adolescence from a fraught stage with fixed characteristics to a dynamic period crucial for normal development. Frances E. Jensen, Sarah-Jayne Blakemore, Antonio Damasio, and others have connected genetics and environment to the remarkable and terrifying aspects of the period such as risk-taking, novelty-seeking, and peer influence. Linda Spear, a developmental psychologist, has examined adolescent brain biology in relation to temperament, and Judy Stamp, an evolutionary biologist, has explored how environments, whether physical or social, shape adolescents' destinies. Psychologist Jeffrey Arnett has popularized the term "emerging adult" and exposed the power of modern culture in shaping the adolescent experience. And, in addition to illuminating this often turbulent time of life for parents and educators, psychologist Laurence Steinberg's work on adolescent neurobiology is being used to question whether younger defendants in criminal cases should be punished as harshly as fully mature adults.

Following in the tradition of these thinkers, but especially inspired by Mead, we use a comparative approach in our research, our teaching, and in this book. However, we push beyond human comparisons to examine the central challenges of adolescents across species. Our focus is not the two-hundred-thousand-year history of *Homo sapiens*, but rather the six-hundred-million-year history of animal life on Earth.

JURASSIC PUBERTY

The terms "adolescence" and "puberty" are sometimes used interchangeably, but although they're related, they're not the same thing. Puberty is the biological process, kicked off by hormones, resulting in an animal's ability to reproduce. Puberty describes strictly physical development—a growth spurt and, among other things, the activation of ovaries and testes to begin the production of eggs and sperm. Great white sharks go through puberty. Crocodiles go through puberty. As do pandas, sloths, and giraffes. Insects go through puberty (it's part of metamorphosis). Every adult Neanderthal went through puberty, as did Lucy, the famous female hominid *Australopithecus afarensis* whose 3.2-million-year-old bones were found in present-day Ethiopia. Dinosaur puberty hit Jane, an adolescent *Tyrannosaurus rex*, sixty-seven million years ago in Montana. She died before she completed it, according to the paleontologists who unearthed her skeleton and gave the young T. rex her name.

While details vary across species, the basic biological sequence of puberty is remarkably similar. The same hormones kick it into high gear in hummingbirds and ostriches, giant anteaters and miniature ponies. Nearly identical hormones get it started in snails and slugs, lobsters and oysters, clams, mussels, and shrimp.

The dazzling array of most life on Earth today erupted 540 million years ago during a period called the Cambrian explosion. But puberty is older even than that. It's part of the life cycle of one of Earth's most ancient life-forms, single-celled protozoa. Protozoa still exist today, and one, *Plasmodium falciparum*, finds its way into human blood by way of a mosquito bite. Once there, the physically immature organism floats harmlessly around the body until it passes through its protozoal puberty and becomes a leading cause of death worldwide: *Plasmodium falciparum* is the parasite that causes malaria.

Despite its sex-specific connotation, puberty exerts its hormonal effects on every organ system in the body. Hearts grow, dramatically increasing cardiovascular performance. Lungs expand in capacity, giving young athletes more endurance (and asthmatics more attacks). Lengthening skeletons provide gangly-limbed pre-adult bodies thrilling new acceleration, but this rapid bone growth is also behind the increased incidence of bone cancers at this age. Child-sized skulls enlarge to adult dimensions, something not only seen in human children but also noted in dinosaurs. Jaws change shape, and so do the teeth within them. In fact, great white sharks are incapable of administering their deadliest bites until after they've gone through puberty.

So puberty is an ancient process of physical transformation. But to attain adulthood, a physically developed young creature must go through a second phase. This one combines body and behavior. It is about learning to think, act, and even feel like a mature member of a group. It's a period of collecting crucial experiences, a time to absorb information from mentors, and test oneself against peers, siblings, and parents.

This phase is adolescence, and it lasts as long as it takes to create a mature adult. In fact, for a species to produce mature adults, as opposed to just physically grown individuals, an adolescence is essential. The quest for maturity through experience is the universal purpose of adolescence in nature.

And the journey can spark astonishing innovations. One of the most famous fossil finds of recent decades is a fish called *Tiktaalik* unearthed by the University of Chicago paleontologist Neil Shubin. These 375-million-year-old creatures bore a clue to our evolutionary past: four small limbs that served as both fins and feet. Those four appendages are evidence of *Tiktaalik*'s pioneering role in one of the most epic stories of life on Earth—the transition from water to land.

Shubin has found that *Tiktaalik* fossils reveal something else. They have been found in a range of sizes, some the length of tennis racquets, others longer than a surfboard. This means something as profound as it is obvious: These ancient fish *grew up*. And during that process, like adolescents today, just-through-puberty *Tiktaalik* individuals would have been especially vulnerable, lacking not only size but experience with predators, with competitors, with sexuality, and with finding food. Vulnerability and inexperience regularly push younger animals into unfamiliar settings. We wrote to Shubin and asked whether he thought it possible that adolescent *Tiktaalik* fish were the ones leading the charge to land. He saw this as plausible and wrote back: "*Tiktaalik* was an animal with big carnivorous adults, so near the top of the food chain, but juvenile stages would be exposed to predation and may have benefited by being partially terrestrial. Likewise, maneuvering on land would be easier in smaller fish, rather than larger ones, at least in incipient stages."

While this remains only a hypothesis, it's consistent with everything we know about the risk-taking and novelty-seeking behavior of adolescents across time and place. Driven by necessity, adolescents explore frontiers. They innovate new ways to survive. And when they do that, they can create the future.

THE "TEENAGE" BRAIN

Among the organs undergoing radical change during puberty and adolescence is the brain. Transitioning "teenage" brains are marvels of upheaval, markedly different from the child's brains they were and the adult brains they will be.

Every brain makes memories, but the teenage brain in particular is storing away huge numbers of them that will shape who we are and how we approach the world for the rest of our lives. Psychologists call this the "reminiscence bump," the especially deep and enduring memories formed during this period (in humans it happens roughly between the ages of fifteen and thirty).

The impulsivity of adolescents, their drive to experiment and seek novelty, and their immature decision-making have been linked to the brain's executive function center, particularly the prefrontal cortex, which matures late in brain development. Adolescents' preference for being with peers and even their conflicts with their parents have also been traced to unique neurobiology in regions of the brain supporting emotion, memory, and reward. So have their mood swings from stratospheric highs to subterranean lows. Vulnerability to substance abuse, self-harming behaviors, and mental illness have also been attributed to the still-developing brain, which doesn't fully finish its transformation until well into a person's late twenties and perhaps even early thirties.

The mysteries of the human teenage brain have been widely chronicled in recent decades, and this research has helped us understand why adolescents behave as they do. Yet this groundbreaking science largely ignores a much bigger revelation: during adolescence, the brains and behaviors of other animals are also going through a massive transformation.

Adolescent birds have a brain region that, like the developing prefrontal cortex in humans, helps young animals gain self-control. The brains of adolescent orcas and dolphins continue to grow after physical and sexual maturity, as do our own. And the changing adolescent brains of other primates and smaller mammals drive tendencies like risk-seeking, sociality, and interest in trying new things. Even adolescent reptiles show unique neurological shifts between juvenile and adult life, as do adolescent fish.

Whether our bodies are covered with skin, scales, or feathers, whether we move by running, flying, swimming, or slithering, we share common biology that builds and shapes our adult selves. This book explores the universality of the period between childhood and adulthood—what we decided to call "wildhood." Looking across the world of animals over hundreds of millions of years of evolutionary time allows us to separate out which aspects of adolescence are unique to a single animal species or human culture, and which are the norm on planet Earth.

THE FOUR CORE LIFE SKILLS

The central insight is this: four fundamental challenges of wildhood are the same for a fruit fly coming of age in the bananas on a kitchen counter, a lion roaring into adulthood on the Serengeti, and a nineteen-year-old balancing work, school, friends, relationships, and other responsibilities. They are:

How to stay safe.

How to navigate social hierarchies.

How to communicate sexually.

How to leave the nest and care for oneself.

Each of these four essential challenges is encountered throughout an animal's life, but adolescence and young adulthood are when they're faced together for the first time, and usually without parental support or protection. The experiences of wildhood build necessary life skills and shape the adult fates of individuals.

Avoiding danger. Finding a place in groups. Learning the rules of attraction. Developing self-sufficiency and purpose. These skills are universal—because they support the survival of young animals moving into the wild world. Learning them is mandatory for a successful life.

Safety. Status. Sex. Self-reliance. The four skills are also at the core of the human experience and the basis of tragedies, comedies, and epic quests.

Millions of things can go wrong for an adolescent animal on the road to adulthood. But when the journey goes well, and a mature adult emerges, it always means the same thing. During its wildhood, that individual faced the four challenges and developed competency in each one. These individuals didn't just grow older; they grew *up*. The journey of wildhood has been undertaken for more than six hundred million years, by countless animals. We believe the ancient legacy of those combined experiences can become a modern atlas for surviving and thriving into adulthood.

COMING OF AGE IN A DIGITAL WORLD

As we'll see, animals develop what, for lack of a better word, we'll call "culture" around transmitting these four life skills to up-and-coming adults. Even within animal species, cultural specifics can vary from region to region and from group to group, just as human cultures have their own endless permutations.

However, one specific area in which humans do in fact stand out from our animal cousins lies in how our teenagers must now traverse two distinct worlds to reach adulthood, one in the real-life communities where they live, the other online.

The four core life skills apply just as much to the internet as they do offline, but these two cultures can be radically different, requiring many modern teens to make two simultaneous journeys to adulthood.

For instance, as we'll explore in Part II, social animals, from fish swimming in the sea to high schoolers rushing to class, must learn to navigate hierarchies of peers. One of the ways they do this is called "association with high-status animals." This term makes immediate sense to anyone who's ever been to school, had a job, or had a social life—you can improve your own status by hanging out with more powerful people. We'll explore the fascinating intricacies of how this works in groups of other animals, but spare a thought for modern human teens and the number of additional hierarchies the internet brings into their lives. If they spend time in multiplayer games or on social media, they're being assessed, sorted, and ranked, sometimes invisibly and sometimes explicitly, alongside an entire universe of other people on those platforms. Imagine the status boost of being praised by a sports or pop star; imagine too the crushing humiliation of being called out by an idol.

Parents and other elders have plenty of experience guiding adolescents and young adults through the real world. But no one has yet aged a full life in the digital world. The four life skills can help sort this new terrain into more easily fathomable categories, because the real-life ones have online correlates: How to stay safe from trolls and predators. How to move through virtual hierarchies. How to express sexuality. How to shape, nurture, and maintain a digital self or identity.

WHY WILDHOOD

When we teach "Coming of Age on Planet Earth," we include an informal poll: Raise your hand if you think you're an adolescent. Next, raise your hand if you consider yourself an adult. Our students are all between eighteen and twenty-three years old, but rarely has a hand shot up immediately or confidently for either question. Often, our students respond with "yes" and "yes"—we're both.

If adolescents don't use the term "adolescent" to describe themselves, what should we call emerging creatures who are fully grown (or almost) but not quite fully grown up? Who are large in size but small in experience and who may be sexually mature but whose brains won't be for many more years?

The term "adolescentia" derives from the Latin word *adolescere*, meaning to grow up, and it appears in medieval texts dating back to the tenth century, describing a religious turning point in the young lives of saints. In North America, the New England Puritans of the mid-1600s considered the age to be a "chusing time," when frivolity was to be left behind and adult employment taken up, but the people in this time were generally called "youth" until the late 1800s, when "adolescents" came into common usage.

Flapper, hipster, bobby-soxer, teenybopper, beatnik, hippie, flower child, punk, b-boy, valley girl, yuppie, Gen Xer—these terms offered ways of talking about young people in specific American cultural contexts throughout the twentieth century. The word "teenager" first appeared in print in 1941 and soon dominated the lexicon. Even today, nearly eighty years later, "teenager" remains the go-to synonym for "adolescent," even as it became scientifically inaccurate when neuroscientists revealed that adolescent brain development starts before thirteen and continues well beyond nineteen. For the past decade or so, "millennial" has neatly covered people in this stage of life, but at this point most millennials have aged out of the adolescent young adult period. "Generation GWoT" is U.S. military parlance for those who have come of age during the Global War on Terrorism. In North America, we've often heard "kids" used as a default term—even by adolescents themselves—but it sounds too young once they're in later high school.

We searched for a better descriptor for both humans and nonhumans in this phase, a word that would cover its ancient commonality. Some terms were too clinical ("pre-adults," "emerging adults," "dispersers").

Some were off-putting or even insulting ("sub-adults," "immatures"). Some were poetic ("fledglings," "deltas," and "elvers," which is the term for adolescent eels). The world's languages held marvels such as the Japanese term *seinenki* (meaning green ones, saplings), or the Russian *lishney cheloveki* (odd people), but we balked at choosing one culture's term over another's.

Our term needed to describe the phase of life in which biology and environment come together to shape mature individuals across all species. It had to be unbounded by a specific age, physiologic sign, or cultural, social, or legal milestone. And it had to capture the vulnerability, excitement, danger, and possibility of this distinct phase of life. We had coined the title of our first book, *Zoobiquity*, by bringing together the Greek root for "animal" with the Latin for "everywhere." For this book, we again created our own term and title. We chose "wild" to capture the unpredictable nature of this life-stage and acknowledge the shared animal roots. And we added the old English suffix "hood," which means both a "state of being" (boyhood, girlhood) and a "group of persons" (neighborhood, sisterhood, knighthood), to indicate membership in the planetwide tribe of adolescents. The phase of life before adulthood, following childhood, across species and evolutionary time, became "wildhood."

A CROSS-DISCIPLINARY APPROACH

The scientific evidence we've assembled and summarized, and which we present here, represents the product of five years of scholarship at UCLA and Harvard. Because our work falls at the intersection of evolutionary biology and medicine, we used research tools from both fields, developing large systematic reviews of comparative adolescence and using the results to create phylogenies. (Systematic reviews are comprehensive, tightly targeted surveys of the world's scientific databases, powered by advances in search technology over the past twenty years. Phylogenies are diagrams of evolutionary relationships among different species, which can be simple family trees or complex computer models containing thousands of data points.) We also conducted fieldwork observing animal adolescents in natural settings and sanctuaries around the world, and interviewed experts in human adolescence, wildlife biology, neurobiology, behavioral ecology, and animal welfare.

We believe our research has important implications for multiple groups, and we've chosen to describe it in ways that can inform both scientific and other audiences. References directly linked to the text are provided as endnotes. An expanded bibliography, including links to our research, source material, and content of interest is available online for readers who parent, teach, study, treat, mentor, coach, or work with adolescents, and most important, for those who are adolescents themselves.

We're writing in early-twenty-first-century America, and our work will reflect that; we don't presume to understand the specifics of every person's adolescent experience. That said, we did have one personal motivation while writing this book. During the whole process, we were parenting adolescent offspring of our own. Kathryn's daughter was thirteen when we started, and Barbara's daughter and son were sixteen and fourteen. All three are older now but being mothers of adolescent humans gave us a practical advantage: we could observe wildhood up close. After field trips to the Arctic Circle, Chengdu, the Gulf of Maine, and North Carolina, we would come home to our own exuberant teens and be reminded of the complex yet fleeting wonder of this age.

A COMMON QUEST

Our office in Harvard's Museum of Comparative Zoology, where we wrote the bulk of this book, has a secret passage that connects it to another world. Up one particular stairwell, if you turn right instead of left, you end up in the Peabody Museum, an institution dedicated to preserving human cultural heritage. Sometimes, immersed in our work, we'd emerge from one world and get lost in the other. On one side, the legacy of comparative zoology, from dinosaur bones to molecular genetics. On the other side, physical objects testifying to millennia of human ingenuity, persistence, collaboration, and love. Both sides—zoology and anthropology, animal and human—a reflection of the diversity of life on our planet.

After crossing this symbolic divide many times, we became as good at identifying the signs of human adolescence in the Peabody's collections as we were at seeing them in animals in the field. We came to feel connection with, almost affection for, these artifacts of growing up. Whether it was a suit of armor from a tiny island in the middle of the Pacific, a youth's golden pendant from fifth-century Mesoamerica, a Lakota courtship blanket, or an Inuit snow shovel, these human touchstones further bridged this unique yet universal phase of life.

As you know from every coming-of-age story you've ever read, youths go on quests. They're kicked out of the house, they escape after a conflict, or they're orphaned, and they head out into the wild world. They're dangerously unprepared, sometimes hilariously, sometimes fatally. On their journeys away from home, they fight off predators and exploiters. They meet friends and learn to identify foes. They might fall in love. And they learn to fend for themselves—finding their own food, making their own homes, and then usually at the end of the story deciding whether to rejoin the community they were born into or reject it and forge a new one of their own.

Our science is told through the real-life coming-of-age stories of four wild animals tracked by biologists over months and years. Our protagonists are not human, but they are all adolescents. Ursula, a king penguin born and raised on South Georgia Island off Antarctica, faces almost-certain death from a monstrous predator on her first day away from her parents. Shrink, a spotted hyena in the Ngorongoro Crater in Tanzania, battles bullies and forms friendships as he navigates the hierarchical hyena version of high school. Salt, a North Atlantic humpback whale born near the Dominican Republic who spends every summer in the Gulf of Maine, confronts sexual desire and learns how to communicate what she wants, and doesn't want, from her partners. And finally, on a harrowing but exhilarating journey away from home, Slavc, a European wolf, nearly starves, drowns, and dies of loneliness as he tries to hunt his own food and find a new community.

We've chosen to tell their stories in a narrative style, which we hope captures the real drama each experienced on the journey from adolescence to adulthood. However, every detail that we provide in these stories is based on and validated by data from GPS, satellite, or radio collar studies, peer-reviewed scientific literature, published reports, and interviews with the investigators involved.

Separated by hundreds of millions of years of evolution, these four wild animals are connected to one another, and to us, through the common experiences, challenges, and ages of wildhood.

Whether experienced in the treacherous waters off Antarctica, the grasslands of Tanzania, a shimmering Caribbean bay, or the Triangle of Death, wildhood extends throughout nature and into our human lives. It shapes and sometimes determines our adult destinies. Wildhood is the common inheritance of all creatures on Earth, an ancient and ongoing legacy ready to be claimed.

PART I

SAFETY

During wildhood, humans and other animals are predator naive. Their inexperience attracts attackers and exploiters who see them as easy prey. Predator training—learning to recognize and deter individuals with violent intentions—may save their lives and prepare them to be more confident adults.

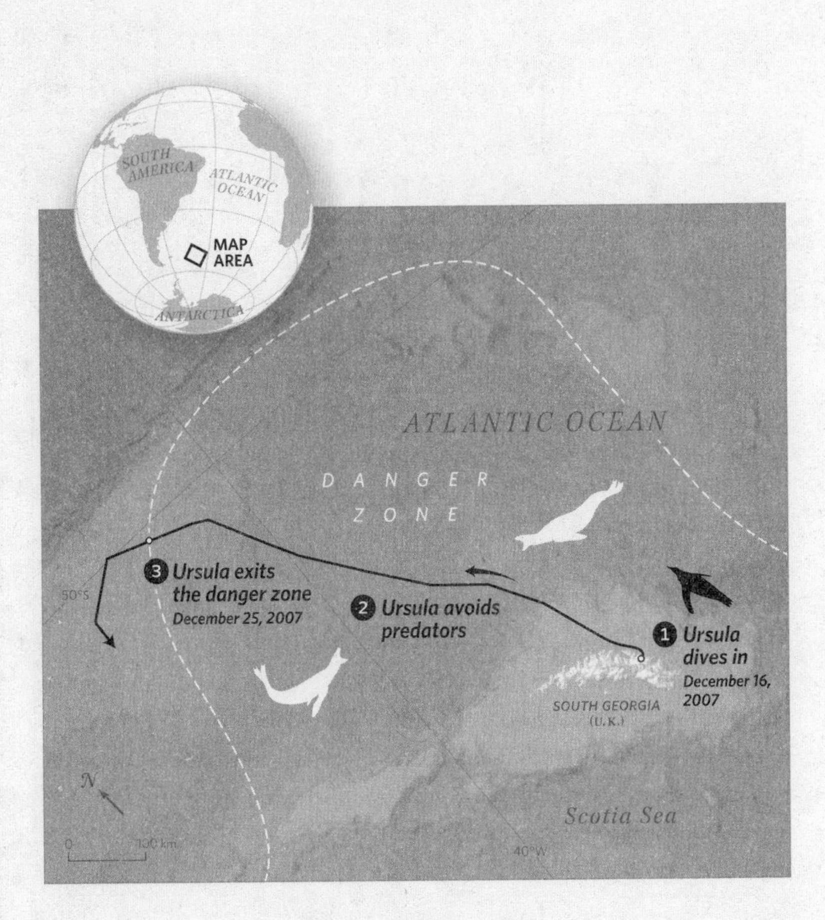

URSULA'S DANGEROUS DAYS

Chapter 1

Dangerous Days

South Georgia Island rises out of the Atlantic Ocean about a thousand miles off Antarctica. If you'd visited there on December 16, 2007, you might have witnessed a defining moment in the life of a young king penguin named Ursula. On that Sunday, Ursula turned away from her parents. She waddled down to the beach with a squawking crowd of her identical-looking peers. Then suddenly she leapt into the frigid water and swam away from home at full speed, without looking back.

Until that moment, Ursula had never ventured more than a hundred yards from where she'd been born. She'd never played in the surf. Not once had she attempted to swim in the open ocean. Ursula had never even fed herself. Up to this point, every meal had been provided by her parents (partially digested and regurgitated straight into her open mouth).

As a fluffy nestling warm under her parents' feathers, Ursula had weathered freezing temperatures and intense winds. Defended by Mom and Dad, she'd survived attacks by skuas, fearsome predatory seabirds that tear apart baby penguins to feed to their own young offspring. Growing up, Ursula, like all king penguins, had a secret language with her parents, unique calls that belonged only to the three of them. For king penguins, parental care lasts a full year and during that time the small family is a tight trio. Mom and Dad care equally for their young, trading off the roles of caregiver, breadwinner, and security guard.

Lately, though, things had changed. Ursula had been shedding the soft brownish down of chick-hood. Sleek black-and-white adult feathers had begun popping through the shaggy patches of her baby plumage. Her squeaky juvenile peeps had deepened into the buzzing honks that make penguin colonies sound like giant, conductorless kazoo orchestras.

Ursula's transformation wasn't just physical. Her behavior too was suddenly different. Overtaken by restlessness, she'd begun wandering farther from her parents. During the day she gathered with other adolescents in chattering penguin gangs. Her edginess has a special scientific name: *zugunruhe*, which is German for "migration anxiety." *Zugunruhe* has been studied in birds, mammals, and even insects that are on the brink of moving away from home territories. Sleeplessness—fueled by shifts in arousing adrenaline and sleep-inducing melatonin—often accompanies *zugunruhe* in animals. A human might describe the feeling of *zugunruhe* with words like "excitement," "dread," and "anticipation."

Until that particular Sunday in December, Ursula's increasing wanderlust had been kept in check by an urge to return each night to the safety of Mom, Dad, and the rest of the rookery. But today was different. Resplendent in her smart new tuxedo, hyped up on adrenaline, and buzzing with her peers, Ursula moved toward the water's edge. Shoulder to shoulder, the jostling adolescents milled, gazing out to sea and glancing back at home. No longer chicks, not quite adults, they paused on the brink of a great unknown.

Like fledgling humans leaving to make their way in the outside world, Ursula faced four great tests. She would quickly have to learn to feed herself and find safe places to rest. She'd need to navigate the social dynamics of her penguin group. She would have to learn to court and communicate with potential mates. And she'd be doing it all without her parents, alone in the middle of the open ocean.

But none of these penguin milestones could happen if Ursula weren't alive. The first great test is to stay safe. Failing this ends a young animal's future before it can even start. Ursula's first challenge was to come face-to-face with death—and survive.

For adolescent penguins dispersing every year from South Georgia Island, the first day away from home is literally sink or swim. Like adolescent animals all over the world, young adult penguins are inexperienced and underprepared. They don't realize predators are dangerous until it's too late. Even if they spot danger, they might not know what to do next. Lacking know-how and unaccompanied by protective parents, adolescents are targets. They're the definition of easy prey.

Ursula's first experience in the water would also be her first encounter with what lay beneath it. And what lay beneath was monstrous. Lurking offshore penguin breeding grounds are predators with jaws so big they can easily swallow a basketball. Picture those massive jaws, lined with teeth like a tiger's, speeding toward a penguin's tennis ball–sized head. That's the maw of one of Earth's elite hunters: leopard seals. A hydrodynamic half-ton of explosive muscle, leopard seals excel in penguin killing. With cool precision, they grab the birds and smack them back and forth on the surface of the water to flay off the feathers. It's a grisly performance worthy of a sushi chef, and leopard seals dispatch ten or more penguins at every meal. Like their feline namesakes, leopard seals are ambush hunters, meaning they hide and wait for prey. Arranging themselves along coastlines like underwater mines, leopard seals skulk along the edges of ice banks, just out of sight. Sometimes they masquerade as flotsam, quietly floating in the waves, the better to surprise their unwary victims. Dispersing adolescent penguins must run this gauntlet of death and come out the other side. If they don't jump in, they can't grow up. But if they don't make it past the leopard seals, as well as the pods of predatory orcas, the first day of the rest of their lives will also be their last. Getting past the danger is a high-stakes test for the penguins who must pass or fail permanently.

If you'd been there to witness this do-or-die moment, you might have noticed that Ursula and two of her peers sported an accessory that distinguished them from their classmates. Stuck to their backs with black tape were tiny transponders, programmed to transmit never-before-gathered information about where penguins go on the day they leave home and in the weeks after. The surprising results would turn out to entirely reframe what biologists knew about penguin behavior. Led by Klemens Pütz, the scientific director of the Zurich-based Antarctic

Research Trust, the multinational investigation included researchers from Europe, Argentina, and the Falkland Islands. Some of the funding came from ecotourists, who as part of their donation got to name the radio-tagged birds.

That's how we know that a penguin named Ursula jumped in the South Polar sea on Sunday, December 16, 2007. The signals from her tracking device pinpointed exactly when she waddled to the beach, pre-plunge. Of the eight penguins that Pütz's team tagged on South Georgia Island that season, three departed that day—Ursula and two others named Tankini and Traudel—along with a crowd of their adolescent peers.

Like high schoolers on graduation night, Ursula and her cohort—the South Georgia Island king penguin class of 2007—were physically grown and ready to leave. But, much like their human counterparts, with little adult experience in the real world they were still behaviorally immature.

Suddenly, they dove. An arch of her back, a sweep of her flippers, and Ursula was speeding straight into the zone of danger. As for her penguin parents—and the biologists tracking her—all they could do was stand by and watch as she swam away.

VULNERABLE BY NATURE

Of the thousands of adolescent king penguins that plunge into predator-patrolled waters every year, many don't make it out alive. Some years survival has been as low as 40 percent. Other years are less deadly, although exact numbers are hard to calculate. No matter what, the first days, weeks, and months after fledging are exceedingly risky for all penguins.

It's sobering to recognize how dangerous life on Earth is for adolescent and young adult animals. In the wild they crash, drown, and starve more often than their adult counterparts. With less experience, they're pushed into jeopardy by older, bigger peers. They're preferentially targeted and killed by predators.

Fortunately, when human adolescents leave home, they don't share the extremely high mortality rate of fledgling penguins. However, adolescent humans do suffer much higher rates of traumatic injury and death compared with adults. A nearly 200 percent increase in mortality is seen between childhood and adolescence in the United States. Almost half of all deaths among adolescents are the unintentional and tragic result of accidents such as motor vehicle collisions, falls, poisonings, and gunfire.

Adolescents drive faster than adults and are generally more reckless. They have the highest rates of criminal behavior and are five times more likely to be the victims of homicide than adults thirty-five or older. Other than toddlers (who stick fingers into sockets) and adults with jobs in electricity-related industries, adolescents have the highest rates of fatal electrocution. Adolescents and young adults fifteen to twenty-four also have the highest rates of death by drowning, other than infants and toddlers under the age of five. Compared with other populations, they're afflicted in great numbers by suicide and by the onset of mental illnesses and addictions. And adolescents are far more likely to binge-drink themselves to intoxication and death than older adults.

Dangers vary by social class and geography, but globally, human adolescents develop half of all new cases of sexually transmitted infections. They're the most vulnerable to sexual assault. Worldwide, the leading cause of death in fifteen- to nineteen-year-old girls continues to be pregnancy-related complications.

Adolescence can be harrowing, but the biology that contributes to the danger and vulnerability also inspires creativity and passion, as Robert Sapolsky, the Stanford neuroscientist and evolutionary biologist, so vividly describes in his book *Behave*:

> Adolescence and early adulthood are the times when someone is most likely to kill, be killed, leave home forever, invent an art form, help overthrow a dictator, ethnically cleanse a village, devote themselves to the needy, become addicted, marry outside their group, transform physics, have hideous fashion taste, break their neck recreationally, commit their life to God, mug an old lady, or be convinced that all of history has converged to make this moment the most consequential,

the most fraught with peril and promise, the most demanding that they get involved and make a difference.

FROM PREDATOR NAIVE TO PREDATOR AWARE

Ursula, of course, didn't know the grim odds before her. Even if she did, perhaps the magical thinking of youth would have made her believe she was chosen for survival. But, in fact, all king penguins are naive when they set off. And we use that word deliberately, without judgment. It's a wildlife biology term for a specific state of development: inexperienced, unsuspecting young animals leaving home for the first time are "predator naive."

For a gazelle, being predator naive means not knowing what a cheetah smells like or how it moves. For adolescent salmon, it means not yet knowing that cod hunt more slowly at night, relying on smell and hearing to find their prey, and that during the day, when they can see, cod strike more quickly. Sea otters are predator naive when they encounter great white sharks for the first time, and predator-naive marmots cavort obliviously outside their burrows, even when coyotes are nearby. For tiny West African Diana monkeys, being predator naive means not yet having the ability to discern the different hunting sounds made by eagles, leopards, and snakes. They cannot predict whether an attack will come from above, below, or around a tree limb.

Predator naive is exactly what human adolescents are too when they enter the world with little experience. They don't recognize what's dangerous. Even when they do, they often don't know what to do about it. This inexperience can be as deadly for human adolescents as it is for young penguins.

A predator-naive teen going off to a party or a young adult moving to a new city won't have literal leopard seals waiting, but the array of dangers they may face are no less lethal: a swerving pickup truck, a drunken hazing ritual, a depressive episode, a predatory adult, or a loaded gun.

It seems tragically counterintuitive that the most vulnerable and underprepared individuals would be thrown into the riskiest possible situations. But facing mortal danger while still maturing is a fact of life for adolescents and young adults across species. It's as true for a young sea turtle that hatches and heads into the ocean without ever meeting its parents as it is for an African elephant that is nurtured for twelve years by its multigenerational, extended family. Animals will ultimately lose parental protection and face the dangerous world on their own. They can't remain predator naive; they must become predator aware if they are to survive. It sets up a paradox for every adolescent: to become experienced you must have experiences. Said another way: to become safe you must take risks. And notably, some risks can't be taken—and their lessons learned—when protective parents are too nearby.

For humans, this paradox underlies a certain terror of parenthood. Parents can't always protect their kids from danger, and sometimes can't even alert them to it. Just as distressing, through their risk-taking, adolescents seem to bring needless danger upon themselves. Whether they're sixth-graders testing thin ice on a pond with their friends or high schoolers masquerading as twenty-two-year-olds to get into a nightclub, adolescents frequently put themselves in danger *deliberately*, to the angst and occasional heartbreak of their parents. The jeopardy they seek out—reckless driving, substance abuse, careless sex—can be baffling to adults. Even when deliberate adolescent risk-taking is of the more mundane variety, like building a bonfire with friends in the woods or sneaking a ride on someone's motorcycle, it's the stuff of parental obsession and late-night, nauseated worrying. It's one thing for a child to be naive to the dangers of the world. It's quite another to know something is dangerous but underestimate the risk and invite it to come closer. Sometimes hilariously, sometimes maddeningly, and sometimes tragically, adolescents don't just stumble into trouble; they voluntarily place themselves squarely in its path.

The behavior seems inexplicable, even contrary to the survival instinct. Taking dangerous risks that could result in death does not seem to make much sense from an evolutionary point of view. And yet, this strange behavior isn't limited to human adolescents. Adolescent risk-taking is seen throughout the animal world. During adolescence, groups of bats taunt predatory owls, and squirrel squads scamper recklessly around rattlesnakes. Not-yet-adult lemurs climb out onto the slimmest branches, and adolescent mountain goats scale the highest ledges. Away from their parents, young adult gazelles saunter up to hungry cheetahs. Adolescent sea otters swim up to great white sharks.

One approach to understanding puzzling behavior is to look for it in other species. Examining the life histories of those animals may then reveal how the "illogical" behavior actually helps them live longer, function better, and have more offspring. For risk-taking, this means first asking: Do other animals take risks during adolescence? And then: How does that adolescent risk-taking help them?

Evolutionary biologists will recognize this approach as an application of Nikolaas Tinbergen's famous "Four Questions." Tinbergen, a Dutch ethologist who won the 1973 Nobel Prize in Physiology or Medicine, believed animal behavior couldn't be fully understood by just explaining its mechanical nuts and bolts or the age in which it occurred. For him, it was always important to look for the behavior across species and determine how it was biologically beneficial. For humans it's helpful to distinguish between the risks teens invite through naivete and the risks they seem to seek out. Both, if survived, can offer future protective benefits. By the end of Part I you'll recognize the distinction. You'll understand why this period of life is so dangerous for all species. And, crucially, you'll understand why taking risks to become safe is not a paradox. It's actually a requirement for adolescent and young adult animals on Earth.

But in order to talk about staying safe, we first must travel to the roots of terror, deep within the ancient connection between mind and body. The story of safety begins with understanding the nature of fear.

Chapter 2

The Nature of Fear

The video shows a roly-poly mother panda, sitting upright, contentedly munching on bamboo. At her feet, sound asleep, snuggles a tiny, adorable baby panda. You watch for eleven seconds, wondering if this is all the video has to offer, when suddenly—*ACHOO!*—the baby sneezes; the mother startles; bamboo goes flying. The rolls of fat on Mom's belly convulse. It's a classic horror movie trick, the jump-scare, but done panda-style.

A second later, everything's back to normal. Baby conks out again. Mother resumes chewing. But unseen, down inside her startled panda heart, the neurochemicals that sparked the electric jolt are being rapidly swept away by her blood. The intense pounding has already been replaced by a calm, regular heartbeat. The mother panda was never in any danger, but the loud sound and sudden movement of her cub's unexpected sneeze jolted her body's fear machinery just the same. That panda startle, laughed at by millions of people around the world on YouTube, is in fact one of the most ancient neural reflexes on planet Earth.

On land and in the sea and sky, animals flinch with fright. The startle response is found not only in humans and other mammals, but in animals with whom we shared common ancestors hundreds of millions of years ago, such as birds, reptiles, fish, and even mollusks, crustaceans, and insects. It's possibly even present in plants. The widespread occurrence of the startle response points to its lifesaving function: alerting an individual that its life is in danger. And it is effective: fast escapes can double or triple an animal's chances of survival.

Flies zoom away from the swatter. Clams snap their shells shut. Crabs scuttle for cover. Clever octopuses have devised a hunting technique that involves triggering the startle reflex in their prey. Positioning themselves on one side of an unsuspecting shrimp, they slowly reach around with one of their arms and tap the shrimp on its opposite side, causing the crustacean to startle-jump right into the octopus's waiting mouth.

Human beings startle at times when the shock isn't even real. Charles Darwin noted in *The Expression of the Emotions in Man and Animals* that "the imagination of something dreadful commonly excites a shudder." Fascinated by the commonality of the fear response across species, Darwin described flinching in orangutans, startling in chimpanzees, recoiling in wild sheep, and jolting in dogs. He intentionally triggered jump-scares in his own infant children. He would make rattling sounds near their faces, noting that "the child blinked its eyes violently every time, and started a little."

Whether you're a person, a panda, or Ursula the penguin escaping a leopard seal, this ancient reflex is triggered automatically when a sight, sound, smell, or memory signals danger. Danger activates an electrical impulse, which fires through neurons, causing muscles to contract and produce a sudden leap, flinch, or twitch.

The physiology of fear involves not only the brain, but also the body's cardiovascular, musculoskeletal, immune, endocrine, and reproductive systems. When fear's powerful whole-body discomfort is paired with an event, place, or individual, animals learn to avoid that stimulus in the future. This is called "fear-conditioning," and it's so powerful that lifelong safety can be learned in a single confrontation. That means if Ursula encounters a leopard seal on her very first swim in the ocean, has a fear response, and lives to tell the tale, she is very likely to pair the negative feeling of fright with the location, sight, smell, and other aspects of her predator. Intense fear is a formidable teacher. Terror's unforgettable lessons are emblazoned into nervous systems and remembered for entire lives.

And, if Ursula survives that first leopard seal run, she's much more likely to survive her second, fourth, and forty-fourth. "As penguins get older, they get more experienced and that makes them safer," Phil Trathan, a senior researcher with the British Antarctic Survey, told us. This is a key point. But of course a near miss only works if it is, in fact, a miss.

ARMORING UP

One day, we were visiting the Peabody anthropology museum and were stopped in our tracks by a menacing human form. It brandished a two-foot-long sword with a blade like nothing we'd ever seen. Although not made of sharpened metal, its potential to rip through skin was just as ominous. It was an inlaid row of sharks' teeth, each about two inches long.

Even more arresting than the shark-tooth sword was the figure's helmet. Made of an entire blowfish, inflated like a balloon, the headgear had spikes extending in every direction. Along with a light-brown vest made of coconut fiber, this was an example of nineteenth-century body armor from the Kiribati on the southern Pacific Gilbert Islands.

The armor was part of an exhibition the Peabody was running at the time called *The Art of War*. As we glanced around the exhibit hall, we saw several other astounding examples of garments that the world's humans have engineered throughout history to defend themselves from one another. There was a nineteenth-century Tlingit hide painted with red-and-black formline art by indigenous people of the Pacific Northwest coast of North America. An eighteenth-century Moro brass helmet and chain mail from the Philippine island of Mindanao. Painted leather and wood armor worn by Lolo, or Yi, warriors from China's Sichuan Province near the Tibetan border.

We took a moment to imagine the individuals who wore these protective clothes. Whether they were adolescents, young adults, or older men, this armor shielded them from one very specific threat: other humans.

Armor design is a window into an era's dangers. World War I was a veritable Cambrian explosion of killing technology, and the development of gas masks and plated steel body armor called "lobster armor" emerged to counter chemical attacks and explosives. More recently, the Kevlar-containing Interceptor Multi-Threat Body Armor System worn by the U.S. armed forces from the late 1990s to the late 2000s was designed to protect against small arms fire and fragmentation from improvised explosive devices.

But weapons on battlefields are not the only potential dangers humans face. By extending the concept, we can see that humans construct external "armors" against many threats, from insect repellent and bed netting to ward off the risk of Lyme disease and malaria, to sunscreen, seat belts, and helmets to protect against skin cancer, car crashes, and bike accidents.

Fear, on the other hand, protects from within. Fear shapes the way animals behave. It triggers responses that have, over hundreds of millions of years, saved lives. Fear is an ancient, protective legacy passed to living creatures over countless generations. And yet, while fear is universal, it's also unique to each individual. No two animals—human or nonhuman—fear exactly the same things in precisely the same way. Each of us has a singular internal armor, tailor-made by our own particular experiences. And much of that internal armor is forged during wildhood, in the stage between childhood and full maturity when adolescents and young adults begin confronting danger on their own.

DEFENSE MECHANISMS

Militaries understand how shields, helmets, and masks protect soldiers from bodily injury. Soldiers' physical protection is their armor. Therapists understand how their patients use internal mental processes to protect themselves from emotional injury. These psychological strategies are their "defense mechanisms."

First conceptualized by turn-of-the-twentieth-century psychoanalysts, defense mechanisms are unconscious mental responses that protect people psychologically from conflicts, tensions, and anxieties. Repression, projection, denial, and rationalization are well-known defense mechanisms that have entered the common language.

Others are less well known. Acting inappropriately friendly to a person you actually detest or insulting a person you have a crush on are examples of a defense mechanism called "reaction formation." Sublimation is another. That's when a person unconsciously funnels aggressive urges into more socially acceptable actions. Channeling hostility and rage into athletic excellence is an example of classic Freudian sublimation.

In the 1940s and '50s, Anna Freud, focusing on adolescence, identified three defense mechanisms she believed emerged during this period to help control heightened sexual urges. They were: intellectualization, repression, and asceticism. Intellectualization is coping with emotional pain by focusing only on the factual aspects of a problem. Repression is hiding socially unacceptable urges or desires from oneself, denying they exist. Asceticism is the channeling of impulses and feelings into rigorous physical practices or self-denial.

Anna Freud and her father Sigmund's ideas are no longer within the mainstream of psychological theory or practice. Yet the defense mechanism lives on in psychology and popular culture as a legacy of their work.

Animal behaviorists don't use the term "psychology" to describe an animal's internal motivations. But they do study the actions that animals take to keep themselves safe from predators. Besides physical defenses like camouflage and claws, horns and thick skin, animals also have behavioral defenses. They can be vigilant, solicit help from others, and alarm call, for example. Taken together, these physical and behavioral protections are called "mechanisms of defense," and we'll explore them more in the next chapter. While Freudians would say that defense mechanisms protect humans from painful feelings, wildlife biologists would say that mechanisms of defense protect animals from existential threats.

Whatever you call these defenses, the responses to emotional and physical dangers learned through experiences in wildhood stay with animals for the rest of their lives.

Some safety knowledge is innate. Wild fish, reptiles, amphibians, birds, and mammals have inborn defenses tailored to the dangers they will face in the wide world. Red-eyed tree frog embryos can perform a neat lifesaving trick. They usually develop over a leisurely seven days before hatching. But if these developing embryos sense the presence of wasps, snakes, or even floods, they can speed-hatch themselves early and swim to safer locations. And the embryos of rainbow fish can detect danger even earlier in gestation. A mere four days after fertilization, the embryos are able to smell when there is a predatory goldfish or perch nearby. They respond to the threat with an increased heart rate, a common response to fear in vertebrate animals.

The safety knowledge an animal lacks at birth it must learn. Safety education continues throughout an animal's life, often intensifying during adolescence. But until that happens, adolescents on their own for the first time, like predator-naive Ursula, can rely only on the limited protection of inborn reflexes, including the startle response, to keep themselves safe.

ISLAND TAMENESS

If dangers in an environment change, an animal may need to remodel its external armor. People have an easier time with this than other animals do—you can take off a bulletproof vest more comfortably than an armadillo can remove his bony back plates. But over time, as threats emerge and recede, physical defenses follow suit, strengthening when necessary, diminishing or disappearing entirely when no longer needed. Similarly, internal armors (defense behaviors) become stronger or weaker in response to what is happening in the environment around the organism.

Island tameness is a fascinating example of this. Animals living on long-isolated islands without predators lose their fear, and with that loss, their antipredator behaviors. When Darwin explored the Galapagos Islands, he noted how easily he could walk up to iguanas and finches, and even ride on giant tortoises. Island-tame animals' fear responses have gone dormant, which is fine if they are living without threat. But if any predators do show up, an island-tame animal is extremely vulnerable.

More broadly, island tameness applies to populations whose predators have died out or been hunted to extinction. Yellowstone elk are a classic example of this kind of non-island island tameness. Wolves were systematically exterminated through the 1800s and 1900s, which allowed the elk to range free, unconcerned about being attacked, all over the national park. When wolves were reintroduced in the 1990s, the elk had to readjust to fear. They had to rebuild and relearn their defenses. This natural experiment in predator-prey relations showed that fear is malleable in island-tame populations and that even once it's gone dormant, it can reemerge when environments change.

Many if not most modern humans live in a state of island tameness. As past threats, such as predatory carnivores, become more and more remote, fear fades away. In some regions of the world, the growing number of parents who don't vaccinate their children may be a uniquely human version of island tameness. The devastating polio and rubella epidemics of the 1950s and '60s are like long-forgotten predators, unremembered and no longer feared. The parents who fear the vaccines more than they fear the diseases leave their children unprotected if the pathogens return. Of course, this practice could change in a heartbeat if the diseases come roaring back. Similarly, an island tameness effect may underlie the relaxation in safe sex practices seen over the past twenty years as the risk of dying of HIV infection has decreased.

Island tameness may help explain even fiscal behavior and economic or political trends. Both individual and institutional investors begin taking greater risks as economic catastrophes fade into a forgotten past.

Rising rates of anxiety in adolescents may also be understood as island tameness. Our animal and human ancestors evolved in environments full of predatory and other existential threats. Powerful fear neurobiology evolved during these dangerous times. Today, many (although not all) humans no longer encounter the kinds of dangers that shaped this neurobiology. What happens when brains and bodies that evolved in environments full of predators and other threats find those dangers removed?

A similar question was posed thirty years ago by a British epidemiologist who noticed a rise in autoimmune diseases like lupus and

Crohn's. David Strachan wondered what happens to immune systems that evolved in environments with many varied pathogens when the world gets cleaner. The "hygiene hypothesis" suggested that human immune systems, unchallenged in overly clean environments, turn inward and begin to attack their own bodies, mistaking normal tissue for pathogens. Might a similar process be driving anxiety in modern adolescents and other individuals?

Lars Svendsen, a Norwegian philosopher at the University of Bergen who studies fear, thinks yes. He believes that many modern humans have a "surplus of consciousness" that gets directed into imagining risks. In relatively safe environments of the modern affluent world, people no longer encounter the physical dangers our ancestors faced. Safer than ever before, with more "brain space" to devote to thinking about risks that don't pan out, we live in a state of what Svendsen calls "permanent fear." Permanent fear, believes Svendsen, isolates individuals and creates anxious, lonely societies because "living a life of fear is incompatible with living a life of happiness."

Unhappiness and anxiety aren't the only adverse consequences of uncontained fear. Fear responses themselves can sometimes, paradoxically, increase danger. When a newly elected Franklin Roosevelt cautioned a turbulent 1933 America that "the only thing we have to fear is fear itself," he might have been giving a lecture on fear to a class studying animal behavior. You don't often hear the second half of the most famous sentence of his presidency, but it perfectly captures the peril of excess fear as the "nameless, unreasoning, unjustified terror which paralyzes needed efforts to convert retreat into advance."

The point is, that while reacting to danger can save a life, it doesn't always come without a cost. Freezing in place can sometimes keep an animal from being detected by predators. Younger animals, especially, rely on stillness (called "tonic immobility") to evade detection. But motionlessness can also delay escape. A scared, overly vigilant animal scanning its surroundings eats less, socializes less, mates less. And showing fear can sometimes actually get an animal killed—as the surprised shrimp discovers on its way into the octopus's mouth. Showing fear can be a tell—an enticing "choose me" signal to watching predators that you might not have what it takes to survive.

The lessons Ursula the penguin learns as a predator-naive adolescent will continue to inform her adult behavior. On the other hand, environments can change, and new dangers emerge. If a freak virus were to wipe out all leopard seals, king penguins like Ursula would likely become island tame within a generation or two. They would relax around coastlines unless another, new predator moved in to occupy the seals' niche. And were this to happen, it would expose a central truth about danger throughout an animal's life: at any age, no matter how experienced, an animal can become predator-naive to new threats all over again.

Chapter 3

Knowing Your Predators

When we last saw the adolescent penguin Ursula, she had just left her parents and was mid-dive, about to speed toward the dangerous domain of the leopard seals. And she was predator naive. No fear conditioning had yet forged her muscle memory into reliable lifesaving behaviors. Her internal armor had not yet been shaped by experience. With no knowledge of the ancient relationship between predator and prey, Ursula couldn't imagine what might happen next.

But we can. Seeing yourself through the eyes of an imaginary hunter can help keep you safe. Let's say you're a cheetah on the African savannah. Driven by the twinge of a hunger pang, you spot a group of gazelles. Here's a possible meal. But you can't chase and eat all of them. You have to pick one. Which will it be? You scan the group, looking for a physically injured gazelle or an unprotected fawn. No such luck. You turn your attention to three adult-sized options. Gazelle A looks good—but is fit and lively, bouncing up and down with energy. Quieter Gazelle B looks like a better choice—but has just spotted you and is now intently watching your every move.

Taking down energetic Gazelle A will require speed and strength. Outsmarting alert Gazelle B without the element of surprise will require skill and a really good tactical plan. Maybe there's another option. Then you spot Gazelle C, a predator-naive, adolescent or young adult. Gazelle C is fully grown but looks willowy. He has learned something about cheetah dangers from his parents, but is predator naive compared to older, more experienced gazelles. To a predator, he may seem unsure of his position in the group. He's neither clustering with the mature gazelles nor nestling with the calves near their mothers. Instead, he's investigating some rustling plants and seems to have no idea you're watching him.

Every time they set out to make a kill, predators must fill out a kind of cost-benefit analysis, a wild spreadsheet. They must calculate the time and energy they can afford to spend in choosing, chasing, and killing. Then they need to weigh that expenditure against how much nutrition the meal will give them. It's what a frugal shopper goes through at a grocery store, deciding how to get the most calories for the fewest dollars. It's what a corporate takeover specialist does to acquire the most vulnerable and valuable companies. Carnivorous predators must estimate how difficult it will be to get a meal. And it turns out that at nature's meat counters all over the world, adolescents are good value.

HUNTING FOR FRESH MEAT AND EASY PREY

Predator-naive animals are attacked and killed by other animals, shot with guns by human hunters, hit by cars, and lured into traps more often than predator-experienced animals. Big in size but meager in experience, unfamiliar with a predator's smells and sounds and easily fooled by camouflage or distraction, predator-naives bumble into dangerous territories. Attempting to escape, they miscalculate their own ability to fight or flee. They crash and drown trying to find their way. Adding to the risk, predators target them, exactly because they're inexperienced, often newly dispersed away from familiar home ranges. Notably, they're unaccompanied by parents.

Off Kodiak, Alaska, for example, swims a kind of killer whale that dispatches its victims in a particularly gruesome way. It bites their throats, rips out their tongues, and tears off their lips. Bigg's orcas, named for a scientist who pioneered orca research methods, specialize in hunting humpback whales—but not just any humpbacks. They go after adolescents who have wandered into dangerous areas with no experienced adults around to protect them. The young humpbacks' lack of sophistication is payday for the Bigg's orcas, which have perfected tracking, attacking, killing, and eating adolescents. They're adolescent hunters.

Scientists studying the predator-prey relationship between cheetahs and an antelope called kudu on game reserves in Southeast Africa found that cheetahs prefer to prey on young adult males. Socially, these kudu adolescents are unstable in their hierarchies, meaning they don't have good backup from other members of their group. Physically, the adolescents are less robust, not as strong, coordinated, and experienced in self-defense as mature males, making them easier to outsmart and outrun.

The question of whether predators prefer to eat adolescents intrigued biologists in Argentina, who examined regurgitated owl pellets for the remnants of tuco-tuco, a South American rodent. The skeletal remains the researchers found were exclusively adolescent. The scientists reported that the owls preferentially hunt adolescents, who travel in more exposed areas. These owls hunt adolescent tuco-tucos for the same reason killer whales hunt adolescent humpbacks and cheetahs hunt adolescent kudu: the cost-benefit analysis always points that way.

Even lowly sardine adolescents aren't safe from predators who specialize in their age group. African penguins (also called jackass penguins) favor adolescent sardines because the young fishes' underdeveloped shoaling skills make them easier to catch. It's worth noting that the penguins who target adolescent sardines are themselves adolescents who aren't strong or skilled enough to hunt with the adults, so they're left behind to grab what they can.

People who hunt deer know that adolescents are especially vulnerable. Out on their own, in unfamiliar terrain, inexperienced yearlings are often first to wander into view and first to be shot. In fact, until about a decade ago, throughout North America, 90 percent of hunted deer were yearlings and young bucks. But at the urging of the Quality Deer Management Association, an advocacy group founded by a wildlife biologist, that has changed. Now most hunters recognize that adolescent bucks need protection and they avoid shooting them. Granting yearlings a reprieve makes for deer populations that are physically and socially healthier.

Humans are perhaps the best predators the world has ever seen, and many animals quickly learn how deadly we can be. The evolutionary biologist Richard Wrangham told us a fascinating tale of poachers in Uganda, who set up snare traps to catch chimpanzees for the bush meat trade. Adolescent chimps, less aware and less experienced, are caught most often; more experienced chimpanzees have learned to scan for the wires and avoid the traps. Chimp babies are safe, because parents protect them.

Whether it's a humpback mother defending her calf from a killer whale, a penguin father going after a predatory skua, or mother hyenas circling their cubs to ward off a lioness raid, parents of many species make sure their calves, chicks, cubs, and other young are protected. Adolescents must often go it alone.

PREDATOR DECEPTION

Deceive or be deceived; eat or be eaten. Prey use deception to avoid capture and death. Playing dead is one effective antipredator strategy. Feigning injury is another, for animals strong enough physically and mentally to lure a predator close. Faking a broken wing, for example, is a deception used by many bird parents to draw predators away from their nesting young.

But deception can be flipped, and predators also use it to trick their prey. They, too, may play dead or hide themselves in ways that prevent their victims from knowing they're there. Deception-assisted predation has been coopted by our species. Throughout history and across cultures, human hunters hide and disguise themselves with camouflage, douse themselves in scents that mask human odor, and mimic the sounds of their prey. Some cunning hunters plan months ahead by planting patches of food like alfalfa, clover, and corn near their favored hunting grounds. Food plots can support populations of animals that might go hungry, but they also draw animals close. The hungriest and most clueless tend to be adolescents and young adults, who fail to recognize a danger their older, more experienced peers can spot easily. Like Hansel and Gretel, young animals are excited by the unexpected feast. Distracted by their hunger, they unwittingly position themselves for the hunter's perfect shot.

Even a kindly fly fisherman–grandpa is in effect a skilled predator practicing the ancient art of predator deception. By mimicking the shape and movement of smaller fish or insects, the lures he casts into the river are disproportionately attractive to young adult, predator-naive fish—the ones that more often take the bait and end up in a frying pan over Grandpa's campfire. Once again, fish who've had some experience are caught less often. There's some evidence that being shy helps young fish avoid lures and also that inborn fish introversion—so-called bait shyness—can protect fish at this age.

The best animal predators can be similarly clever with their lethal intent. Sharks, for example, learn to approach prey with the sun behind them. Backlit, they're harder to detect. Crocodiles hide themselves in watering holes, lying very still with only their nostrils exposed. Pharaoh cuttlefish can change their colors and alter their movements to appear to be harmless hermit crabs. Because their prey aren't afraid of hermit crabs, the cuttlefish can get closer and increase the chances of killing their victim. Lacking experience, adolescent and young adult animals are more likely to fall for these deceptions, with catastrophic consequences.

While carnivores aren't generally a concern for modern humans, to the terror of parents everywhere, adolescents can be targeted for abduction by predators using everything from force to deceptive techniques.

A ten-year analysis by the National Center for Missing and Exploited Children (NCMEC) looked at nearly ten thousand incidents of attempted abductions on infants, toddlers, young children, and teenagers (eighteen and younger) in the United States between 2005 and 2014. It showed that abductors tailor their methods based on the predator-naivete of their targets. For example, the study found that for both the youngest and oldest children, the mostly male predators had to use force, including weapons, to carry out their abductions. This was because when going after the younger victims, abductors often had to overcome a protective parent. Similarly, when attacking older children, who likely recognized what was happening, force was needed to overcome the teens' size and developing street smarts. For these older adolescents (sixteen to eighteen), abductors also chose to target them in secluded areas—parking structures, hiking trails—far from help. Just as unfamiliar environments put animals at increased risk of predation, unknown, out-of-the-way places cut down on human adolescents' ability to escape or call for help.

In contrast, for the predator-naive victims in between, aged roughly eight to fifteen, would-be abductors didn't need to use much force or even travel to secluded areas for the ambush. Instead, they generally used a very different ploy: verbal persuasion. So inexperienced was this age group that they didn't need to be threatened with force or separated from the group first. Predator-naive middle school children could be lured with sweet talk.

Most of these abductions occurred when the developing adolescents were walking to and from school. Offenders offered rides and candy or drinks. They asked their victims for help finding a pet or a person. Frequently offenders pretended they themselves had a child who was missing. Twenty percent of offenders got to their victims just by giving them compliments. Three percent asked for directions. Sometimes, like pharaoh cuttlefish, they took on a nonthreatening appearance, mimicking authority figures like doctors, nurses, or police officers. All these strategies help predators avoid detection and save them time, energy, and potential exposure.

According to the study, the predator-naive students in the middle of the age range—between eight and fifteen—were especially easy prey, because they were less likely to cry out. Perhaps the signature fatal bite of the adolescent-hunting Bigg's orcas (the ones that rip out their victims' lips and throats) serves a grisly secondary function of muting any calls for help. Silence, it seems, can be deadly for targeted adolescents across species.

Fortunately, forced abduction is fairly rare in the United States. However, statistics on other crimes back up what we know about the danger of being predator naive for adolescents and young adults. Sex traffickers, for example, prefer victims who are still uncertain about and unstable in the adult world. As a former sex trafficker told an investigative reporter in a 2017 documentary called *Selling Girls*, "I don't waste my time with confident women. Pimps seek out people who are too naive to notice something isn't right." Not only do these sexual exploiters use the ecological techniques of animal predators, but they also recognize who their most uncomplicated victims will be.

During wildhood, safety is compromised by naivete and inexperience. Adolescent animals do learn a panoply of antipredation behaviors that, with practice, improve their chances against predators. We'll dive into those in later chapters. But wildhood is also marked by another nearly universal vulnerability, one that's less bloody but also often deadly. Adolescents, whether they're human or nonhuman, enter a world that is not only waiting to prey upon them. It also just doesn't like them very much.

ACT YOUR AGE (ACTUALLY, DON'T)

One curious feature of adolescence is that it makes the people going through it feel as though they're the first and only ones ever to have done so. But for every wave of adolescents feeling unique, there's a corresponding generation of adults feeling exasperated by the excesses of youth. There's a quote often attributed to Mark Twain: "When a child turns thirteen," he is said to have quipped, you should "put him in a barrel and feed him through a hole in the lid." And when they turn sixteen? "Plug the hole!" Twain advised. Though the words are almost certainly apocryphal, their popularity reveals some clang of truth that keeps them circulating.

Youth is prized in many fields of human endeavor, from music to sports, writing to acting. But a closer look at the shared experience of adolescence reveals that the flip side of obsession with youth is a simmering ambivalence about them. Adults often have intolerance, disdain, and sometimes even a particular kind of loathing for adolescents. There's even a word for it: "ephebiphobia," the fear or dislike of youth.

At its most benign, ephebiphobia is simply patronizing headshaking, embodied in the cliches "youth is wasted on the young" and "kids today . . ." That's the tone of Aristotle's oft-quoted complaint that youth "would rather always do noble deeds than useful ones; all their mistakes are in doing things excessively and vehemently. They overdo everything—they love too much, hate too much."

It's easy to hear some affection in Aristotle's words, but more committed ephebiphobes go beyond fond scolding. A magnetic sonic device called the Mosquito targets the biological vulnerability of teenagers across the United Kingdom and elsewhere. The Mosquito works by emitting an ultra-high-frequency blast (around 19 to 20 kilohertz) that teens can hear but older adults have generally lost the ability to. Placed in parks and near shops, anti-loitering devices like the Mosquito make areas intolerable for young people and encourage them to leave. They're an impersonal, electronic "get off my lawn" signal.

Other ephebiphobes may intend harm. Like predators, they view adolescents as easy targets. In many countries around the world, ephebiphobia is institutionalized and hides in the predatory practices of societal establishments from banks and hospitals to sports franchises and the military.

Financial bodies prey on adolescents' unsophistication about money and their supposed impulsivity—or perhaps it's just inexperience—in spending it. Credit card companies often deliberately focus their marketing on pre-adults as young as high school students. In the United States, 10 percent of college students graduate with more than $10,000 of credit card debt. The gaming and gambling industries also target adolescents, who have six times the pathological gambling rate of adults.

Because of their physical potential and youthful idealism, adolescents are co-opted in other arenas as well. College sports programs use adolescents' desire for education to make billions off their bodies. On police forces, rookie cops may be assigned the most dangerous beats. Agence France-Presse reported in 2015 that in Beijing, inexperience and financial need propelled seventeen- to twenty-four-year-olds into contract firefighting jobs that lacked professional safety training, making these adolescents disproportionately vulnerable to death and injury.

Exploitation of youth is apparent in the forced conscription of adolescent and child soldiers, which happens around the world today just as it has in the past. The youngest, poorest soldiers in Roman legions were adolescents, and these Velite and Hastati troops, inexperienced and lacking the best weaponry, were placed in the most dangerous battle positions, and suffered the highest casualties. In the eighteenth century, thousands of adolescent boys, some as young as eleven and twelve, were recruited to become ships' boys for the British Royal Navy. Without resources or social status, these adolescents had few options other than to join. Jim Hawkins, the young protagonist in Robert Louis Stevenson's *Treasure Island*, becomes a ship's boy at the age of thirteen after his father's death. His coming-of-age transformation from naive child to capable young adult paints a much rosier picture than the typical outcome for these exploited adolescents. Another Stevenson classic, *Kidnapped*, was inspired by the real-life story of Peter Williamson, a thirteen-year-old from Scotland who in 1743 was lured onto a ship with seventy other boys and sold into a seven-year term of servitude in Philadelphia.

Thirteen was shown to be a key age in a study of another form of exploitation: recruitment by street gangs. Gang leaders seek out younger adolescents, exploiting their need for acceptance with promises of enhanced connection to a group or cause. Drug dealers target them too, knowing how effective their products are in falsely raising the status—or the feeling of status—in their teen customers.

Like deer-hunting advocates imposing protections for vulnerable yearling deer, societies do sometimes put specific protections in place for adolescents. In 1988, the R.J. Reynolds tobacco company rolled out a campaign to sell Camel cigarettes to adolescents, targeting them with a cartoon character named Joe Camel. Despite Joe's skeevy vibe and the near-instant outrage from public health and parent groups, Joe Camel slung cigarettes for almost nine years until the promotion was finally pulled, perhaps in response to pressure from the Federal Trade Commission. In 2018, e-cigarette use, also called vaping, had increased 900 percent among U.S. high school students compared with 2011. The Centers for Disease Control reported that among high school students, e-cigarette use had leapt from 220,000 students in 2011 to 3.05 million students in 2018. The Federal Drug Administration warned retailers of penalties for selling vaping devices to minors, and laws to limit the sale of flavored e-cigarettes marketed to teenagers started to emerge from state and federal legislatures.

ADULTOCENTRISM

Sometimes, ephebiphobia isn't so much fear or hatred of adolescents, but an unwillingness to see them at all. Or worse, seeing a big body and assuming it belongs to a mature adult. "Adultocentrism" is the tendency for pre-adult phases of the life cycle to be diminished in value or overlooked altogether. An Italian biologist, Alessandro Minelli, believes it is impeding scientific progress. He calls on scientists to "grant similar status to all life stages" for a profoundly important reason: that "understanding of non-adult phases of life could reframe how biologists understand evolution."

Adultocentrism creeps into the decision-making of scientists and physicians, who should know better. The result: unintentional discrimination against sick adolescents. One of the most vulnerable groups in the world, adolescents and young adults with cancer have worse survival and higher relapse rates than children and adults with many similar cancers. Lack of access to cancer specialty centers is a contributing factor. In the United States, adolescents and young adults are the fastest-growing medically uninsured group, so even once cancer is diagnosed, they are less likely to be treated by experts at national research centers.

But the primary driver of the tragic disparity in cancer survival in this group: adolescents and young adults have some of the lowest levels of participation in lifesaving clinical trials in the world. Most clinical trials exclude patients under eighteen. Feeling too old for pediatric trials, and too young for those aimed at adults, adolescents find themselves in cancer no-man's-land, what pediatric oncologist Joshua Schiffman describes as "the wild west" of cancer referral.

There are many explanations given for the exclusion of adolescents and young adults, but a simple truth underlies it: adolescents don't conform to the "model" subjects of adult or pediatric research. Their differences muddle the results. (Not too many decades ago, this kind of thinking was applied to a different "nonconforming" group: women. Female reproductive cycles were said to complicate the research. As a result, much of the medical investigation of the past century exclusively focused on—and benefited—men.)

The tendency to see a big body and assume there should be adult accountability can even sway medical ethics committees making decisions about who will receive precious organ transplants. Adolescents have been denied transplants because the committees think they won't stick to the pre- and post-op rules.

WHEN ADULTS DO THE WRONG THING

Ephebiphobia refers to a fear of human adolescents, but some people unwittingly extend that intolerance to the youth of many species. For example, avian adolescence varies in timing, but among Indian ring-necked parakeets it occurs between four months and a year of age. For owners of these beautiful, brightly colored pet birds, the behavioral change from compliant chick to hissing, biting, defiant adolescent can seem to happen overnight. Other pet birds begin incessantly singing or talking loudly when they hit adolescence. Some become more territorial and aggressive or want nothing to do anymore with their owners. Even less welcome to a human showing off his prize pet might be the sexual changes brought on by bird adolescence. They include feather plucking, screaming, and, yes, masturbation.

Although all of this behavior is developmentally normal, some owners aren't prepared and can't tolerate it. They stop engaging with their pets or give them away. It even happens with dogs, the most popular pet in the United States. If you've ever known a canine going through adolescence, you know how annoying they can be. They chew shoes and furniture, frolic hyperactively, bark and growl inappropriately, and run away at the park when it's time to go home. Applying Aristotle, they love too much, and they hate too much.

Adolescence is the period at which pet dogs are most likely to be banished to a backyard, left alone for hours tied to a stake, or simply abandoned on the street. The odds of a dog being surrendered to a shelter skyrocket at this age because of behavioral issues.

More than half of all dogs in American shelters are adolescents between the ages of five months and three years: after puppyhood and before doghood. Since a majority of all surrendered dogs are euthanized, this means that simply being an adolescent can count as a fatal condition. Experts note that these dogs are simply "young adults exhibiting potentially resolvable behavior problems that owners are ill-equipped to handle." And since 96 percent of surrendered dogs have no obedience training whatsoever, their owners never really gave them a chance to succeed.

But while some humans mistreat young animals due to ignorance or indifference, adults from animals' own species can be downright exploitative. Wild, older songbirds native to northern Europe and Asia have been seen by researchers in Finland to restrict younger birds' access to food by intimidation or force. Exploiting the weaker position of the adolescent birds has a double benefit for the older ones. First, the dominants are better nourished. Second, keeping the younger birds hungry makes the dominants safer. When feeding birds spot a predator, they flee to hide in the bushes. The birds cannot be 100 percent sure when the predator has left, but they can't wait too long to resume feeding, or they won't get enough to eat. Hungry birds are the first to venture back out of the safe hiding place. The better-fed birds can afford to hold back. Not only do the subordinates smoke out any predators, but they may also fill the belly of those predators, further protecting the dominant birds.

These subordinate, adolescent birds aren't necessarily impulsive risk-takers by nature. They are taking risks, but it's because they have no other choice. Like animals of all ages, when adolescents can't get the resources they need to survive, they become more desperate. This sets them up to be exploited. And as we know from stories of runaway and neglected children, the risks they have to take to survive can be severe and heartbreaking.

PUPPY LICENSE

Notably, young animals of many species have a special status, a leeway period granted by older members of the group. This reprieve from the hierarchy is called "puppy license" by the behaviorists who see it in dogs, but it's a feature of family dynamics in a range of species, including primates, who have their own "monkey license." Older animals will overlook, or gently correct, an inappropriate display of dominance as long as the offender is young enough not to know better. Puppy license also covers play: older dogs seem to enjoy puppy playfulness, and may encourage young dogs by wrestling more gently, growling more softly, and sometimes letting the puppies win.

As soon as that young dog hits a certain point in adolescence, however, its puppy license expires. Behaviors that were lightheartedly tolerated just a few days before are now met with adult pushback. Although the dog is still young and may lack experience, it is challenged and treated like an adult. In the human world and in the dog world, as juveniles mature into wildhood and their puppy licenses expire, a tolerant world becomes irritated and intolerant. The new adolescent finds itself in an unfamiliar realm of annoyance, aggression, and even victimization. No more immunity, no more second chances. It's called growing up.

Some modern adolescents do enjoy a suspension of adult responsibilities. Families may provide economic support; the legal system may forgive some infractions; embarrassing mistakes can be explained away as youthful indiscretion. The importance of puppy license for humans was recognized by the developmental psychologist Erik Erikson and anthropologist Margaret Mead. They believed adolescence should carry a "psychosocial moratorium." What they meant was that during adolescence, different roles and behaviors should be allowed to be sampled without the consequences and obligations of adult life.

THE TREACHEROUSNESS OF UNFAMILIAR TERRAIN

Penguin license may be different from puppy license, given the different social structures of mammals and birds. But king penguin parents are tolerant of the begging and squawking of their adolescent offspring. Once they leave home, however, those youth are generally without parents. Ursula had never been in the open ocean before, and suddenly finding herself in unfamiliar terrain worsened her odds of survival, whether leopard seals chased her or not. New landscapes are risky for inexperienced adolescents and young adults.

Imagine it's a chilly autumn morning in a Pennsylvania forest, dark and cold. An adolescent white-tailed deer startles awake. His head is heavy with his growing antlers, still soft with fuzz. It's a transformative moment for him: his first-ever morning waking up alone, without his mother by his side. The young buck has spent the past year and a half moving through the forest with his mom, learning which routes to take and which to avoid. When she flagged her white tail in warning, he would run or freeze. If she paused and rotated her ears to double-check a sound, he would too, learning to listen for snapping twigs. When she sampled the air to detect ominous odors, he sniffed alongside her. Under her guidance, he steered clear of coyotes, cars, and hunters and learned which healthy plants to eat and which dangerous or low-nutrition foods to ignore.

But no longer. The day before, driven by the instinctive wanderlust of the *zugunruhe* behavior, and maybe even a push from Mom, this deer had dispersed, wandering five miles away from the home where he was born. Starting today, all the responsibilities of adulthood are on him. Drawing on what he's learned from his mother, and coupling it with trial and error, this young buck will need to anticipate and avoid danger all on his own. Find his own food. Figure out where to sleep at night. If this young buck could read, he might relate to a scene in *Charlotte's Web* in which Wilbur the pig is about to depart the familiar farm with its warm slops and friendly faces. On the brink of walking away, Wilbur suddenly turns and heads back to his barn. Settling into his straw bed, he muses, "I'm really too young to go out into the world alone," and curls into one more night's sleep in the safe home he knows.

Joe Hamilton, the wildlife biologist who founded the Quality Deer Management Association, described for us the challenges young deer face and why they need protection when they're out on their own for the first time.

> Their mothers are kicking them out, and they're sort of bouncing around like a ball in a pinball machine trying to set up their home range in an area that's not stepping on the toes of others . . . sometimes two or three or four or five miles away from where they were born. . . . They're thrust into habitats they're not familiar with. They don't know the lay of the land.
>
> These young bucks get into a lot of trouble that way. They're having to move around a good bit to get familiar with it. That increases their chances of bumping into predators, bobcats and coyotes, for example.
>
> It's like a young person going into a new town. They're going to get into more trouble than they can handle, usually, until they get a little bit of experience and they learn the ropes: where to go, where not to go . . .

Hamilton added, "If you see deer activity in the early fall in the daytime, it's almost always young bucks. They're just curious. They're trying to set up home and they're more vulnerable to hunters. They're more vulnerable to predators . . . Down here [in South Carolina], if they're swimming creeks or small rivers, often they learn the hard way about alligators."

Learning the hard way works—as long as you live to tell the tale. As adolescents' knowledge of danger grows, they start to learn something very important: dangerous things aren't necessarily dangerous all the time. Motor vehicles are the leading cause of death for human adolescents, for example, but they're also a part of daily life and usually follow certain predictable rules.

WHEN DANGER IS LESS DANGEROUS

In the wild, predators are sometimes very dangerous and at other times remarkably harmless. First of all, they're not always hunting—they can't, and don't, hunt 24/7. They're also on more of a schedule than you might imagine. Some specialize in dawn/dusk raids; others hunt only during certain times of the year or in certain weather or light conditions. And, sometimes, when they've just eaten, they really aren't dangerous at all.

Take, for example, a rattlesnake that has just feasted on a California ground squirrel. Belly full, that snake physically hasn't room for any more squirrels and is probably not motivated to hunt. Experienced ground squirrels develop the ability to tell the difference between a sated snake and a hungry one. They are more vigilant around snakes looking for a meal and more relaxed when they sense that snakes are fed.

Moose can differentiate between wolves that are hungry and wolves that aren't. Garter snakes can tell when hawks are hunting and when they're just flying by. If she makes it past enough leopard seals, Ursula will come to learn that her major predator pauses hunting at midday for a couple of hours' rest.

Learning how to coexist with hungry animal neighbors has been a fact of life for hundreds of millions of years on our planet. And some rules of engagement have evolved that direct the behavior of both predator and prey. For a predator, knowing these rules is the difference between eating and going hungry. For prey animals, understanding them can be the difference between life and death.

That makes understanding both sides of the interaction a key part of adolescent learning about safety and danger. King penguins like Ursula become skilled predators themselves although they're also hunted aggressively by leopard seals. They can power themselves after fish and krill with the precision of targeted missiles.

Besides getting to know a leopard seal's daily hunting rhythms, Ursula will be learning an ancient secret of predator behavior. It's called the "Predator's Sequence," and it's a series of predictable offensive moves that all predators use to successfully hunt and kill other animals. Knowing the Predator's Sequence is like getting a peek inside the other team's playbook. It may give the hunted a lifesaving heads-up about what the hunter will do next.

THE PREDATOR'S SEQUENCE

When leopard seals pursue penguins, they play out the same series of steps as a cheetah taking down a gazelle or a hawk swooping onto a field mouse. It's the same sequence a T. rex ran to nab a duckbill dinosaur. Even ladybugs follow the same steps to hunt aphids. Human hunters follow the Predator's Sequence when they recreationally prey on pheasants and deer.

For the hunter, the Predator's Sequence is straightforward, just four easy steps: *Detect. Assess. Attack. Kill.* Day after day, kill after kill, predators must go through these four steps in that exact order, perfectly or nearly perfectly every time. Detect. Assess. Attack. Kill.

The choreography of killing is like classical ballet. Each step is precise, and each phrase of steps leads into the next. As in ballet, the components can be broken down and anticipated. Carnivores must practice their technique as much as any ambitious young dancer, but the strict structure of their routine doesn't vary.

If predators' roles are simple, the prey's part in this macabre and high-stakes pas de deux is even simpler. It can be conveyed in a single imperative: *Terminate the interaction as soon as possible.*

But that straightforward directive has a complicated reality with infinite variations. Predators follow the choreography strictly: detect, assess, attack, kill. Prey must improvise—change rhythm, syncopate, freeze in midstep, sometimes simultaneously.

Prey can undercut their hunters' ability to carry out the Predator's Sequence by underreacting and fooling expectations, or overreacting and messing up their stride. Like all great improvisers, prey aren't just making it up as they go along. The best improvising—and the best predator-evasion, whether in the savannah, sea, or sky—comes from paying close attention in the moment, after hours spent practicing skills. The more you rehearse movements and sequences, the better you are when it's your turn onstage. One other secret of great improvisation and of great prey-evasion is this: learn from the pros. The best improvisers and the safest prey watch and learn techniques from older, better, more experienced members of their community.

If you look closely at the Predator's Sequence, you'll notice that the first half of the interaction tends to favor the prey. The second half favors the predator. That means that if the prey can avoid pre-assault detection and assessment, they are much more likely to avoid attack and death. It seems obvious, but it has crucial wisdom for all predator-naive adolescents and even for adults facing new fears. A good general strategy is to avoid the first two steps: detection and assessment.

The second half of the Predator's Sequence—attack and kill—is the much scarier scenario. And it's much harder to repel a predator once an attack is under way. In fact, the antipredation strategies you probably know best—fight or flight—are called "behaviors of last resort" by antipredation experts describing them in wild animals. Being less experienced, slower, weaker, less confident, and sometimes still lacking their full armamentarium of adult physical defenses (tusks, claws, spines), adolescents are at even more of a disadvantage than adult prey if the Predator's Sequence tips into the second half. There are many, many behaviors wild animals undertake to protect themselves long before they get to last resort.

PREDATOR'S SEQUENCE STEP 1: DETECT

The first trick for not getting attacked and eaten is to avoid being spotted in the first place. Animals have evolved astonishing behaviors and physical characteristics that allow them to avoid detection. That's because predators are searching for them with a range of ingenious tools. Their hunting eyes and ears can perceive ranges that ours can't. Some predators smell or taste air and water, using a skill called "chemo-sensing." They feel for currents that might betray the movement—and reveal the location—of their dinner-to-be. And many have senses we humans either don't have or don't use. Sharks perceive electromagnetic currents through specialized skin cells; bats use ultrasonic hearing to "see" in the dark.

Against this array of detection technology, prey aren't helpless. Pretend for a moment that you're an adolescent *Molossus molossus* bat, also known as the velvety free-tailed bat. It's dusk, and you're flying around looking for dinner. Bats are the only mammals that can fly, and *Molossus molossus* is the fastest of them all. In fact, it's one of the fastest mammals on Earth. While its maneuverability is poor, it hunts over large distances.

At the moment, you're cruising along, listening for moths and beetles to grab and eat. But as you're hunting for your own meal, a barn owl is hunting too. And she eats bats. Unbeknownst to you, she spots your group. Whether you know it or not, your predator has just completed the first step of the Predator's Sequence. You've been detected.

Rewind this image for a moment. Animals can stop the Predator's Sequence before it even starts, by preventing being detected. One of the most obvious ways to do this is to hide. Hiding is more effective if you can remain absolutely still. To avoid detection by predators that can track the beating hearts of their terrified prey, some animals have evolved responses that slam the brakes on heart muscles. This quells the sounds, stifles the movements, and throws off any predator that might be listening in. It's called a vagal response, and it is an ancient cardiac trick we share with other mammals, birds, reptiles, and fish. You've felt it hundreds of times. It's that feeling of nausea during moments of terror—a near miss with a speeding bus, or realizing you've posted a reputation-annihilating message. That watery feeling in your gut and throat is produced by a sudden downshift in your nervous system related to the lifesaving ability of animals to slow their hearts to hide from predators.

Besides hiding, animals can improve their safety through vigilance. Too little can mean the end of an animal's life. But too much can effectively paralyze the animal, keeping it from feeding, socializing, mating, and accomplishing the other necessities of living. So maturing animals have to learn how to balance too much vigilance with too little. One way is to join a group and trade watch shifts with other animals. More eyes looking out for predators is one of the reasons group living evolved.

In fact, as groups get larger, animals get safer. For one thing, their individual risk is diluted by all the others in the group since a predator can't eat them all at once. But the predator's attack success also goes down because of a phenomenon called the "confusion effect." It is hard, literally confusing, for a predator to track one individual in a group of matching creatures. Try keeping your eye on a single bird or fish while watching a flock of starlings or a school of sardines. It's really hard to do. Identically uniformed football players on a field, indistinguishable T-shirt-clad dancers in a drill team, even a pile of perfectly ripe oranges at the grocery store can elicit a confusion effect when you try to pick just one. This is extremely protective for the individuals in that group—especially perhaps for inexperienced younger animals trying to hide within the group as they're gaining the strength and experience to defend themselves.

The confusion effect has an equally powerful flip side: the "oddity effect." Because it's so hard to track an individual within a group of seeming replicas, any little thing that makes an individual stand out will catch the eye of a predator. A misshapen fin. A wing of a different color. An immature behavior or vocalization that calls attention to itself. A body that's bigger and taller, or smaller and shorter, than those of the rest of the group. A smell that's off or distinctive from the others.

Because predators can more easily target group members that stand out, the oddity effect means that animals are safer—especially as younger adolescents with immature defenses—when they adopt the look and behavior of the majority. Oddity endangers animals from birds to fish to mammals, including, of course, humans.

To study the oddity effect, a wildlife biologist in Tanzania in the 1960s painted the horns of some wildebeest white. He then released them into groups of other wildebeest that had unpainted horns. With their conspicuous headgear, the white-painted wildebeest were targeted and attacked by hyenas. The biologist controlled for other factors and found that it was the conspicuousness, the oddness that attracted the predators.

In a different study, scientists dyed some minnows blue and found that when they swam with a group of black fish, the blue ones were caught first by predators.

Similarly, a study of the social ostracism of albino catfish found that they faced higher predation. But the scientists noticed another trend: the albinos weren't only eaten more; they were also consistently shunned from their groups. Not only were the albinos already at higher risk of being killed because they looked different, but they were also pushed out by their own groups and deprived of the benefit of safety in numbers.

Group shunning points out an absolutely fascinating aspect of the oddity effect. Investigators believe the albino catfish were "rejected" by their peers because the odd fish increased predatory danger for the whole group. Without the presence of the odd fish, the group as a whole stood a better chance of confounding a predator with the confusion effect. But the presence of the different-looking fish attracted a predator's gaze not only to it as an individual but also to the group as a whole.

The oddity effect helps explain why fish prefer shoaling with other fish that look like them and why birds of a feather literally do flock together. Behavioral similarity—swimming or flying with uniform speed, dexterity, and angle—reduces the risk of being preyed upon.

Being odd is dangerous for prey animals, and part of the work of avoiding danger is not sticking out. We humans don't live in herds or flocks. And dying in the jaws of a predatory animal is a vanishingly rare occurrence in our modern societies. But when we are together with others in groups, some of our responses are strikingly similar to those of animals living collectively in schools, flocks, or herds. Humans streaming into football stadiums and wildebeest crossing over narrow river straits follow similar patterns of movement. Group decision-making by flocks of birds, schools of fish, swarms of bees, and crowds of humans follows common patterns.

The tendency of some humans to group with others they resemble may be explained by culture or a hardwired preference for kin. But contributing to this may also be echoes from the oddity effect, an ancient animal instinct to avoid the notice of those who mean you harm.

The oddity effect in animals may also underlie a behavior seen in humans called appearance-based bullying. Common during adolescence, particularly during the earlier middle school years, appearance-based bullying involves shunning an individual who looks or acts odd. While the danger to the group wouldn't be predation, an odd-looking individual might attract unwanted attention or jeopardize the status level of the group. One of the ways adolescent animals demonstrate their inexperience is by looking different, by not having the awareness or ability to blend in with the group. As a fourteen-year-old boy once admitted to us, a secret for surviving middle school is: "just don't be weird."

Blending in, not standing out, slouching to appear smaller, avoiding eye contact (covering up with a hoodie or hair), these are all ways that humans, especially adolescents, hide within their groups. They may be ways of avoiding being chosen as targets. Knowing that could make a parent feel some compassion when their ninth-grader begs for the branded sneakers, T-shirts, or jeans that everyone else has.

Hiding, vigilance, the confusion effect—these behaviors all help animals avoid detection by predators. They can avoid detection in one other way that seems self-evident: by not going into the killing zone in the first place. Avoiding places where predators seeking easy targets hang out or have struck before, whether that's a particular section of a river, park, club, or campus, is a simple but powerful safety strategy.

But avoiding all risk is impossible, and so when they're forced (or choose) to be in dangerous areas, inexperienced animals should, according to UCLA evolutionary biologist and expert on animal fear Daniel Blumstein, "overestimate risk, limit their exposure, and be very careful." That takes us to the second step of the Predator's Sequence.

PREDATOR'S SEQUENCE STEP 2: ASSESS

Let's go back to the bat and the owl. Remember in this scenario you are a *Molossus molossus* bat, one of the world's fastest mammals but nonetheless vulnerable to predation by owls. Your group has just been detected by a barn owl. Now the Predator's Sequence moves on to step two: Assess.

As you fly along, the owl is evaluating you and your peers. She judges, weighs, calculates, and gauges your bodies and behaviors. You're a buffet of bats, but she can't eat all of you at once; how will she choose just one? This is a crucial early lesson for the predator-naive animal. The second step of the predatory sequence, how predators choose their prey, is full of possibility and heartbreak. Across animal groups, hunters look for easy targets, impaired individuals that can't or won't put up a fight, such as the young, the off-guard, the unaware, and the undefended.

This is a golden piece of knowledge for adolescents and their parents. As we noted before, a predator will calculate the costs and benefits of every attack before she invests. A biologist would say this barn owl is assessing your profitability.

Anything you can do to communicate to your predator that the final two steps of the predatory sequence—attack and kill—will be costly can encourage her to choose a different individual. It could even save your whole group by convincing her to move out of your area altogether.

Animals communicate their costliness all the time with a fascinating suite of behaviors called "signals of unprofitability." Signals of unprofitability send one specific message to a would-be predator: you will bankrupt your reserves of energy and waste valuable time by coming after me.

The attack step of the Predator's Sequence relies heavily on speed, stealth, and surprise. Especially for ambush predators, those that hunt by hiding and waiting, the element of surprise is perhaps the most important weapon they possess. But even coursing hunters, like wolves and orcas, which chase down their prey, benefit greatly if their victims never saw them coming. Breaking the advantage of surprise is an extremely powerful way for victim-prey to fend off an attack. And communicating to a predator that you're aware of its presence is something that reliably encourages predators to hunt elsewhere.

Signaling "I've spotted you, so you've lost your element of surprise" can be simple. For example, when they detect rattlesnakes, California ground squirrels stand up on their hind legs. Brown hares assume a similar posture to signal to lurking red foxes that they've been seen. An alert posture can be enough to convince the snake or fox to move along and find a less wary victim.

The "I know you're there" signal can be vocal, and quite complex, as well. Diana monkeys in Cote d'Ivoire, for example, have been recorded using different alarm calls depending on which of several predators is tracking them. If a Diana monkey spots a leopard or a crowned hawk eagle, she or he will release a long-distance call that both warns other monkeys and lets the cat or bird know that the surprise is over. Alarm calls are learned vocal behaviors, and animals learn them most intensely during wildhood.

A different signal of unprofitability goes a step further than simply letting the predator know he or she has been spotted. "Quality advertisement" signals tell a would-be attacker that the prey is in excellent physical condition and will be really hard to run down and kill. In other words, the prey communicates, "Come after me and you're going to waste your own energy and still have an empty belly. Better choose someone else."

Throwing an "I'm strong and not scared" signal of unprofitability is essentially what adolescent psychologist Laurence Steinberg recommends for human adolescents dealing with bullies. While being targeted by a predator versus a bully is different in many ways, the strategy Steinberg recommends in *You and Your Adolescent* is familiar: "if possible, look the bully in the eye and walk by him without engaging him."

Like detection signals, quality advertisements can be simple. In the presence of snakes, kangaroo rats start drumming their big feet. When they hear the sound, snakes give up their stalking. Spotted skunks stamp their front feet. Before kangaroo rats and skunks have learned how to foot-drum, learning that intensifies in adolescence, they're at a disadvantage.

Another unprofitability signal, called "stotting," combines detection ("I see you") and quality advertisement ("I'm strong and healthy and can outrun or outsmart you"). The classic example of stotting is seen in Thomson's gazelles. Picture a long-legged, tan gazelle with a black-and-white stripe on each flank and two graceful ridged horns spiraling up from its head. Have you ever seen one of these gazelles seem to spring straight up and then bounce, stiff-legged, like it's on a pogo stick, across the savannah? This strange gait, called "pronking," seems to serve no other purpose than as a wonderful way for Thomson's gazelles to signal to cheetahs that they are not worth pursuing. It's a demonstration of excess energy, of youthful exuberance—and that can be very off-putting to a middle-aged predator who's just looking for a light lunch.

Skylarks, for another example, sing loud and complex songs when they escape peregrine falcons. Their unique escape song lasts thirteen seconds, and they sing it only at the start of their grueling ascent two hundred feet into the sky—never when they're just sitting on a perch. Why would a skylark choose the moment of escape, when it needs every bit of energy and focus, to start singing—and to start singing as loudly and as perfectly as it can? It's because falcons usually give up the chase when they hear the song. If they don't hear the song, or when it's sung badly, they're more likely to keep up the pursuit. That's because "only a very fit skylark can afford to sing while being chased by a predator," said the biologist who studied this phenomenon. Adolescent skylarks are at a disadvantage here because they're less physically strong than fully mature adults and they've had less time to practice and perfect the song. If you're a skylark, you're better off singing beautifully, but if you can't sing, you should crouch and hide rather than mounting an aerial escape.

The skylark's stotting song is like the exhilarating extra speed Usain Bolt lays on around meter seventy of his hundred-meter dashes. Who in his right mind would continue pursuing Usain Bolt when he makes a hard task seem easy? Another example of stotting occurs in African antelopes called klipspringers, which are primarily targeted by jackals. Like skylarks, when a klipspringer senses its predator nearby, it starts to sing. But klipspringers aren't soloists. They live in monogamously mated pairs, and their stotting signal to predators is a duet. Alarm duetting shows a predator that these two antelopes are fit, have great aerobic capacity, and have backup. The predator should look elsewhere. The behavior protects the two antelopes and secondarily bonds them.

Klipspringer mates tend to stay close throughout their lives, and they begin learning and practicing the songs in adolescence.

Pronking, defensive singing, alarm duetting—these are all ways that wild animals communicate with their predators in order to remain safe themselves. Adolescent animals stot to project a confidence they might not yet have, essentially faking it till they make it.

For humans, stotting means signaling that pursuing you will be more trouble than a predator wants to go to. This can be as literal as walking with a scary guard dog, posting a notice that your home is alarmed, or flashing a weapon. The shields and intimidating helmets in the Peabody's *Art of War* exhibit had elements of stotting to them—the wardrobe of combat is not simply for physical defense. Stotting for adult humans can mean lawyering up or associating with powerful groups. Computer encryption is an especially modern form of stotting; most hackers say that when they encounter significant encryption they go elsewhere. Locks on doors, bars on windows, even if not foolproof, signal to would-be robbers that this is a harder target.

There are many signals of unprofitability that adolescent humans can use to stay safe, such as downplaying adolescent gawkiness or giggliness, making oneself appear larger or older, even pulling out a phone that could be used to place an alarm call. Dampening the startle response by becoming unflappable and seemingly fearless—all these signal unprofitability to would-be attackers. A swaggering group of adolescent boys may annoy or intimidate an adult, but those adolescents might be keeping their own fear in check. However clumsily or over-expressively they're doing it, part of their behavior may be self-protective, a human version of the animal urge to stot.

PREDATOR'S SEQUENCE STEP 3: ATTACK

Bat-you has been unlucky enough to have been detected, then assessed, and chosen. As a bat with tremendous speed, you still have a fighting chance. But you are now in the more dangerous second half of the Predator's Sequence, and the hunter has the upper hand. As prey, this third phase, Attack, puts you in last-resort mode.

Managing this part of the Predator's Sequence requires strength, size, and brains—from both parties. Hunters need to be physically capable of wrestling or incapacitating their chosen individuals. Their physical skills need to be matched with mental stamina to overcome the profound energy of another animal fighting for its life.

The hungry barn owl swoops in. It's likely you won't hear a thing. Barn owls are one of Earth's quietest predators, with splendid, angel-like wings that are remarkably maneuverable, with specialized super-soft feathers that cut sound. Your attacker's silent movements are smooth and undulant overhead, not jerky and flappy and loud, like other birds of prey. Her disc-shaped face functions like a satellite receiver. Working within that concave dish, her ears, one placed higher than the other, pull in sounds that create a 3-D image of you and your location in her mind's eye. Once the owl's audio equipment is locked on you, it's very hard to shake her.

As prey, your mandate is to terminate the interaction as soon as possible. But what to do now? Sensing the imminent attack, your bat heartbeat accelerates, powering your muscles for one of two famous fight-or-flight responses: a physically vigorous counterattack on the owl or a speedy and successful getaway. But a third possible response exists for prey species, including mammals as well as birds, reptiles, and fish: a sudden, dramatic slowing of your heart rate, leading in essence to a faint. This seemingly paradoxical cardiac reflex suddenly reduces the blood pumping to your brain, causing your body to become motionless. For predators who track their prey by their noises and movements, complete stillness becomes a perfect audio camouflage. Human beings retain this fear-triggered slowing of the heart, and it underlies the most common type of fainting in adolescents and young adults.

Now the attack is under way; the owl is racing toward you. Will your heart rev up so you can fight the owl or fly away? Will your heart plunge and send you into a freeze or faint? We'll find out your fate in a moment, but the odds are generally against you. Barn owls have impressive kill rates, around 85 percent. On the other hand, most predators have much lower odds of success. Tigers make only one kill for every twenty attempts. Polar bears do a bit better, with one for every ten. Leopards and lions are much more successful, but still make only one kill for every three or four tries. These facts alone could comfort any creature that finds itself the object of a predator. As Yogi Berra liked to say, it ain't over till it's over.

PREDATOR'S SEQUENCE STEP 4: KILL

To end the Predator's Sequence, the hunter has to kill her victim. This requires a practiced technique. A swift kill minimizes energy waste, as well as the chances that her victim will escape or that other animals will come to the rescue. It also protects the attacker herself from getting hurt if the animal fights back. For prey, the kill phase pretty much means game-over, which is why nature videos showing young animals escaping with their lives can be popular—they're rare.

The barn owl clamps on to your bat body with her talons. Normally her next move would be to squeeze the life out of you with her claws as she flies somewhere to pick you apart with her beak. But this time, bat-you gets lucky. Your speeding heart may, as it did during the attack, suddenly decelerate, causing your brain to lose its blood supply and your body to go limp. In response the owl releases her grip slightly, and suddenly your heart revs. Making use of every bit of your hundred-mile-per-hour flying ability, you speed off to safety. You're wounded but have gained crucial knowledge about how barn owls hunt.

Being disproportionately targeted during the predatory sequence is part of the vulnerability of wildhood. But nature has endowed her young creatures with ways to stay safe. Some they're born with. Some come from being part of a group. And some, like stotting and other signals of unprofitability and quality advertisement, can be learned.

But these essential lessons about your predator's moves, abilities, and weaknesses are only available to young bats like you who venture out of protected places. A bat, or any other animal, can't deeply learn to understand danger by hiding somewhere. And that brings us to the most thrilling—and most dangerous—moment of emerging as a young adult into a new and exciting world: facing your fears.

Chapter 4

The Self-Confident Fish

Ursula has little choice about how she leaves home. For thousands of generations, adolescent king penguins have simply jumped naively into the deep end and relied on trial and error to make it through the leopard seal gauntlet. Trial-and-error learning is a potent teaching tool. Gaining the confidence to face threats by yourself is a crucial part of wildhood: learning how to be safe in the world is a necessary step on the way to becoming an adult animal. And much of that learning can't happen without trying and sometimes failing.

But for an individual, failing only works as a teaching tool if it doesn't ruin the rest of your life. Surviving physically is essential. In the human world, social dangers often stand in for physical dangers, and people can get seriously harmed through blows from crime, drugs, and even misuse or missteps with social media. Failing reputationally or legally can ruin your life, so blind trial-and-error learning in those areas can be dangerous too. Modern human adolescents, like their animal counterparts, need to be able to try and sometimes to fail. But today's high-stakes testing and the potential for indelible damage through social media have made this kind of learning riskier to try.

Parents who are queasy about throwing their emerging adolescents into dangerous situations might be relieved to know that many animal parents are cautious too. Gazelles and monkeys don't just leave their adolescents to the cheetahs and the snakes. In the wild, trial-and-error learning sometimes begins with parents or other trusted adults showing the way at first.

As noted by the biologists Bennett G. Galef, Jr., and Kevin N. Laland: "A naive young animal, newly recruited to a population and faced with . . . challenges, would be well advised to take advantage of opportunities provided by interaction with adults of its species."

PREDATOR AVOIDANCE TRAINING

Learning that predators are a fact of life can begin from the first moments of a young animal's life. Although Ursula's parents couldn't teach her how to deal with a leopard seal, they likely did demonstrate proper reactions to attacking seabirds.

Some parent-provided training is just leading by example. Offspring observe how their parents deal with a particular danger and absorb the lessons. Ring-tailed lemurs, for example, carry their young on their backs into stink battles—the turf fights they have with rival gangs of lemurs that are waged by pumping scent from specialized glands at the enemy. From a young age, those lemurs understand the smells, sounds, sights, and motion of this (admittedly nonpredatory, although conflict-laden) adult encounter.

When a wolf attacks a mother bison and she weaves and lunges to protect her calf, the calf is observing the sights, sounds, and smells of the predator as well as the moves her mother uses in response—all while still under her mother's protection. Bison mothers also learn from terrifying—and especially traumatic—experiences with predators. It makes them fiercer. Bison mothers whose calves were killed by wolves were 500 percent more vigilant with future calves than mothers whose offspring hadn't been attacked. The tragic experience made them safer, more informed mothers.

ALARM CALLING

When animals spot a predator, they may produce sounds known as "alarm calls." Throughout the animal kingdom, alarm calling serves a triple function: It warns other community members. It summons help. And it puts a predator on alert that he has been spotted. As we've seen, nothing ruins a predator's attack plans like taking away the element of surprise.

Offspring listen to their parents' alarm calls, learning to discern among the dangers they announce. Nestling Japanese *Parus major* birds learn different alarm calls for a jungle crow and a rat snake. If their mom calls out, "Look out; it's a jungle crow!" the nestlings crouch in place. If she signals, "Hey, it's a snake!" they flee the nest.

Of course, young animals are also learning from their elders how to make alarm calls themselves, in order to scare off the predator and alert others that they need help. The NCMEC study on child and adolescent abductions, discussed earlier, showed that a successful tactic for preventing abductions was screaming and making noise. The study found that nearby adults were vastly more likely to take notice and help when youngsters vocalized—and that those incidents were much more likely to end in arrests of the offenders.

However, as adolescent animals are learning the intricacies of different warning sounds and when to make them, they can raise false alarms. Because of this, adults often pay little attention or even ignore the calls of adolescent and young adult animals. Adult sea otters, for example, don't budge when adolescents alarm, but they respond when another adult raises an alarm about an incoming great white shark. Just like human parents, animal parents have to ascertain when to rush to their children's aid and when to understand that they are simply practicing the noises that may someday save their lives.

MOBBING

Alarm calling to danger is a reasonable defensive strategy. But in the wild, sometimes the best defense is a strong offense. If you've ever seen a group of songbirds dive-bombing a cat, crows attacking a hawk, ground squirrels staring down a snake, or even a bunch of dogs barking together contagiously when the doorbell rings, you've witnessed a very effective antipredator strategy: mobbing. When a group of animals pulls together and makes a big commotion to intimidate a predator, they are mobbing. Mobbing involves the whole community. It's loud. The birds, primates, and other mammals who do it scream and shout and squawk and bellow at the intruder. They run and dive-bomb and charge the predator. Mobbing is not subtle.

Sometimes the predator gets hurt or killed by the mob, and sometimes the predator fights back. Sometimes the predator attacks anyway and individuals in the mob are hurt or killed. But the most common outcome of mobbing is that the predator gives up and slinks away to look for easier prey elsewhere.

Mobbing is a very effective signal of detection—nothing ruins a predator's element of surprise like a bunch of angry adults yelling and screaming and running at him just as he was assessing the profitability of attacking their offspring. And it's a great signal of unprofitability. Only the hungriest or hardiest predator would continue an attack against a group of grown animals when it could find an unattended adolescent instead.

If you ever get a view of animal mobbing, look closely at the individual mobbers. Most include some adolescents and younger adults. Mobbing is on-the-job training with real-world danger. Being allowed into the group defense ritual gives adolescents and young adults valuable, hands-on experience identifying predators and taking action against them. So mobbing is performed for intimidation and safety, but also serves a teaching function. And studies of fish, birds, and mammals have shown that young adult animals given the opportunity to mob with their parents and older group members have higher survival rates than their inexperienced peers.

Mobbing is an important tool for young humans to learn too. When people gather together to protest authority, they are mobbing. Gandhi's Salt March, the storming of the Bastille during the French Revolution, the Selma to Montgomery marches of 1965, and the Estonian Singing Revolutions of 1986–1991 are all examples of vulnerable individuals who together became a force to be reckoned with. In fact, freedom of assembly is, in essence, the right to mob. And whether an adolescent meerkat is mobbing a cobra or a teenager is marching on Washington, if they are doing it alongside parents and grandparents they are learning something crucial: how competent adults effectively confront powerful foes.

PEER PRESSURE AND RISK-TAKING

When Atlantic salmon hit adolescence at two years old, they must undertake a harrowing journey. These so-called smolt leave their familiar home rivers and streams and travel hundreds of miles to the ocean. And they have to do so while not getting attacked and eaten by a progression of ravenous predators lining their river route. Under the water lurk cod and eel. From the air swoop birds of prey, in particular one called a merganser, whose beak stabs like a javelin. From the riverbank, bears swipe at the salmon with their claws.

If they do survive the river journey, they'll encounter new predators hungrily awaiting their arrival in the open seas—larger striped bass, cod, sharks, toothed whales. For up to four years the young salmon will grow and mature in the ocean. During this time and when they return to their native rivers to breed at the end of their lives, they are especially vulnerable to one predator, smarter and deadlier than all the others combined. If you enjoy lox or salmon steaks, you are that predator. Salmon are huge business.

Salmon farmers make their livings by helping salmon avoid natural predators and live long enough to become enticingly large targets for our species to catch, sell, and eat. In learning how to protect salmon from natural predators, these farmers and the scientists who work with them have learned some interesting things about how young salmon in the wild defend themselves.

A common farming practice is to scoop just-hatched, wild baby salmon from a river and move them into tanks for a couple of years until they're the age when they would normally start their adolescent voyage. At that point, salmon farmers dump the smolt back into the river and, crossing their fingers that they'll safely return, allow them to swim off downstream to the ocean.

There's a problem with this, though. Farm-raised fish are 100 percent predator naive. They've never seen a cod or eel in their lives. They've never sensed the shadow of a merganser overhead moments before its deadly bill stabs the surface. They've never felt the ripple of a bear claw slicing through the water.

And so it's probably not surprising that farm-raised, predator-naive adolescent salmon making their ancestral journey downriver die at significantly higher rates than the smolt who were raised in the wild. Hand-raised fish, the hothouse flowers of the aquatic world, are in fact so easy to catch, so obviously unable to handle life in the wild, that the predatory cod, eels, birds, and bears have learned to wait for them when they're released into the water every year.

Swedish and Norwegian scientists decided to see whether hatchery-raised salmon could be trained to do better against predators in wild environments. They divided a group of salmon smolt into three groups. Group 1 was placed in tanks containing freely swimming predatory cod. These smolt were left to fend for themselves, with "direct experience of the hunting predator," in the words of the scientists.

Group 2 was also placed in a tank with cod, but with a crucial difference: a transparent net ran down the center of the tank, with the cod sequestered on one side of it. The young salmon could see their deadly foes. They could smell and hear them. They could feel, literally on their skin, how the cods' movements rippled and folded the water around them. They had the chance to pick up on the cods' daily hunting rhythms. But the smolt were in no danger of actually being harmed. The cod couldn't attack them through the net.

Group 3 was held in happy innocence in regular fishery tanks, with no predators of any kind.

Over the course of the experiment, the scientists observed a number of responses that the young fish had to the predator cod. Groups 1 and 2, the fish housed with cod, kept their distance from the predators. Whether young salmon accidentally approached a cod or the cod approached them, a universal response was for the salmon to move away. (The fate of the predator-naive salmon of Group 3 wasn't good—more about this in a moment.)

The "cod-smart" salmon from Groups 1 and 2 exhibited three getaway strategies.

One response was to swim away from the threat as fast as they could, with no regard for their surroundings. The scientists called this flailing form of panicked swimming "wobbling." The movement pushed the fish to the surface of the water, leaving them exposed and imperiled. Wobbling is a telltale sign of immature, inexperienced fish.

Other fish in Groups 1 and 2 had a completely different reaction. Instead of flailing at the top of the water, they dove to the bottom and stayed still. This behavior is called "freezing," and during the Scandinavian study, a number of the fish engaged in it every time they were attacked by cod.

When they weren't wobbling or freezing, the young fish exhibited a fascinating third option: they banded together with their peers. Sensing danger, these fish would suddenly all point their heads in the same direction, decrease the distance between their bodies, and coordinate their movements as precisely as a drill team. This swimming behavior, called "shoaling," appears to be completely instinctive. Other studies have shown that many fish are born with this reflex. But there's a crucial secret to successful shoaling: although the behavior is instinctive, it can't be sparked unless a fish practices with other fish. Fish that are raised apart from other fish—as well as fish raised without predators to trigger the instinct—don't develop this fundamental, lifesaving skill set.

Alone, like a single hand clapping, a solitary fish couldn't do it. But the smolt in the study didn't need to be in a big group to get the safety training—and the future benefit—of shoaling. Just a single other fish ignited the shoaling behavior.

The U.S. Navy undertakes similar training when it hones this instinct in aeronautic teams like the Blue Angels. Individual pilots can practice on their own all they want and be the best in their class on the simulator.

But how do they learn to fly in formation under real circumstances? They have to do it for real, and they have to do it with other fighter pilots.

Group shoaling is widespread throughout fish species. As they get the hang of it, they start learning ever more complicated combinations that can be so distinctive and elaborate scientists have named them. Fish do the hourglass, the skitter, the Trafalgar, the jigger-pole, and the doodle-sock.

The urge to move together in synchrony in response to danger is also seen in birds: clouds of starlings create their swooping patterns in response to predatory raptors in the vicinity. Mammals form herds and stampedes in response to nearby dangers. Dolphins traveling in tightly coordinated groups are more likely to win fights with rival dolphins.

Humans share with other animals some incredible examples of coordinated physiology. Researchers tracking the heartbeats of singers in a chorus found that their cardiac rhythms became synchronized. Similar physiologic synching has been demonstrated in studies of dancing partners, soccer teammates, even patients and their therapists. Being part of a group literally changes the physiology of the individual members—transforming them into a new, and often more effective, collective organism.

THE POWER IN GROUPS

We can't ask dolphins and starlings what it feels like to be a single individual schooling and flocking through water or sky. But we can ask people. Does moving in unison produce an emotional reaction in humans?

Anthropologists at UCLA led by Dan Fessler, an expert on the evolution of behavior and culture, wanted to test the idea that moving in a coordinated way with other people could change the way those people felt. In 2013, they recruited ninety-six undergraduate men and set them a simple task: walk an eight-hundred-foot route around UCLA's Pauley Pavilion with another man. Half the group was instructed to march in lockstep, matching their stride with their partner's. The other half got no specific instructions. They were simply told to accompany their partner. All the pairs were told not to talk during the walk.

Unbeknownst to the participants, their walking partners had secret agendas. They were fellow UCLA male undergrads, but they were also members of the anthropology department studying with Fessler.

When the men returned from their short walks, Fessler and his team showed them a photo of a mature man with an angry expression. The experimental subjects were then asked to guess how tall and muscular the man was.

Fessler and the team were testing the men's sense of formidability. And they got clear and interesting results. The men who'd walked in unison with their partners judged the man in the photo to be smaller, less muscular, and less imposing, whereas those who'd walked without coordinated purpose found him more physically intimidating. It wasn't just the presence of a companion. It was engaging in coordinated movement with someone else that produced the greater feeling of strength.

"The ability to move in unison indicates that one is part of an effective fighting alliance," said Fessler. "That's no accident. In order for individuals to be synchronized, they have to be motivated to coordinate their behavior—they have to be paying attention to what one another are doing, and they have to be skilled and competent. A deep part of our brain registers this connection."

Fessler's work has revealed other effects of synchronized movement. Not only do men feel stronger and more invincible while moving in unison, but their opponents also *perceive* them to be more formidable.

Another of Fessler's findings was somewhat more troubling, though not entirely surprising. The powerful confidence arising from being part of a synchronized group cuts both ways. The group strength and protective sense of invulnerability could also tip into abuse of that power. Moving in coordinated ways with other men—such as riot police marching on protestors, and vice versa—can make men more likely to use violence. In Fessler's words, walking in sync may make men more likely to think, "Yeah, we could take that guy."

Adolescent salmon learn to shoal defensively, in order to protect themselves from predators. The Scandinavian salmon study didn't look at whether shoaling also increased aggressiveness in the fish, the way it can with groups of men moving together. But this is where the salmon experiment gets even more interesting. Remember that in this study were three groups of salmon: Group 1 had direct experience of the hunting cod; Group 2 were in the water with the cod but protected by a net; and Group 3 had no experience of hunters at all. It turned out that Group 1 were the best shoalers when tested in freshwater and salt water. These were the fish that had had direct contact with predators, and they were better able to band together with other fish. Not only did predator-exposure teach them about staying safe from the thing that could kill them, but it also improved their social functioning.

In a refreshing departure from academic language, the Scandinavian researchers described these young fish as having "self-confidence." The fish who'd had direct experience with the thing that could kill them had more of it. "Positive outcomes from threatening and dangerous situations," they found, greatly improved the fishes' future safety as adults.

The predator-experienced fish in Group 1—cod with no net—were far and away the most self-confident. Although some did get eaten, the ones that survived were the quickest to work with their peers for group defense, and the safest overall.

Group 2, the net-protected salmon, showed a limited ability to shoal in freshwater and zero tendency to shoal in seawater. These predator-aware fish—having had some experience but protected by the net—showed only the beginnings of safe antipredator behavior outside the tank.

The completely predator-naive fish, however, Group 3, who'd never encountered cod, were in big trouble and experienced the worst outcomes when they were returned to the river. These sheltered individuals had lived their equivalent of tween and early-teen years blissfully unaware of the ravenous fish, birds, and bears that specifically target them. Without experience, these predator-naives overreacted and responded inappropriately. They "wobbled" more in the water than predator-experienced fish. Or they reacted not at all, a nonresponse the scientists concluded was a form of physiological stress, a kind of inert freak-out we might call deer in the headlights or even a panic attack, which made them easy targets for the cod.

This story has two central lessons. One, to become safe, an animal must encounter danger. The smolt with no predator experience fared the worst. The smolt who'd gotten at least a look at the cod through the net fared better. But by far, the fish who'd experienced actual encounters, even near misses—who'd felt danger in their bones and muscles—were best prepared for safer salmon adult life.

The other lesson is to not be isolated as an adolescent. Peers can help peers build that feeling of confidence. They literally activate in one another the lifesaving skills of working as a team. They offer one another opportunities to practice this skill. Isolation may keep a young animal safe temporarily. But animals growing up without peers cannot learn the safety skills they need to function in the real world. And as we'll see, this finding holds true for young animals beyond fish.

The lessons learned or not learned by the salmon in the study make for a powerful message. Young animals who spend time with their peers get safer. The reason is very simple: because they can glean information about opportunities and threats by watching their peers' successes and mistakes. Although they can cause worry and grief to the adults in their lives, the risks that adolescent animals take together are some of the most useful and important experiences they'll have.

Chapter 5

School for Survival

Whether Ursula was among the first penguins to jump into the icy water that fateful December morning isn't known. But groups of penguins typically wait at the water's edge, watching to see what happens to the first penguins who dive in. If a leopard seal appears to attack their peers they hold back, delaying their entry. Penguins, like most animals, learn a great deal about staying safe from the other penguins they spend time with. Social learning with peers is one of the most powerful educational tools on the planet.

Adolescent guppies from a river in Trinidad were taken out of their safe, low-predation river and placed in one teeming with predators. As you might expect, the predator-inexperienced guppies fared poorly compared with predator-experienced, river-smart fish. However, when the naive fish were given a chance to watch their more experienced peers confronting or evading predators for a while, they got wiser. Before long, the fish who had never had direct experience with predators demonstrated better anti-predator behaviors too. These guppies made themselves safer by watching, hanging out with, and in effect learning from more experienced peers.

Animals that have the chance to learn from more experienced peers also begin communicating better about danger. They move from responding individually to the call of a parent to communicating with one another. Of course, friends can also provide valuable examples of what *not* to do. Watching bad things happen to its cohorts provides a fish, bird, or mammal with lessons it can't get anywhere else.

In April 2017, a group of students admitted to the 2021 Harvard class-to-be received an email. The admissions office had identified them as having traded "sexually explicit memes and messages that sometimes targeted minority groups" on a Facebook messaging site. At least ten of these students had their Harvard acceptances revoked.

Some might say these aspiring college students had been naive in addition to insensitive. But as accomplished graduating high school seniors in 2017, they were likely well versed in the real-world consequences of online harassment. They did it anyway.

While their decisions to post didn't kill them, their online behavior changed their lives forever. But it didn't just change their own lives. It served as a public example to aspiring college applicants and other adolescent users of social media who heard about the story. Watching the worst happen can be a grim but powerful lesson.

The most dramatic and often tragic way adolescent animals learn from their peers' negative examples is by being on the scene when a group member gets killed. Immediately following a fatal crash involving a friend or classmate, adolescents drive more carefully, according to therapists who specialize in this age group. Though one might wish they never had to face this kind of loss, witnessing or even hearing about a teenage peer whose life has been cut short can serve as a lesson for predator-naive humans, too, whether it's learning to be safe around cars or fires or thinking twice about using alcohol or drugs.

One study of starlings showed that observing a peer struggle with an owl taught adolescents to avoid them. And even just watching their peers be scared—without the presence of a predator—sped up adolescent fishes' ability to recognize threats. Other senses can add to the lessons. A combination of scent molecules called "schrekstoff" is released from the torn skin and scales of injured fish. When they're close enough in the water, fish can literally smell a doomed peer's fear and tragic end. It is from other fish that they learn what could kill them.

It's understandable for human parents to worry about the influence of peers on their offspring. Peer pressure sometimes does lead to regrettable and dangerous decisions. But social learning from peers can also teach adolescents valuable lessons they can't obtain elsewhere.

Parents can't impart some lessons for a simple reason—they're too old. Their age and good sense make them far less likely to go off and do something stupid that their kids can witness, be scared by, and vow to do the opposite of. This is particularly relevant for modern parents whose adolescent children are exploring digital worlds that didn't even exist when they were coming of age. And it makes peer learning all the more important. Even for naturally conscientious, risk-averse adolescents—ones who avoid riding motorcycles, say, or don't impulsively post on social media—witnessing the real consequences of misguided behavior can serve to confirm their aversion and continue to keep them safe.

Peer pressure is often cast as a frightening and negative influence in the lives of adolescent and young adult humans. But reframed through a different lens, one of animal social learning, it begins to shift from a cautionary concern to a universal behavior, an invaluable strategy for teaching adolescent animals to be safe.

PREDATOR INSPECTION

In the wild, brushes with death are an unavoidable reality. Given that danger is ubiquitous, why do adolescents, who are already naive, inexperienced, and vulnerable, increase their jeopardy by taking unnecessary risks?

Simply answered: in order to become safer adults. Many animal adolescents are biologically primed to move toward—even sometimes seek out—danger as a way of learning what it is and how to avoid it. The term of art for this counterintuitive behavior is both clinical and evocative. It's called "predator inspection."

Like human adolescents who underestimate risk, animal adolescents lack the experience to recognize and assess threat. Predator inspection is one way of getting that experience.

Remember the *Molossus molossus* bats chased by barn owls? When those bats realize they've been detected by predators, they send out distress calls. Upon hearing alarm calls, most bats self-protectively fly away. But not the adolescents and young adults. They fly straight *toward* the danger.

This behavior was observed by scientists at the Smithsonian Tropical Research Station at Barro Colorado Island, Panama, who played recordings of bat distress calls and saw what happened. The adolescent and young adult bats that flew toward the danger, the scientists concluded, weren't doing it to help. This wasn't altruism. The adolescent bats were embarking on inspection flights. And the object of those inspection flights was to gather information about their biggest lethal threat—something so dangerous it made adult bats cry out in fear and warning—their owl predators.

With predator inspection, peer influence can play a key role. And grouping with peers who are more experienced may be the ultimate safety learning strategy. One study of juvenile minnows inspecting a simulated predatory pike fish showed that by themselves, minnow individuals weren't willing to get close. But as part of a group they moved in and were able to get a better look.

When those fish, with their fresh-gained knowledge of pike, returned to the larger group, they were changed. They were more aware, and more vigilant. Their feeding behaviors went from indiscriminate to more cautious, and they became more physically active. They were no longer predator naive. Experienced, they came back to a group of inexperienced fish, and that's where something even more fascinating happened: the naive fish started to act like the inspector fish. Even though they hadn't seen the predators with their own eyes, the innocent fish acquired knowledge from the returning inspectors. They benefited from spending time with the fish who had taken the risks and had survived.

Predator inspection behavior has been studied in fish, birds, and many hoofed animals. Slender Thomson's gazelles skip up to hungry cheetahs, inquisitive meerkats gather in a cobra's strike zone, or California sea otters roll up to great white sharks. While there are differences in how it's done, three things about predator inspection are true for all the species that do it. First, adolescent and young adult animals have a greater propensity to engage in predator inspection behavior. Second, predator inspection is dangerous, and inspectors sometimes die. Not surprisingly, adolescent inspectors are at the highest risk. In one study, adolescent Thompson's gazelles who moved toward a cheetah to inspect were attacked and killed in 1 out of 417 approaches. Adult gazelles, on the other hand, were more than ten times safer, eaten only once every five thousand times that they approached cheetahs. Third, and finally, while it is clearly dangerous, provided they survive it, predator inspection makes animals safer in the long run.

The occurrence of predator inspection in so many species suggests a function and purpose for it. If it only resulted in death, it would soon disappear as a behavior. One reason it may improve safety is that when prey animals approach their predators, it often causes them to retreat or leave the area. But for young, nonhuman animals, predator inspection has an additional and special function: it equips them with crucial information about dangers in their environments. It gives them direct experience with a predator. It is an engineered near miss.

If learning about the dangers of the real world underlies the animal behavior of predator inspection, then human adolescent attraction to adult activities, even before they are equipped to handle the dangers, may also be driven by a human version of predator inspection. Without understanding the adaptive benefits of predator inspection, we cannot see its functional roots in the natural world. And we cannot fully understand why human adolescents do it. Not all adolescent risk-taking is a rebellious attempt at individuation. And not all risk-taking is to be avoided at all costs.

Approaching and inspecting danger, even the very predators they've been warned about, could be one way of understanding why teens get fake IDs and sneak out in the middle of the night to go to bars and nightclubs. Like their adolescent animal ancestors, they may seek to encounter the very predators from whom their parents and communities are trying to protect them. Horror films have been surging at the US box office, and the average audience age is younger than for other genres of film.

Adolescent morbid fascination with and even attraction to the macabre and violent—whether it's terrifying true-crime stories or the death-defying plunges of a roller coaster—may be, in part, an expression of predator inspection and social learning in modern humans. Interest in scary media has been understood by social scientists in more anthropological terms, as a modern coming-of-age experience. Youth can demonstrate their ability to maintain their composure and handle simulated danger, displaying adultlike self-control to their peers. Coming-of-age rituals from around the world have included similar demonstrations of courage. Young Sumatran Mentawaian women undergo the painful process of having their teeth sharpened, and Amazonian Satere-Mawi adolescents wear a glove packed with thousands of painfully stinging *Hormiga veinticuatro* ants for ten minutes. Not flinching in the face of fear is part of the process.

But adolescent attraction to scary media may also have connections to antipredator behavior. Learning to stifle a startle reflex in order to avoid sparking the Predator's Sequence could have protective benefits. But a more powerful possibility is that fascination with crime, gore, and adult content in general is driven by an innate adolescent instinct to learn about the dangers of the world.

Whether on a computer or a movie screen, through an earbud, or on the page, young adults can come face-to-face with the leading fears of contemporary humans—serial killers, mass murderers, climate catastrophes, drug addiction, terrorism. Adolescent attraction to immersive video games, too, may be connected to the reasons adolescent animals are drawn to cheetahs and owls. Although simulated, these games give contemporary teens close-up exposure to death by gunshot, bombing, torture, needle, and high-speed road crash.

Predator inspection allows an individual to learn about something dangerous without actually experiencing it. Adolescent interest in the realities of the adult world may be less a loss of innocence, and more a gaining of lifesaving safety knowledge.

BEYOND SAFETY: HOW TO SURVIVE

"The juvenile male was lean, in poor condition, and had minor injuries. Its hindquarters were covered in stalked barnacles, which suggest that it had spent a long period at sea." So reads the description of a wayward adolescent leopard seal, one of hundreds of leopard seals that failed to survive the launch from their childhood home around Antarctica in September 2006.

Before they start terrorizing penguins like Ursula and her peers, leopard seals begin life as soft, defenseless, cuddly seal pups. Like other mammals, they nurse and snuggle with their mothers, getting protection and life lessons at Mom's warm and furry side. A little later, these young seal pups grow into predator-naive, inexperienced, vulnerable adolescents. Leopard seal adolescents don't know the four life skills either at first. It's not easy to be a big-bodied, inexperienced, vulnerable leopard seal trying to learn how to stay safe, navigate hierarchies, communicate sexually, and fend for him- or herself. And like the penguins, bats, gazelles, and white-tailed deer that succumb to predators, the leopard seals, owls, cheetahs, foxes, and wolves that hunt them are in lethal danger during inexperienced adolescent wildhood, with their big bodies and small know-how.

The fates of leopard seal adolescents point out an interesting ecological fact. Although predation can be devastating and deadly, the true threat stalking most animals during wildhood is starvation. A leopard seal that doesn't learn to hunt will not live through adolescence. Neither will Ursula or any other penguin in the wild. Adolescent predators and prey share this common foe. As we'll see in Part IV, fending off starvation determines survival at every level from the immediate to the future, from their first independent meal to their last. How well they learn to forage and hunt during wildhood is crucial.

Hungry animals take more risks. They venture out of hiding places driven by it. They are forced to eat lower-quality food if they don't know where to find better options (or are kept from them by stronger, older group members). They can also ignorantly eat poisonous food. And the hungriest animals are often adolescents.

After all this talk of predators, you might be surprised to find out that the penguin study that tracked Ursula wasn't focused on how penguins avoid leopard seals. In fact, it was designed to study how and where dispersing young penguins learn to feed themselves. As the scientists noted, dispersal is "a period when individuals may be susceptible to increased mortality given their naive foraging behavior." Klemens Pütz and his international research team concluded that "inexperienced king penguins . . . develop their foraging skills progressively over time." In other words, they have to learn how to feed themselves. And interestingly, for penguins, that learning happens with their peers, without adults around. Ursula and Tankini and Traudel and the rest of the adolescents went off on their own to learn how to eat, honing their skills among same-age peers before having to compete for resources against mature adults. Pütz told us, "They really have to learn; they have to develop diving behavior. They have to develop foraging and physiologic skills like holding their breath and surfacing."

Leopard seals may hunt penguins, but penguins are fantastic hunters themselves. Still, getting good at nabbing fish and scooping up krill takes long months of practice. Becoming a successful, mature hunter who can provide good food for oneself and eventually others does not happen overnight.

EVERY ENDING IS A NEW BEGINNING

There's no way of knowing what Ursula's parents felt when she took the plunge. Bill Fraser, an ecologist and penguin expert, says the penguin parents he's observed barely even look up when their offspring go.

But take a moment to consider that, just like a human child venturing away from home for the first time, there was a moment when Ursula was living with Mom and Dad, receiving food literally beak-to-beak, and huddling close to them for protection from the cold and from ravenous seabirds. Then came the moment when she jumped in the water and swam away. Something happened in between. As far as we know, Ursula's parents didn't have a penguin version of pride watching her depart, something like, "Look at her go; she's got this!" Her parents didn't turn a critical eye on her diving technique, like a human parent critiquing their student driver, "You should've tucked a little closer on that turn." And they certainly didn't stand at the edge of the water, like human empty nesters waving goodbye to their kid—at once hopeful and bereft—and gaze at Ursula's feathered shoulders moving toward the water's edge, shrinking smaller and smaller, feeling their parental hearts constricting as a voice in their heads silently begged: *be safe, be safe, be safe.*

A painful truth about parenting is that once offspring have moved beyond their watchful and protective presence, mothers and fathers have little control over the fate of their young. Parents do not have the power, either mental or physical, to protect them forever. And this truth is further sharpened by the irony that in order for adolescents and young adults to learn to be safe from danger, they must plunge toward it. In fact, protecting them too long from learning about predation, danger, and death could actually be the worst mistake an animal or human parent can make.

Being raised in an overly protective environment deprives young animals of learning the skills they need to feel safe as adults. Something is lost when opportunities to experience danger are taken away. Growing up will be dangerous, but both parents and children have to let it unfold. The alternative is ending up a predator-naive adult, which can be even worse.

For human parents, the tension between continuing to offer care and protection and encouraging independence by withdrawing support can be confusing at best and catastrophic at worst. The competing parental impulses to over- or under-protect are rooted in universal, juxtaposed facts of adolescent children's limited experience and grown bodies. There's a lesson that human parents can extract from animal life, which is this: no one knows the exact amount of protection to provide in every situation. Rather, the ideal amount must be calibrated to an individual's strengths and weaknesses, and to the local environmental conditions.

For example, in very dangerous, predator-dense environments, animal parents often provide ongoing protection to young offspring, even young adults. Under circumstances where resources are scarce, some animal parents may continue to provide land and food for their adult offspring. On the other hand, when environments are safer, or hold more resources, prolonged protection and prolonged provision are not only unnecessary, but can hinder the development of the young adult into an animal optimally positioned to take advantage of that rich environment. (We'll talk more about extended parental care and cultivating self-reliance in Part IV.)

The South Georgia Island king penguin class of 2007 didn't have a graduation speaker. Ursula's parents couldn't send a care package or text her encouraging words. But if they could have, there are a number of things they or any parent should want their predator-naive offspring to know.

First: You're spectacular, and young, and youth has a magnetic pull that will attract attention. This makes you easy prey.

Second: You're naive and unprotected. Inexperience kills, and it's most potently deadly whenever you're in a new environment.

Third: You have options. You can avoid danger by overestimating risk, staying away from predator-prone areas, learning as much as you can about potential predators, and carrying yourself well. Stot and use signals of unprofitability to encourage potential predators to move on.

And finally: Have friends. There's safety in numbers, and you can learn a lot from what they do right, and even from what they mess up.

On that December Sunday in 2007, Ursula took a leap, and then she swam for her life. Did she survive her first day? Pütz knew the answer: yes, Ursula made it, and so did Tankini and Traudel. Although statistically a third of their cohorts would have died, all three got past the leopard seals on their first try. From his electronic tracking data, Pütz could see where Ursula went over the next three months. She swam straight south, toward an area off Antarctica where food is plentiful. She swam around ten kilometers a day, traveling with groups of other adolescents while together they learned to hunt fish and krill.

After three months, her transmitter's signal went dead. Pütz doesn't know exactly what happened to her but said it was likely that the radio-tag had just fallen off. Ursula, no longer predator-naive, was well launched on her coming-of-age journey. Swimming with friends, learning to eat, king penguins typically spend four or five years exploring the Southern Ocean gaining experience before settling down to breed.

For adolescent penguins, and young adults of any species, growing up looms like the open ocean, filled with excitement and danger. They are biologically primed to underestimate the risk and even impulsively take chances. If they don't dive in, they'll never gain the experiences they need to survive as mature adults. But experience can be costly. This is the paradox of safety in wildhood. Inexperience can be fatal. But gaining the experience necessary to be safe can also be deadly. In a dangerous world in which inexperience can cost you everything, a wise approach could save your life: learn as much as you can from parents, from peers, and, crucially, from the environment. Then, when the time is right: leap.

PART II

STATUS

Humans and other animals must learn to navigate status hierarchies, which often favor privileged creatures. Learning the rules of groups in wildhood will determine whether they eat or go hungry, are safe or in danger, and are tolerated, shunned, isolated, or accepted.

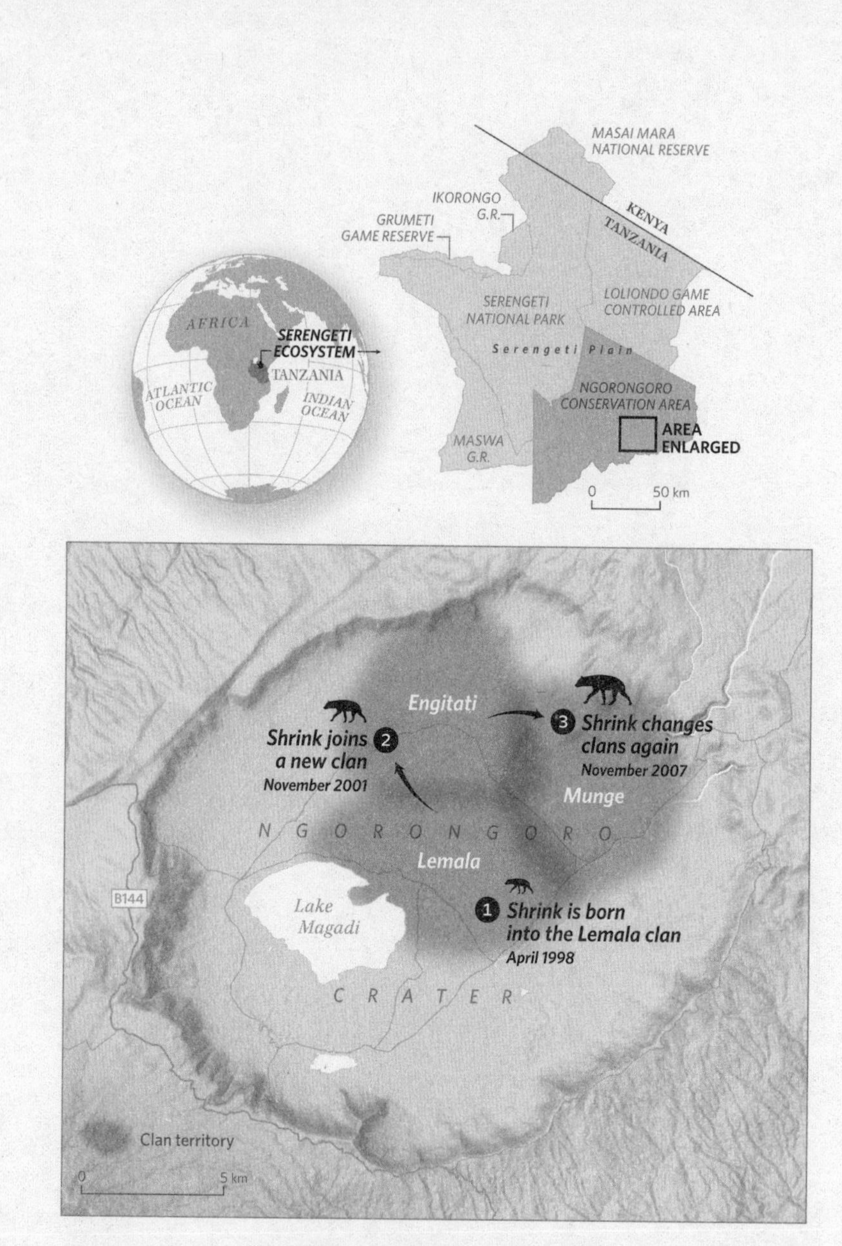

SHRINK TAKES ON PRIVILEGE

Chapter 6

The Age of Assessment

Looking like a Labrador puppy crossed with a koala cub, Shrink, with his shiny black fur and bright eyes, was born identical to all baby hyenas—except for one thing. He had a funny right ear. It had a little bend in it that folded the outer edge into a heart shape. Shrink's ear gave him a rakish, distinctive look. It made him special.

Objectively, however, Shrink wasn't at all special. He was born the lowest-ranking cub to a low-ranking mother in a clan of spotted hyenas in Tanzania's Ngorongoro Crater in 1998. Shrink should have entered the world in obscurity, struggled his whole life at the bottom of hyena society, and died with no one knowing a thing about him. But things turned out differently. Strong-willed, charismatic, even creative, Shrink steered his life in a different direction, against tremendous odds, within one of the strictest and most aggressively enforced status hierarchies on the planet.

As a male, Shrink arrived on Earth at the bottom. Females rule most hyena clans, and high-ranking daughters inherit their mothers' social status. Males inherit their mothers' ranks as well, but they lose their birthright if they immigrate to other clans. In hyena society, the lowest-ranking members are always outsider males.

Hyenas come into the world ready to fight for their place in the group. Of all the carnivores, only hyenas are born with their eyes already wide open, the better to see the competition. Their teeth are fully erupted at birth, ready to bite. From his first breath, Shrink was equipped for combat, and that was lucky for him because the moment he emerged, his twin sister was waiting for him. With the advantage of having been born first, she had been sucking down mother's milk as fast as she could to fortify herself for battle. She attacked Shrink immediately, and they fought and clawed for the best position at their mother's breast.

Shrink's story came to us via Oliver Höner, a Swiss-Brazilian biologist with the Leibniz Institute for Zoo and Wildlife Research in Berlin. For more than two decades, Höner and his zoologist colleagues have spent much of each year in Tanzania, conducting field research on hyena social behavior. Their Spotted Hyena Project, unlike other studies of animal behavior, is noninvasive, meaning they don't handle or immobilize the wild animals in order to track their behaviors and interactions. They use only detailed field observation, and no electronic tagging. They do use technology, including DNA testing and video recording, but the observational approach means that the scientists are keenly aware of the hyenas as individuals within groups. They scrutinize them physically, mapping spot patterns, scars, and ear notches in order to identify who's who in the field. And they log the animals' personality traits and how those traits evolve over days, years, life cycles, and even generations. Their trove of data, started in 1996, now covers several thousand individual hyenas.

Höner told us that when Shrink was born in April 1998, his clan was ruled by a beautiful young queen named Mafuta. Queen Mafuta had ascended to power suddenly after her mother's unexpected death in a lion attack. Although she wasn't yet fully grown and most of her clan members were stronger and more experienced, Mafuta grabbed her chance. Decisive and magnetic, Mafuta had an instinct for power. Fortified by her mother's legacy, she teamed up with her older sister, and with other close relatives, they were able to keep all the other clan members in check. As Mafuta assumed her rightful place on the throne, the rest of the clan arranged themselves into the hierarchy beneath her.

As queen, Mafuta got the first of everything. When the clan brought down a wildebeest, she was first to fill her belly, while lower-ranking hyenas were lucky to get a few mouthfuls of meat. Mafuta took the most favorable sleeping positions under the acacia trees, sheltered from wind and rain and safer from intruders; it was her privilege, like that of all high-ranking animals, to get more and better sleep. She received protection from her powerful sister, and her nieces, plus other clan warriors that backed their coalition. Queens also receive the first choice of mates and, when it's time to give birth, the first choice of den locations.

Shrink's mother, Beba, on the other hand, was at the opposite end of the clan hierarchy. Beba had been at the bottom her whole life. Picked on by higher-ranking females and their offspring, pushed to the edges of social groups, forced to eat last when the group made a kill, Beba knew her place and mostly tried to mind her own business while scrounging around for the bare essentials of food, safety, and shelter.

In the spring of 1998, Höner could see that Beba was more stressed than usual. She was pregnant with twins—one of them the infant that would become Shrink. By coincidence, it so happened that Queen Mafuta was pregnant, too, with her first royal cub. This cub would be in line to inherit the throne, and with the fortune of its mother's high rank, would be guaranteed advantages other hyenas wouldn't. The lives of these two about-to-emerge infant hyenas—one born of a queen, the other of a pauper—would come to intersect in ways that would alter Shrink's adolescent life and destiny.

RULING THE ROOST

It was 1901 and a six-year-old boy was playing in the backyard of the Oslo cottage that his parents had rented for the summer. The house came with a flock of chickens, and the boy, who was bright, sheltered, and intuitive, spent every day observing them. He gave each bird a name. He memorized their quirks and affiliations. Leaving the chickens at the end of the season was difficult for the sensitive boy, and he thought about them all winter.

The next spring, the boy begged his mother for a flock of his own. Perhaps she was indulging her only child or just wanted to keep him occupied during the long Norwegian summer days. Or maybe she was seizing the chance to kindle an interest in science or instill a sense of responsibility. Whatever the reason, she granted his wish. The boy passed another summer tending to a flock.

The following summer, the boy cared for more chickens, and again the next, until after a few years, he'd spent hundreds of hours observing the birds. With a precocious attention to detail, he chronicled the food they ate, and in what quantities, and meticulously logged the eggs they laid. He recorded daily weather patterns and tried to discern how it affected the hens. But what most fascinated him, what he loved above all, was mapping how the birds related to one another. On page after page, he drew complex triangles and diagrams of how the chickens rotated through their hierarchies. Day by day he noted who was sick and who was well and what that meant for group stability and group strife.

What the boy, at age ten, had noticed and would later name was the Pecking Order. It would be many years before Thorleif Schjelderup-Ebbe, at twenty-eight, would publish his findings formally in a 1922 paper for the German psychology journal *Zeitschrift für Psychologie*. But "Weitere beiträge zur sozial-und individual psychologie des haushuhns" ("Contributions to the social and individual psychology of the chicken") even today informs the basis of how we understand the ancient and powerful construction of rank and status in living creatures. The natural, inborn process that chickens use to arrange themselves in a hierarchy—observed by a ten-year-old boy—is the same one that stratifies groups of animals from elephants and raccoons to fish, reptiles, and, of course, birds. This ranking process is constantly at work in human groups too—and never more intensely than during wildhood.

Wildhood is when much of an individual's future status is learned and solidified. How young animals are ranked and rated during this period will shape their place in the world and their sense of belonging for the rest of their lives. Some of what they'll be judged on they can do nothing about: they're born with—or into—it. Some of their status they can learn or cultivate, or, in rare cases, change.

All animals, including adolescents, assess one another for size, strength, and attractiveness. They evaluate age, health, and reproductive prospects. Physical abilities like swimming, flying, and fighting are contested and flaunted. On their way to joining the ranks of mature adults, they also cannily assess the power of families, friends, and rivals. They're accepted into or spurned from the social groups that will govern their future opportunities. The pressures on an animal to perform at this time of life are immense. And with good reason, because the stakes are too.

Across species, coming of age means entering an age of assessment.

THE GRAVITATIONAL PULL OF STATUS

Status. Hierarchy. Position. Standing. Class. Station. Prestige. Many just call it popularity, and some schoolkids today bluntly but accurately call it "relevance." Whatever word you use for it, social rank—an individual's place within a group—is a powerful shaper of personal identity.

For nonhuman animals, social rank may be less about personal identity than it is for people. But it nonetheless profoundly influences how individuals exist in their worlds. Social rank may determine whether an animal eats or starves, has offspring or remains barren, is protected from danger or cast out to the wolves. Animals will suffer pain, forgo food, give up sex, and betray others just to ensure they're not left out or driven from a group. You might say that for social animals, status is like gravity. It's powerful and inescapable. It's invisible. It exerts an omnipresent force, and it molds how a creature moves through the world and behaves around others.

In nature, the lower an animal slips in group rank, the worse his or her life becomes. Higher-ranking animals have greater access to food, territory, and other resources. Failing to learn how to strategically recruit allies and navigate enemies, pay attention to peers or ignore bystanders, can cost an animal potential resources, homes, and mates.

The highest-ranking rooster in the coop, for example, has the privilege of announcing the breaking of the dawn—he crows first and until he does, his competitive subordinates must suppress the urge. Dominant female hamsters prevent the embryos of subordinates from implanting. High-ranking crayfish claim thermally perfect 75-degree locations, sending subordinates to waters that are too warm or too cold. The top-ranking homing pigeon claims the highest perch. And top-ranking fish swim near the front of the school, where the water is high in oxygen and low in fish feces. Low-ranking fish at the back of the school have the exact opposite experience.

It isn't only a matter of comfort. Getting sorted to the bottom can be a life sentence—and sometimes even a death sentence. High-ranking animals enjoy privileged safety positions in their flocks and herds, so they're less likely to be attacked, captured, and eaten by predators. High-ranking grey mullet fish occupy positions inside the school, far from the dangerous outside edges from which predators pick off their meals. Lower-ranking fish are pushed toward the "domain of danger," often, but not always, at the perimeter. Lower-ranking animals in general spend more time vigilant, scanning for predators. They therefore get less sleep, and the sleep they do get is poorer quality. High rank enhances an animal's safety; low rank pushes him or her into greater risk.

Group living offers benefits to animals. With more eyes to scan the environment, safety in numbers takes predator danger off individuals. Sharing resources and information makes work more efficient and keeps them better fed. Groups allow younger members to learn and grow before taking on responsibilities. But when individuals come together, having recognizable social structures and rules can help reduce conflict. Hierarchies can also keep animal groups organized and more productive.

Higher-status members of hierarchies have first dibs on food, territory, mates, and safe havens. And they will vigorously defend their position and privilege. Recognizing one's place in the group is crucial to staying alive and brain systems alert animals to second-by-second shifts in their rising or falling social positions. Neurochemical messages nudge an animal to adjust its behavior in response to the social stew simmering around it. As far as we know, nonhuman animals experience these neurochemical "status signals" as noxious, pleasurable, or something in between. But humans register these same neurochemical status signals as emotions. In fact, our emotional lives derive from this status-detection physiology, an inherited legacy from our status-conscious animal ancestors for whom shifts in social position could mean opportunity or the end.

Animals that don't see the full complexity of a social hierarchy may miss opportunities to better their position. But those that don't understand their place may be attacked, injured, killed, or driven from the group. Social animals observe and evaluate every tiny detail of daily social life, scanning not only for opportunities to rise in status, but also to spot and prevent one catastrophic thing from happening: status descent.

Rapidly detecting status descent is fundamental to survival.

THE PECKING ORDER OF ANGELS

Centuries before young Thorleif noted pecking orders in his chickens, European theologians were ranking the angels in heaven. They sketched elaborate hierarchies (literally *hieros* = sacred and *arkhia* = rule) ranging from stern seraphim and cherubim at the top, down to lowly but good-natured archangels and angels at the bottom. Among the privileges of the seraphim was the honor of sitting closest to God's throne. Low-ranking angels, on the other hand, had to spend their time tending to the far less exalted affairs of humans. Hierarchies, defined, are organizational systems in which individuals are ranked as superior or subordinate to others.

Schjelderup-Ebbe noticed that pecking orders established themselves very quickly. He saw that when new chickens were introduced, a new order rippled through the group almost invisibly, until every hen knew her place. After the seconds of disruption while the hierarchy was in flux, the flock calmed down into a (seemingly) peaceable and functional group. The term "pecking order" was literal: chickens used their beaks to maintain the hierarchy. The top chicken could peck any other hen in the flock. Her next-in-line, right below her, could peck every other hen except the top chicken above her. The third in line could peck everyone except the top two chickens . . . and so on down the line.

A range of different hierarchy types is seen in animals. They can be despotic or coalitional, triangular, stable, or flexible. In humans and many other species hierarchies are often linear. We humans share a deep-seated, inborn ability to figure out where we rank, and where we fit in. As Marc Bekoff, a scholar of animal behavior, describes it: "we social animals . . . are hard-wired to sort ourselves into hierarchies in which someone is at the top and someone is at the bottom and the rest of the group is arrayed between those two points."

Before we continue looking at how status shapes the lives of animals, it helps to understand two words that might seem interchangeable, and are often used that way, but that social scientists and animal behaviorists differentiate. They are: "rank" and "status."

An animal's rank is its absolute position in a group, measured as objectively as possible. Status, in contrast, isn't an objective measure. It's the *perception* of rank. Status depends on what other group members think and decide. Status and rank may sometimes be the same, but they can be different. An example from human life would be a family that is widely believed to have millions whose actual net worth is far less. Their rank (how much money they have) is lower than their status (public perception of their financial standing).

Every animal in a group has its own rank and status. It's part of the tremendous diversity within herds, flocks, and schools, even if it's not initially apparent to the untrained human eye.

THE DIVERSE HERD

Visualize a flock of starlings—the swirling clouds of birds that gather at sunset in massive formations containing thousands. To our eyes, they may seem as indistinguishable as peanuts in a bag. But every member of that flock is an individual, different from every other bird. Some are male; some are female. Some have been mature adults for several years; some are experiencing their first year in the flock with the big body of an adolescent but the inexperience of one too. Just as people come in different sizes, starlings can be stocky or wiry, tall or short, stiff or loose-jointed, calm or high-strung.

The differences don't stop there. Beyond age, sex, and size, each bird has a different personal profile encompassing fighting experience, athletic ability, physical attractiveness, and sensitivity. Libido varies; some have huge sexual appetites while others are more laid-back. Animal behaviorists now routinely measure personality traits such as boldness and shyness—even in animals as uncelebrated as cockroaches and pigeons. And, of course, a bird's parental status, social ties, birth order, and life experience all combine to make it unique. What you're seeing as the sun sets over that field and the birds start flocking is a speeding, flowing organization chart of starling hierarchy, where every individual has a place but not all the places are equal.

It's the same for every shoal of anchovies, every stampede of caribou, every exaltation of larks, and every troop of bonobos. Each individual has a place. And the social energy that goes into determining those positions is also what holds together the hierarchy.

Animals can deduce status relationships among unknowns using an ability called transitive rank inference (TRI), essentially, "If A is ranked higher than B and B is ranked higher than C, then A ranks higher than C." For hyenas it means if Shrink loses a fight to his sister and then watches her lose to a new hyena, then Shrink can assume that he ranks lower than the new hyena too. He knows where he stands without actually having to fight the new hyena.

Transitive rank inference is a behavioral shortcut that minimizes direct conflict, keeps the peace, and lowers the risk of injury. For chickens, TRI means the top hen doesn't have to deliver an actual peck every time she wants to assert dominance over another. A thrust of her head, a cluck, a feather ruffle can come to substitute for an actual blood-drawing lunge. The sparrow at your bird feeder who sends the other sparrows flying with an almost imperceptible feather flick; or a bossy blue whale whose intimidating presence on the breeding ground shuts down the reproductive systems of her younger sisters and cousins; or the dominant cat who keeps her subordinates edgy and on the jump with a slight narrowing of her eyes—dominant animals know how to send status signals and subordinates learn what they mean.

Mammals, birds, and fish all can use TRI to determine their position in a group. Finding this ability in species separated by hundreds of millions of years of evolutionary time suggests that perceiving social relationships quickly is an ancient and crucial life skill for social animals. During wildhood, as adolescents' social brain systems strengthen, transitive rank inference is one of the ways they learn their place in a group. What used to be play among juveniles transforms into a series of contests: of strength, skill, and perseverance. Primed with highly sensitive neurocircuitry that registers every compliment as acceptance and clocks every slight as rejection, human adolescents pay close attention to where they fall in the hierarchy. The gravity of social status doesn't just influence how they act; it shapes how they feel. Changes in status—whether in real life or on a screen—can induce euphoria, despair, and every feeling in between in adolescents and young adults.

According to public health sources, the twenty-first century so far has been an era of widespread loneliness and disconnection, especially for adolescents. Anxiety and depression have joined smoking and poor nutrition as urgent health concerns around the world. Parents and educators point to high-stakes testing and other academic assessments in an adolescent's school life. Psychiatrists point to genetics, hormones, and neurochemical changes in the brain. Economists and lawmakers cite geopolitics and global recessions. Everyone blames social media. All these factors can dial up stress and mental anguish at any age. But we believe the roots of adolescent and young adult angst—from simple mood swings to more serious depressive episodes—can be found in the ancient circuitry that powers animal hierarchies.

Connecting animal hierarchies to personal emotion helps us understand why finding a place in a group, especially for adolescents, feels so miserable when it goes wrong—and so jubilant when it goes well. Obsession with status, it turns out, is entirely natural. And hierarchy formation, with the status-seeking that drives it, isn't a game you can opt out of. Given that, it's better to learn the rules.

BORN ON THIRD BASE

Spotted hyenas in the Ngorongoro Crater are pregnant for around a hundred days. When it was time to give birth, Beba, like all hyena mothers, went off to a secluded spot. There, on the fringes of the territory, far from the group, Beba began her labor. With twins, her delivery was high risk. But because she'd given birth before, the process went better than it could have. Shrink and his sister were born without any injury to Beba.

Around the same time, a few kilometers away in a private den, Queen Mafuta was also in labor. Hers was likely much harder than Beba's, because this was her first pregnancy. Hyenas' birth canals are exceptionally narrow and so inelastic that their firstborn often dies of asphyxiation. But the royal cub was born healthy—and male. Höner and his research team named the new princeling "Meregesh."

Thanks to the privilege he'd been born into, Meregesh was destined for a good life. Like other high-born cubs, he would enjoy the benefits of better prenatal nutrition. His mother the queen was able to make a lot of milk, thanks to her priority access to food. In addition to quantity, her royal milk was also better quality. Each suck of the energy-rich, performance-enhancing elixir gave Meregesh an advantage over less-privileged hyenas. Because they receive greater quantities of superior milk, higher-born hyenas grow faster than lower-borns. This means they nurse for less time—nine or so months compared with up to two years. Better nourished, higher-borns survive their first year of life more often than lower-borns.

Hyenas aren't the only animals given a head start by nutritional endowments from their parents early in life. Within litters of Eurasian lynx, newborn cubs who monopolize access to their mother's milk go on to become dominants. High-ranking canary mothers anoint their eggs with more testosterone than subordinate canary mothers do, granting their chicks an inherited competitive advantage. In the yolks of fish eggs, hormone levels vary too, affecting the social status of newly hatched fry. Among minnows, dominant females lay eggs with higher testosterone, which means their newborn offspring will emerge possibly more primed for dominant status than those of subordinate minnows.

Across the crater in Beba's den, Shrink was facing yet another disadvantage. Beba's milk was thinner and less nutritious than the milk Mafuta made for Prince Meregesh. Beba also had much less of it and she was feeding twins. So, when Shrink arrived, he had just about every odd stacked against him. He was male. He was second-born. His mother was low-ranking. Her milk was inferior and scarce. And Shrink himself was small and weak from poor prenatal nutrition. Within seconds he was set upon by his slightly older, slightly better-fed twin sister. In Shrink's first encounter with a hierarchy, a subgroup with only two members, he was already at the bottom.

Chapter 7

The Rules of Groups

One morning when he was around two weeks old, Shrink awoke to the feeling of his mother's teeth on the nape of his neck. He squirmed but she held him fast as she wriggled him out of the only home he'd known in his young life. Shrink bounced along, carried by Beba, under the still-dark sky. When she finally came to a stop and dropped him to the ground, Shrink could sense movement around him. He could tell that his twin sister was there, but so were other, unfamiliar hyenas. He may have heard whimpering and growls. His mother left him there and trotted away.

Shrink didn't know it, but this was his first day of the next phase of hyena development: life in the communal den. After two or three weeks alone with their mothers and any siblings, spotted hyena cubs are brought to a single, shared den along with every other hyena cub in the clan, from the princesses and princelings on down to the offspring of the lowest-ranking adults. Shrink and his sister were there and so was Meregesh, son of Queen Mafuta. In this communal hyena daycare, the young animals' social lives expand, and they begin to understand their place in the wider order of the clan.

In the communal den, the young hyenas spend long days and nights together without adult supervision. Their mothers visit their cubs regularly once or twice a day to nurse, but the cubs are generally left alone to battle and bully, play and range freely.

For Beba, although relieved of the constant demand of her hungry and undernourished twins, this new phase was a different kind of struggle.

Already forced to live at the margins of her group, pushed to pour her scanty resources into finding food and defending herself, Beba couldn't visit the hyena daycare as often as the other mothers could. When she did, she couldn't provide for her cubs as lavishly. She came regularly to nurse, but barely had enough milk for one hungry hyena, much less two. Parched and starving, Shrink was often shoved out of the way by his sister when their mother came to visit.

Queen Mafuta, on the other hand, visited several times a day. She was overflowing with milk and Meregesh ate copiously. Mafuta also brought Meregesh extra treats of meat, something the lower-status mothers couldn't.

For Shrink, his twin sister, Prince Meregesh, and their other cub-peers, the first few days in the communal den were probably intimidating, according to field scientists who've studied this period. After being left by their mothers, hyena cubs are tentative and easily spooked. They cower at anything that moves—even stalks of grass blowing in the wind or stray insects walking by. But soon, they begin challenging and attacking instead of submitting to the things that disturb them, and that includes their fellow cubs. They begin play-fighting, and in these early encounters the young hyenas rack up wins and losses based on a set of ranking characteristics consistent across animal groups.

HOW HIERARCHIES HAPPEN

How the status and ranks of individuals are determined within groups varies across species. But some characteristics are widespread in nature. The following are some of the common criteria that influence where an individual lands in a hierarchy.

Size

Physical size is a major predictor of status rank in many animal societies. From birds to fish, crustaceans to mammals, and even in some spiders, being larger contributes to higher rank. For some animals, size is not as important; for a female hyena, family ties and social networks mean much more than her body mass. For male hyenas, size is less important than kinship, social networks, and age.

Age

Getting older is a status booster in many animals. Age confers higher rank in feral ponies, African elephants, mountain goats, meerkats, chimps, bottlenose dolphins, and humans. For growing animals, size and age are connected. Older siblings usually dominate younger siblings, at least until a certain age. Gap years and redshirting, which allow humans to gain a year before competing in the next stage of academics or sports, may have their origins in this age-based bump seen across the animal kingdom. Seniority—hanging around long enough for competition to leave or die and then replacing them as the next-oldest—also improves opportunities to inherit territory. And the extra time allows younger individuals to learn crucial life skills by watching older, more experienced animals. Shrink's youth was another strike against him. A male hyena's age and status are tightly linked. Males have to wait years to work their way up the status ladder, unless their friends and alliances can give them a leg up.

Grooming

Attractiveness—even physical beauty—can raise status among human beings. But animals, too, have what Darwin called "a taste for the beautiful." Displays of splendor in male animals have been viewed primarily as advertisements of desirable genetics and access to resources, aimed at choosy females. For flamingos, as an example, bright orange-pink feathers billboard a diet high in healthy carotenes. Looking at those vibrant feathers, a potential mate would know that this male had fine genetics and feasted regularly on the best shrimp. A paler gray bird would suggest the opposite. Of course, access to good food can be an environmental issue, well out of the control or agency of the animal individual.

Some of the same physical features that draw in mates are also status symbols in same-sex hierarchies. A black swan's elaborately curled wing feathers may help him attract females, but the ruffle also signals high social status to other males.

Another element that seems to enhance attractiveness, and raise status, is grooming. The best-groomed birds, fish, and primates tend to be the highest-ranking. They also tend to be the healthiest, physically. Receiving grooming from other animals in one's group is a benefit of high rank. Schjelderup-Ebbe recognized the grooming gap between birds who sported "bright, sleek, beautiful, and clean plumage" and those at the low end of the pecking order with "rumpled, disordered plumage, often with dirt hanging to it."

Lower-ranked individuals may physically groom higher-ranking animals in exchange for resources such as protection and food, and also to boost their own status through association. Watching who grooms whom in groups of fish, birds, and mammals can help observers get a good sense of status relationships. It doesn't take much to spot this same trend among humans and to realize how much grooming is associated with status. Humans can use words as a kind of social grooming. Compliments trigger neurochemical responses that are similar to the ones produced by physical grooming. Humans can use praise and flattery the way animals use plucking, rubbing, and nibbling to curry favor of dominants in the group. Extending this concept of "spoken grooming" to social media, one can discern hierarchies from who does the posting and who does the liking.

Like other well-groomed, high-ranking animals, top hyenas are less blemished than others—perhaps in part because lower-ranking animals don't dare attack them. But high-ranking hyenas also have superior immune systems, receive more grooming from others, and thus have lower parasite loads.

Sex

Finally, an animal's sex (what has been called gender in humans) factors into status in some fish, reptiles, birds, and mammals. Females are the dominant sex in some species, males in others. Schools of colorful tropical clownfish form hierarchies with a dominant female always at the top. The benefits for the fish in the highest position are great. So much so that if the female clownfish dies and the position opens up, subordinate males will transform themselves into females in order to compete for a shot at occupying the top spot. It takes them about forty days to transition from male to female, a process that involves doubling in size and converting testicular tissue to ovarian.

Many factors contribute to why males or females dominate certain animal societies. While biology plays a role, environmental conditions—food availability, predator density—are also important.

Shrink could do nothing about his genetic inheritances—his sex, age, birth order, size, attractiveness, or parents.

But he still had a chance. Animal behavior, some of it innate but some that can be learned and adapted to their advantage, plays a considerable role in shaping an animal's status. For Shrink, these tried-and-true behavioral techniques turned out to be his key to survival.

ASSOCIATION WITH HIGH-STATUS ANIMALS

Spotted hyenas like Shrink spend a great deal of time with their blood relatives. But when they aren't with kin, they prefer hanging around with animals at their social rank or higher.

Many primates, including several species of baboons, macaques, vervet monkeys, and of course humans prefer high-ranking social companions to low-ranking individuals. Status strongly influences affiliations and friendships among horses and cattle in herds. High-ranking dairy cows choose stalls next to one another and follow closely behind other members of their cow clique when walking in lines. An animal's rank makes it more appealing as an alliance partner and sometimes even as a sex partner. Male bison, for example, may show little interest in a middle- or low-ranking female, preferring to mate with the high-ranking females in the herd. And since higher-ranking animals are often in close physical proximity to one another, just standing, grazing, or lounging next to admired members of the herd may be a status booster.

Association with high-status animals—and the reflected glory of advertising those friends—may be the biological drive behind posting a selfie from an in-crowd party or displaying a shelf full of photos with politicians and celebrities. Letters of recommendation, name-dropping in conversation, and lunching with popular students or colleagues accomplish similar purposes. The power of association to raise status also likely underlies the attraction to high-profile companies, prestigious schools, winning sports teams, and elite branches of military and public service.

SIGNS OF STATUS

Besides hanging out with the popular crowd, some animals also carry props to show off their status. Luxe fur, glamorous feathers, splendidly elaborate horns, inconveniently long tails—these all attest to the riches backing up the animals who own them. Energy and time must be devoted to maintaining plush pelage and beautiful skin and dragging around other wild accessories like enormous antler racks. These animal social signifiers, called "status badges" by biologists, tell the animal's peers, "I am special." Flaunting status badges shows off an individual's genetics, social networks, and grooming resources. Knowing the money and time humans expend on propping up our status, perhaps it's not surprising to learn that animals without resources can sometimes fake their way up a hierarchy by flashing faux status badges. A fiddler crab that loses his one massive claw in a fight will plunge down the hierarchy. He can grow a new one, but it's lighter weight and not as effective in battle. Still, waving his ersatz claw, he can fool the other crabs into thinking his fake weapon is the real thing. Unless he's called out, and forced to fight with his mock cudgel, he can usually claw his way back up the hierarchy.

In a display case in the Mesoamerican hall at Harvard's Peabody Museum, next to an elaborately carved jade head and a ceramic jaguar bowl, sits a gold pendant about the size of your thumbnail. We spotted it one day as we were searching the museum for artifacts that indicated high social rank. This particular case held items treasured by elite members of the Late Classic Maya period, a time of hereditary kings, grand public architecture, and advancements in astronomical thought.

Mayans, who lived around one to four thousand years ago, were as status-conscious, just as we are today, and as Shrink's clan was in 1998. And the pendant carries clues to that. Carved into the gold is the figure of an adolescent young man in profile, his head topped by an enormous headdress radiating like a sun. Headdresses like these were decorated to resemble revered animals such as jaguars or falcons and often adorned with quetzel feathers. Mayan headdresses were high-status items, and commoners were forbidden from wearing them. Around the youth's waist is a decorative shield with an ornamental stone ax head called a *hacha*, thought to have been used in a Classic Maya ceremonial ballgame called *pitz. Pitz* players were usually elites, and the games were watched by thousands of people in arena-like stadiums. Like many universities today, Mayan culture prized athletes, particularly athletes who could combine brains and beauty. All of these symbols indicate this adolescent's honored place in his society. This young man was like a Heisman Trophy winner of the eighth century, and the pendant with his likeness would likely have been used as an offering at an elite burial.

Anthropologist and archaeologist Stephen Houston, an expert on Maya civilization, suggests in his 2018 book *The Gifted Passage* that adolescent males, perhaps because they were potentially heirs, held positions of notably high status in Mayan society. Images of adolescent males are widespread on many Mayan ceramics, hieroglyphics, and murals. High-status elites enjoyed perks in Mayan society that included living in large houses in central locations, wearing fashionable clothing and accessories, and regularly eating meat and drinking chocolate—a rare treat for commoners.

For all the extra food, comforts, and luxury items, Mayan nobility came with a distinct disadvantage: rules governing warfare among groups of Mayan clans often required royals to be the first into battle.

Similarly, in fights with other clans, and in defending the group from lions, hyena queens must be first to offer their lives for the cause. Death by lion is one of the commonest ends to the reigns of hyena alphas. In these moments of drama and violence, the clear succession of hyena hierarchies can be very useful to the group. Because everyone knows their place in the hierarchy so well, the leader-in-waiting, the second-highest-ranking princess or prince, can step in immediately upon the death of the alpha, and transition is seamless. Höner and other hyena experts have observed clans accept a transfer of power, even during the gore of battle, when a female is killed and her daughter steps in to take her place before the fight has even been won.

THE GESTURES AND SOUNDS OF STATUS

Because life is potentially physically and socially more dangerous for them, subordinate animals are more edgy, vigilant, and nervous. A low-status wolf betrays his rank with darting eyes and gestures of submission. His shoulders slump. He bows his head and licks his lips. In contrast, the movements of higher-ranking wolves tend to be more purposeful. Motionless and unblinking, they're more likely to make bold or hostile movements, such as chasing other group members and lunging with open mouths.

Starting very early, as young as four weeks, Shrink would have been learning the body language of hyena status. Higher-ranking individuals like Meregesh would learn to keep their tails upright and ears cocked, whereas lower-ranking individuals would be expected to keep their tails between their legs, ears backward, teeth bared, and head downward. Performing these behaviors in ritualized greeting ceremonies confirms status relationships and strengthens friendships.

Studies of dominance gestures in humans show similar differences between the relaxed posture and steady eye gazes of dominants and superfluous body movements and darting eyes of subordinates. Higher-ranking humans tend to show their status in verbal language that is quicker, more confident, and more enunciated. They interrupt more too.

Frans de Waal, the author and primatologist, describes in *Our Inner Ape* how the human voice reveals status cues that may seem subtle but are powerfully, if intuitively, understood. The pitch of your voice, he writes, is an "unconscious social instrument" that betrays your position in the hierarchy. Everyone has a personal pitch, but, "in the course of a conversation, people tend to converge." They settle on a single pitch, de Waal explains, "and it is always the lower status person who does the adjusting." As de Waal writes, the *Larry King Live* talk show provided a demonstration of this effect. "The host, Larry King, would adjust his timbre to that of high-ranking guests, like Mike Wallace or Elizabeth Taylor. Low-ranking guests, on the other hand, would adjust their timbre to that of King."

Younger adolescents with their higher-pitched voices may often find themselves vocally jockeying to be heard. And older adolescent males may notice their status in the home or classroom changing along with their deepening voices.

Spotted hyenas are famous for a particular vocalization: their high-pitched, staccato hoot-laugh, or giggle, that gives them the nickname "laughing hyenas." Although this distinctive sound has long been attributed to all hyenas, in fact it's a marker of lower status, made by subordinate animals who are communicating with higher-ranking members of the group. A team of Berkeley psychobiologists presenting to an acoustics conference in 2008 reported that giggles are made by "distressed, or submissive, animals in situations where they are both excited and conflicted between approaching and leaving the situation. For example . . . submissive individuals at a kill waiting their turn while being chased away by higher ranking animals."

While the characteristic giggle is made mostly by lower-ranking hyenas, all hyenas produce many different sounds. One of them, the so-called whoop, is a loud, long-distance communication call that starts low and swoops up and down in pitch. Every hyena's whoop is distinctive, and scientists like Höner can learn to identify individuals by them. Notably, the Berkeley team also reported findings that "immigrant males who are approaching a new clan produce a high number of whoops as to carefully advertise their arrival into a group that could potentially reject them."

Shrink was going to use every tactic he could muster, including vocalizations. Besides high-status associates, status badges, body language, and voice, there was another asset he was developing as he progressed through adolescence. It's called the Social Brain Network.

THE SOCIAL BRAIN NETWORK

Discerning one's place in hierarchies is critical for humans and other social species. Specialized brain cells and regions dedicated to social awareness and function are found in fish, reptiles, birds, and mammals. Collectively, these systems are called the Social Brain Network (SBN).

In mammals, the SBN is housed in six separate but linked regions of the brain. Picture one of those maps in the back of an airline's in-flight magazine, showing where in the world the airline flies. Glowing hubs and arching lines indicate planes landing and taking off all over the globe. Imagine your brain as the world and your social brain—the SBN—consisting of six of those hubs, connected and communicating at all times. The six hubs bring together visual input, stored social memories, fear associations, hormonal information, coping behaviors, and logical decision-making.

Anytime you're with or thinking about other people, your Social Brain Network is active. It helps you process facial expressions, make sense of body language, assess others' emotional states, and interpret tones of voice. SBNs allow you to read a room, make a sale, know when to walk away, and know when to run. It's hard to overstate the importance of this neural network to daily human life: abnormalities in SBN development have been linked to brain and social deficit conditions, including autism spectrum disorder. Brain injury may impair the SBN's ability to regulate social function. Inappropriate laughter, public displays of sexual behavior, reduced empathy, and out-of-character temper tantrums have all been seen in patients with brain tumors or injuries affecting regions of the SBN.

Our ability to understand and connect not only with other humans but also with cats and dogs, birds and horses, points to our common ancestry as social animals. A dog's social brain helps it decode where it ranks at the dog park and possibly within its human family too. Recent studies of dog-human communication have shown that the same regions in the social brains of both are activated when hearing emotional voices of either species. The ability of an experienced equestrienne to sense the emotional state of her horse seems to be reciprocal. Evidence that horses can read their riders, too, suggests interaction between horse and human social brains.

STATUS MAPPING

The social brains of babies sit primed, waiting and ready, to guide them through the social world they have just entered. Within months of birth human babies smile socially at, stare at, and study other babies. By six months they differentiate and prefer certain people in their lives, and by nine months they want to participate in activities with others. By a year they connect dominance to power and begin accurately differentiating between dominants and subordinates, and by two years status ranks emerge among playing toddlers. From the outcomes of these interactions, the earliest ranked, linear hierarchies of their young lives are established.

At playdates and on playgrounds for years to come, the ranking and sorting will continue. As toddler cartographers, they construct mental status maps featuring their peers and themselves. By four they can perceive which peers have high status, and they show a clear preference for spending time with them. They're interested in what dominant peers are up to and spend a disproportionate amount of time watching them. Preferential interest in watching high-status individuals is a trait that continues through adult life—this trait can explain the enduring appeal of gossip tabloids and paparazzi. Adult humans and rhesus macaques have this in common: monkeys will pass up the sweet juice they love for an opportunity to watch a high-status monkey on a screen. In contrast, when shown the activities of low-status monkeys, the watchers have minimal interest. The scientists have to bribe them with extra juice just to get them to pay attention even for a short while.

As humans and animals enter wildhood, developing social competence becomes mission critical. And at no other time in an individual's life is the SBN more active than in adolescence. Sarah-Jayne Blakemore, an author and University College London neuroscientist, has used imaging and other methods to show the major influence that social peers begin to have on decision-making and risk-taking in adolescents. Compared with adults and children, she writes, "adolescents are more sociable, form more complex and hierarchical peer relationships and are more sensitive to acceptance and rejection by peers." Laurence Steinberg, an author and Temple University psychologist, suggests that both the immature cognitive control of the adolescent brain and its increased sensitivity to reward may play roles in the power of peer influence in this age group. As he and his research colleagues reported, "Peer relations are never more salient than in adolescence."

Walking into a cafeteria, classroom, party, or workplace, an adolescent's Social Brain Network, which started wiring itself in infancy, is zinging with input. It's coordinating the six involved brain regions, while gauging the social landscape. The preoptic area pulls in visuals, clocking or deploying up-down glances for maximum assessment. The midbrain contributes memories of slights and snubs; the amygdala (the brain's fear center) flashes emotion and feelings of panic or dread. While the hypothalamus signals the release of stress hormones like cortisol or soothing hormones like oxytocin, the lateral septum promotes active stress coping behaviors. Keeping it all together is the prefrontal cortex, which decides and judges, moderates and plans the next move. It's busy up there, with those six regions zapping messages back and forth, making sure the individual understands the hierarchy and what he or she needs to do to cope within it. And this happens every day, all day long, every time any animal with an SBN—from humans to fish—encounters any other animal with an SBN, even other family members.

Of all the processes of broader brain reorganization that happen during wildhood, the sharpening of the SBN is one of the most crucial. The experiences adolescents have while their SBNs are calibrating often stay with them for the rest of their lives. A nearly universal ability to recall intensely humiliating or exhilarating moments of adolescence points to this. And one's perception of rank in adolescence may also be internalized. The adult brain that navigates friendships, business, politics, and social interactions is built during these sensitive years, and the adult may retain a mental map of her position in earlier adolescent social hierarchies. By late adolescence, SBN development is nearly fully complete. Like an eye in the sky, it will guide the individual through social terrain for the rest of her life.

WHY IT GETS BETTER

Dominance hierarchies are common among many groups of animals. They form and are regulated through aggression, violence, or the threat of force. Dominance hierarchies are a part of human history and modern life. They have been used to control large groups of people, like whole countries, or a single individual, like a spouse. Dictatorships, military occupations, prison societies, and physically abusive relationships are examples.

But in our species, status may rise based on an individual's excellence at something else, something less brute than physical force. If a group values a skill, an attribute, some know-how, or another quality, the person who possesses it can gather what is called "prestige." A person is said to have prestige if the deference shown to him or her is freely given without the threat of force. MacArthur Geniuses, Academy Award winners, YouTube stars, Malala Yousafzai, Yo-Yo Ma, J. K. Rowling, and your favorite Olympic athlete are all "prestigious." Their high status is based on group admiration for their scientific, artistic, humanitarian, or athletic abilities and contributions. Prestige doesn't have to involve celebrity or wealth. The marksman who consistently makes the shot; the home baker who brings the best brownies; the third-grader with the best bottle-flipping skills; the lawyer who wins a lot of cases; the fertility doctor with the highest pregnancy rate; the uncle who can always soothe the crying baby—humans value many forms of prestige.

In human hierarchies, dominance and prestige often interact, and, as has been shown time and again throughout history, both can be used for power and control. But for adolescents, understanding the difference can be revelatory, because at a key moment in adolescent development, the balance shifts. Popularity criteria for elementary, middle, and early high school hierarchies often fall outside individual control: Size. Attractiveness. Age. Athletic ability. Parental wealth. But, in mid-adolescence, a surge in competence-based hierarchies (prestige) occurs. A process called "niche picking" emerges, with students finding groups in which their particular skills and traits are valued. Their status, therefore, can rise. Competence may take the form of ability (musical, academic) or a high level of knowledge about a shared interest (politics, eccentric film, fashion, sports, video games).

Prestige hierarchies based on these competencies are often welcomed by students lacking in classic high school popularity characteristics. Moving beyond the tyranny of popularity hierarchies to find groups in which one's strengths translate into higher status underlies the "it gets better" advice given to adolescents struggling through this period of life.

Prestige hierarchies based on valued abilities also demonstrate the power of environment to affect status. Circumstances shift and qualities once of little value rise in importance. One generation's nerds are the next generation's app designers and computer programmers.

Baby hyenas, like baby humans, are born with SBNs primed to navigate complex and ferocious social terrain. And that was a very good thing for Shrink. As it turned out, with the deck stacked against him in so many ways, Shrink's greatest strength would be his social savvy.

Chapter 8

Privileged Creatures

The Ngorongoro Crater is the remnant of a monster volcano that blew open and then collapsed in on itself around three million years ago in what is now Tanzania. Ngorongoro, appropriately, means "big hole." Today, grasses carpet the flat caldera floor, supporting lush vegetation and an array of animals nourished by the fertile area and its river system.

For Shrink in 1998, however, this was no verdant playing field. In the communal den he was having a rough time. Every interaction with other cubs was a battle. When he tried to play with them, they turned on him. Smaller, younger, and male, Shrink was picked on by everyone. Even the mothers of other cubs would take a swipe.

MATERNAL RANK INHERITANCE

In communal dens, hyena cub hierarchy reshuffles after a few months of group living. At first, hierarchies are based on the usual markers of animal status—age, size, looks, sex. But by the age of four months, a near-linear hierarchy appears, with the cubs of the highest-ranking mother at the top and all the other little hyenas ranked below them. And that new hierarchy disregards age, size, sex, and looks. Instead it perfectly mirrors the hierarchy ranks of the cubs' mothers. The cubs of the highest-ranking female hyena sit at the top, with the offspring of the next highest below them, and the next next highest below them, and so on down the line. Meregesh, as the son of the queen, was at the top of the hierarchy, and Shrink was at the very bottom.

The social reshuffling among hyenas is a consequence of an ancient and powerful force called "maternal rank inheritance." An example of the "silver spoon effect," maternal rank inheritance guarantees that the sons and daughters of high-status mothers receive that high rank—and all the privileges associated with it—as a birthright. It may be surprising that an individual's position in a wild-animal hierarchy could be based on family connections, not on winning competitions of physical and competitive ability. Nepotism seems so human.

But with a moment of thought it ceases to be surprising. Having offspring that survive and go on to reproduce is the definition of success in evolutionary terms. It is natural for parents to want their young to inherit the advantages that would increase the chances of this happening. Maternal rank inheritance is an insurance policy for high-ranking animal parents whose offspring might not rise to the top based on merit.

Hyena clans are not meritocracies. When a hyena cub is born, he or she automatically receives the rank just below its mother's. The entire group knows this and shifts down one notch to make room for the young offspring, a social convention like adult travelers stepping aside to allow the children of first-class ticket holders to board the plane ahead of them.

Rank inheritance isn't limited to hyenas, and it isn't limited to mothers. From red deer to snow monkeys, the lucky heirs of high-ranking parents find themselves with desirable positions in the hierarchy, which are literally unearned. For these privileged creatures—sperm whales, domestic swine, wild spider monkeys, and many more—status rank is more than a perk; it's an entitlement. Then it becomes a way of life.

The offspring of the high-ranking and well-connected also inherit the social connections of their parents. They benefit from these established adult networks. Young adult birds, fish, and mammals, having grown up in the swirl of their parents' friends, go on to socialize within those extended social webs and even frequently interbreed with offspring of their parents' associates.

Among species, notably birds, where parenting roles are more equal, a father's rank boosts the status of his offspring. However, in mammalian life, where the care for young falls disproportionately on mothers, paternal rank has far less influence on offspring dominance than maternal rank. With chimpanzees living in Gombe National Park in Tanzania, researchers looking at conflicts between juveniles found that "offspring were more likely to win if their mother outranked their opponent's mother." For chimps, it seems, a playground taunt is more likely to be "my mother can beat up your mother" than "my father can beat up your father."

MATERNAL INTERVENTION

All hyena mothers try to interfere on behalf of their cubs, but the dominant mothers of high-ranking cubs are the most successful. Lower-ranking hyenas, like Shrink's mother, Beba, try to resolve conflicts by blocking the antagonist with their bodies. Or they create diversions, hoping to distract the group to bring the conflict to an end without resolving it. These mothers turn out to be less effective, in part because their tactics are more conciliatory. It's not that the Bebas of the hyena world are unaware that their cubs' social status is in danger—they just go about their interventions less aggressively, and thus less effectively, perhaps because of the ways they themselves were taught to handle conflict as cubs. It might also be because if they intervened more aggressively on behalf of their cubs, they'd be punished by higher-ranking adults.

High-ranking mothers, on the other hand, just go for it—zooming straight in on their kids' rivals with direct physical attacks. These displays demonstrate to the group the offspring's superiority. They also show the cubs how to wield power, how aggressive behavior is performed.

After working as a team a few times, with mother leading the way and cub feeling the elevated mood of winning, the mothers step back. Then, like Cinderella's stepsisters trying out their mother's cruelty, the younger offspring begin initiating attacks themselves. They choose targets that are easy to beat, those without parents or other allies around, and little hope of fighting back.

Maternal intervention doesn't always guarantee success. A hyena expert at Michigan State University, Kay E. Holekamp, told us that sometimes young females are born into high rank but "lack the chops" to take charge. In these cases, despite their mothers' efforts, the animals don't go on to inherit their mothers' positions. But that doesn't mean they have to leave the group or tumble to the bottom of the hierarchy. These individuals often live comfortably in the middle, free from energetically costly dominance displays but also benefiting from group kills and protection against predators.

As the mothers observe their offspring stepping in, they begin to pull away, applying backup muscle only when necessary. This is also the time when the mothers begin recruiting other adults to fight on behalf of their adolescent offspring. An unsupported young hyena trying to go up against a rival cub who has backup from a strong mother and the mother's adult coalition usually doesn't stand a chance. That challenger will likely experience the "loser effect"—a process in which social losing begets more social losing. Eventually a losing cub stops even trying to challenge a higher-ranking cub. We'll examine the loser effect more closely in the next chapter.

Having mastered bullying her own peers and felt the support of her mother's adult coalitions behind her, the high-born hyena female faces her next lesson: learning how to assert herself with—or some might say, to bully—lower-ranking adults. Again the mother leads the way, encouraging and supporting her rising hyena daughter to pick fights. Over time, these lower-ranking adults (and the rest of the clan observing the interactions) come to learn that despite their age, seniority, and size, they occupy a rank below that of the privileged daughter. Eventually the high-ranking mother doesn't even have to be present for other group members to recognize and respect the daughter's position.

A vivid primate example of this kind of training behavior comes from Cambridge behaviorist Tim Clutton-Brock, who describes a low-ranking female adult Sri Lankan macaque out in the forest foraging for fruit. She was stuffing the food in her cheeks when suddenly a younger and smaller juvenile, the daughter of a high-ranking mother, plopped herself down and reached for the older monkey's lower lip. Grabbing it, the younger monkey stretched the older monkey's lip until she could reach in and pull out the partially chewed fruit. The subordinate adult relinquished her bounty without protest, recognizing that she must defer to this juvenile from a dominant matriline or she would be punished. And making sure the whole encounter went down as planned was the dominant mother, who sat fifty meters away, watching from a tree.

Maternal interventions aren't found only in mammal societies. And they aren't all aggressive or physically violent. Primate, bird, and fish parents give gifts to adults in their communities that are positioned to help their offspring. These gifts can be literal, in the form of food. Or they can be behavioral, often in the form of grooming. Lower-status cleaner wrasses, a kind of fish that lives in coral reefs and eats parasites off other fish, and lower-status baboons will nibble or pluck at higher-ranking animals, and the overture can solidify or elevate their offspring's social position.

Maternal rank inheritance and intervention are a bitter pill for an adolescent hyena like Shrink. It's one that members of other species, including our own, are forced to swallow throughout their lives. No matter how good, smart, strong, or prepared you are, if you find yourself up against offspring of a high-ranking individual, you can be in for a really hard battle. The daughters of high-ranking hyenas (like the offspring of nature's other power moms) have advantages you might not be able to see: They've had better nutrition, and probably have stronger immune systems. They may be more aggressive—or accustomed to getting their way—because they've been trained to demand what they want. They've had more opportunities and more protection from their mistakes. And often, their parents have explicitly trained them from a young age how to be bullies and how not to lose.

Privileged creatures can be found among fish, reptiles, birds, and mammals. They're born with advantages and primed to behave a certain way. Wildhood is when these behaviors kick in, and the nature of each individual's adolescence is greatly influenced by the advantages he or she inherits.

In human societies, infancy and young childhood can be periods of relative class blindness. But as children enter adolescence, class, rank, status, and position become more apparent. One of the great challenges of adolescence is entering an adult world that is rigged against less well-born animals. Understanding privilege in the natural world, from parental rank inheritance and beyond, is crucial to understanding it in our own.

TERRITORY INHERITANCE

Privilege in the animal world can also come in the form of territory inheritance, which offers a powerful leg up for any lucky animal or human. Like human dynasts, advantaged Eurasian beavers born to parents with safe, resource-rich real estate will inherit that territory, including all dams and other structures on it, when their parents die. Juvenile pikas, red foxes, and scrub jays also inherit their parents' territories upon their death. And self-sacrificing red squirrel mothers bequeath their full territories to their adolescent offspring and then set off on middle-aged journeys to find other places to live. If the bequeathal comes before the offspring is ready to defend the area, the gift often comes with ongoing protection by one or the other parent until the offspring is skilled enough to defend it on her or his own.

Level playing fields don't exist in nature, and the ancient roots of animal privilege are everywhere. There are privileged bison, privileged birds, and privileged bears, privileged insects in first-class sections of the swarm and privileged oysters tucked into cozier, safer, more comfortable beds. In flower fields grow privileged tulips, the offspring of powerful flower "parents," planted in sunnier or wetter parts of the field. At the base of trees, deep in the woods, grow privileged truffles, with lives of ease that their kin and other nearby fungi might resent or covet if they could.

Privilege exerts its influence even at the microscopic level. Some individual cells receive greater advantages than others. Within a malignant tumor the size of a pencil eraser, for example, billions of cells compete with one another for resources. Like swallows in a flock, each cell in the tumor is an individual with unique strengths and weaknesses. Some have better access to blood supplies and exploit that access to replicate greedily. Some have safer positions at the center of the tumor out of the range of chemo or immunotherapies. Some were stressed early in their development. Others had easy starts. One hypothesis for why cancers metastasize, in fact, is that, deprived of resources, the more desperate cells leave their originating tumor—their homeland, so to speak—to try their luck in better, greener pastures elsewhere. Or, hounded by predatory T-cells, they race off on journeys of the body to set up residences in regions far from the original cancer.

And, as is the case within our species, some entire animal populations have privilege over other groups based solely on where they were born. Perhaps even more than parentage, environment is often destiny. A ragtag group of unexceptional hyenas in a lush, food-rich, lion-scarce environment will likely do very well compared with a clan composed of the strongest and most clever hyenas living in the midst of a drought, famine, or poaching zone.

Even in his disadvantaged position, Shrink was privileged compared with his hyena peers outside of Ngorongoro Crater. In most hyena clans, when twins are born, one so restricts access to food that the other dies of starvation. Not in the Ngorongoro Crater, however. During many months of the year, the Ngorongoro teems with food—one study counted a whopping 219 possible prey animals per square kilometer. The impoverished Serengeti hyenas nearby have to fight over a mere 3.3 possible prey items per square kilometer. And survivability is linked to this environment-determined privilege. In the Serengeti, both twins rarely live. In the Ngorongoro, they almost always do. Even triplets, which are rarer than twins, survive in the Ngorongoro, according to Höner's research. Shrink was privileged, in a way, to be born there.

PUTTING PRIVILEGE IN PERSPECTIVE

The animal roots of human privilege are ubiquitous, hiding in plain sight if we know how to look for them. From an evolutionary standpoint it makes sense. Animal parents want their offspring to be advantaged with better access to the resources and safety that will increase their chances of surviving and reproducing. Layers of unearned advantages for some animals—and the absence of these for others—influence their every interaction.

In human societies that consider themselves to be merit-based, adolescents may be told that good things will happen to those who work the hardest. College legacy admissions, access to internships and jobs, and introductions to powerful networks are observable ways that privilege furthers itself. But globally, many enter worlds in which the circumstances of their young lives—their health, environment, and family; their wealth, race, and gender—are far more powerful in shaping their future destinies than merit. But the influence of privilege on the lives of adolescents goes far beyond that. Privilege determines whether an adolescent lives in poverty, has clean water, and is physically safe. Privilege determines whether an adolescent has access to reproductive health care. Privilege shapes educational and occupational opportunities.

The ubiquitous influence of privilege in the natural world, from the life histories of individual cancer cells to those of wild animals, is certainly not an endorsement of or justification for using it to stratify human social classes. On the contrary, as uncomfortable and even depressing as it may be to encounter at first, the ancient animal origins of privilege contain important lessons for modern humans.

There are species for whom rank is believed to be fully determined by winning fights. But victorious animals often come to contests preloaded with generations and layers of advantages. Uninformed observers may see what looks like a fair fight between animal equals. But privilege is omnipresent, shaping outcomes. Recognizing its invisible power in nature can help us understand its oversized influence on the lives of young humans. The frequent contests and assessments of wildhood may appear to be based on merit. But the primal, embedded nature of privilege paints a more complicated picture.

To build airplanes that fly, humans needed to understand the laws of gravity. To develop antibiotics to fight infection, we had to study the strategies of pathogens. If we are to be successful in building equitable societies, exposing and understanding privilege in the natural world is crucial.

As powerful as privilege is in nature, it does not always determine the destinies of adolescents. As we've seen, when environments change through weather, disease outbreaks, or other unexpected events, the disadvantages of some can become advantages. For adolescents and young adults facing privilege-based obstacles, making that change happen rather than waiting for it can rearrange their circumstances. For some, changing environments may be a good idea.

Oliver Höner told us that lower-ranking spotted hyenas can leave a social situation and remake themselves elsewhere more easily than higher-ranking hyenas can. In fact, their early-life adversity may make them more behaviorally flexible than higher-ranking hyenas. Studies of meerkats and feral guinea pigs also indicate that some adversity during adolescence actually promotes innovation and tenacity. With hyenas, because food is more consistently available to the higher-ranking females, they usually stay at home; they're less motivated to look elsewhere. Lower-ranking females and males, on the other hand, go on more excursions outside the home range, and they're the first to learn when a new territory has opened up. As Höner explained:

"We've seen low-ranking females being very successful after they have moved to a vacant area. It doesn't happen a lot in intact ecosystems that areas get vacant, but it can happen, be it due to disease outbreaks or other forces. In one case in Kenya it was because a whole clan was poisoned by poachers, so it was a whole territory that was completely empty. And low-ranking females moved in and lived a very happy life. Much happier than had they stayed in their original clan."

While place of birth, resources, and family connections play powerful roles in the lives of all animals, they do not always determine destiny. Difficult early circumstances are not always fatal and may build strength.

New alliances can be formed. Contests can be won. Environments can change. And young individuals with some motivation, knowledge of how privilege works, and a little bit of luck can shape their lives to be very different from the ones they were born into.

Of course, that's not always easy, as Shrink soon learned.

Chapter 9

The Pain of Social Descent

In the United States in the 1950s, five people became severely depressed: a widow, a retired policeman, a business executive, a housewife, and a college professor. Experiencing depression in 1950s America was certainly not unusual. But these five people shared a medical condition that had nothing to do with their mental health. When they fell into depression, all five were being treated for high blood pressure. And all five were receiving a drug called reserpine. Reserpine lowers blood pressure by lowering levels of neurotransmitters called monoamines. But those lowered levels of monoamines also seemed to be lowering the moods of the five patients. The *New England Journal of Medicine* issue reporting these cases also noted that when the patients stopped taking reserpine their depressions lifted and their normal moods were restored. This study helped spark the extraordinarily influential, though only partially accurate, monoamine hypothesis: depression is caused by—or at least driven by—low levels of monoamines.

In the six decades since the study, this connection has been studied and refined, but the basics remain that, while human depression with all its complexities can't be reduced to the work of a single group of molecules, the monoamines clearly play an important role in shaping human moods. And the most well-known of the monoamines? Serotonin, the neurotransmitter whose levels are manipulated by SSRI antidepressant drugs such as Prozac, Celexa, Lexapro, Paxil, and Zoloft. People on SSRIs today are taking them based on years of evidence that in humans, raising levels of serotonin in certain parts of the brain can sometimes improve mood.

Now consider a different piece of knowledge, from the world of animal behavior. When a lobster is born, as a tiny free-swimming larva, it neither looks nor behaves like the giant clawed fighter that it may one day be. But within three months it transforms into a tiny juvenile version of its adult form. For the next several years juveniles hide themselves as they get older and bigger. By about six to eight years they're close to adult-sized adolescents. At this point, just like hyenas and humans, lobsters begin sorting themselves into hierarchies. And like chickens setting up a pecking order, wild lobster hierarchies are established with very few fights. By watching the behavior and smelling the urine of other lobsters, individuals can determine and remember who's above them and who's below. High-ranking lobsters evict subordinates from their burrows, attacking them with leg grabs and antennae locks. Lower-ranking lobsters retreat with submissive tail flips. Lobsters, whose ancient ancestors appeared on Earth during a period of massive global wildfires, have been skirmishing for status for 360 million years.

But a certain substance has the power to change all that. Scientists studying rank relationships among these crustaceans found that if lower-ranking Norwegian lobsters were given this substance they showed fewer of their characteristic low-ranking behaviors. They were less likely to retreat when challenged. They were more willing to fight, which is uncommon in subordinate lobsters. And they began to posture and pose the way dominant lobsters do, most notably with the classic lobster "meral spread"—a threat display in which they raise up the front half of their bodies and wave their claws menacingly. In other words, although nothing in their environments had changed except the substance they had been given, they began acting and appearing as if they were no longer low-ranking.

A similar study in crayfish showed the same thing: when subordinate crayfish were given this substance, they no longer withdrew from threats or fights, behavior that suggested their status had risen. They didn't have to actually fight and win. Their postures, poses, and behaviors were enough to establish their dominance. And, like the lobsters, the crayfishes' peers began treating them as if their status had gone up. Rank perception (status) became rank reality. The effect has also been seen in fish and mammals: treated with this substance, lower-ranking animals acted higher-ranking, and because of that, their cohorts began treating them that way.

The substance, of course, is serotonin. In animals, serotonin plays a central role in brain activity related to social rank, specifically the rise and fall in status. It's the same central role it plays in the rising and falling of human moods. Putting these two ideas together suggests an important connection between the work of animal behaviorists and that of human psychiatrists: a link between the regulation of mood and an animal's status.

HELPLESS AND HOPELESS

As we've seen, social descent is a ubiquitous experience for social animals. No one stays on top forever. And we've seen how brain systems like the Social Brain Network and transitive rank inference detect changes in status and send neurochemical messages—status signals—that nudge animals toward behaviors that improve their chances of survival. But what do these signals "feel" like? Nonhuman animals can't tell us. But scientists observing the behavior of low-ranking animals suggest that if these animals could talk, they would say it doesn't feel good at all.

In the early twentieth century, Thorleif Schjelderup-Ebbe, with a liberal combination of anthropomorphism and objective scrutiny, described dominant birds fallen from positions of "unlimited authority" as "deeply depressed in spirit, humble, with drooping wings and heads in the dust." The dethroned birds were "overcome with paralysis although one cannot detect any physical injury." Schjelderup-Ebbe further noted that the severity of this reaction was worsened if the bird had been an "absolute ruler for a long time," and added that if the social descent was extreme "it is nearly always fatal."

Other ornithologists have confirmed this observation. V. C. Wynne-Edwards, an English zoologist of the twentieth century, noted that

Scottish red grouse who suffer low status after failing in contests with others for territory sometimes "mope and die." If they were human, these birds would be diagnosed as depressed. The trigger for their depression? Social descent.

Forty years ago, a Belgian ornithologist who was also a psychiatrist described similarities in the behaviors of his patients and the birds he loved to study. Albert Demaret noticed that birds with territory proudly strutted about. It reminded him of the swaggering he'd observed in his human patients in high spirits. His depressed patients, on the other hand, behaved more like birds furtively lurking on another's territory. They skulked and minced, and kept quiet, notably limiting their singing.

We can't ask individual birds how they feel when they lose coveted, protected positions at the head of the flock and are pushed to its dangerous edges—any more than we can ask fish, lizards, or nonhuman mammals.

But we can ask people. Insults, humiliation, financial losses, romantic disappointments—all of these status-reducing experiences lower our moods. They make us sad. Even just thinking about a potentially cringeworthy comment or situation can be enough to cause a surge of distress. In extreme cases of status descent, the pain may be so severe that some people consider extreme measures, including substance abuse and self-harm to get relief from the agony.

The emotional experiences of human life may be unique to our species, but having an emotional brain is not. Many of the brain processes and chemicals that drive human feelings are also found in the brain reward systems that we share with many other species. These systems work with a classic carrot-stick dynamic. Simplistically, feelings of pleasure reward behaviors that promote survival. Our bodies unleash elixirs of neurochemicals, such as dopamine, serotonin, oxytocin, and endorphins, to tell us, "Good job! You just did something right. Keep doing that and there are more good feelings where that came from."

On the flip side, drops in mood are created by a shower of noxious neurochemicals, such as cortisol and adrenaline. The unpleasant sensations created by this bath are made even worse by the withdrawal of pleasure-producing neurotransmitters. What it feels like to other animals we don't and may never know. But in our species, we call it decreased mood, even sadness. The chemical reprimand motivates animals to change their behaviors in ways that will restore and elevate their position.

Rising in status improves animals' chances of survival. When animals rise in status, they are chemically rewarded. Rising in status feels great.

Plummeting in status, on the other hand, reduces animals' chances of survival. When animals fall in status, they are chemically punished. Status descent feels terrible.

Newer research on status and serotonin in anole lizards, bluebanded gobies, lobsters, crayfish, rainbow trout, and more that connect serotonin and status suggest this possibility: Serotonin levels don't *control* an animal's mood. Serotonin, along with other neurotransmitters, signals a shift in an animal's *status.*

The status-mood connection is a powerful lens for interpreting adolescent and young adult behavior, mood swings, anxiety, and depression. Public humiliation and other forms of significant status descent may even increase vulnerability to suicide. Losing status rank is painful—literally. So is spending your young life at the bottom of social hierarchies.

During wildhood, the combination of adolescents' increased sensitivity to social status and intensified experience of social distress may set them up for depression. Social pain is excruciating and not something to be trivialized. It may therefore be not only insensitive, but perhaps even ignorant to ask an adolescent why they care so much about what others think. After all, it has been the job and preoccupation of every social animal adolescent, from humans, to hyenas, to lobsters, to scan its world for clues about its status and to pay close attention to what it learns. And to intensely feel—sometimes with joy and sometimes with pain—when that status shifts.

SOCIAL PAIN

Underlying those spiraling negative feelings that accompany status descent is what we call "social pain," a phenomenon that has been studied extensively by UCLA neuroscientist Naomi Eisenberger. Her work connects the physical and emotional pain of being excluded.

In one study, her team imaged the brains of adolescents while they played an internet-based game that simulated social exclusion. Eisenberger's findings suggest not only that the neural pathways involved with physical pain and social pain are shared, but also that exclusion is especially painful for adolescents. In other words, there's a reason that kids may do things their status-oblivious parents can't explain: it hurts so much to be left out.

Eisenberger has also linked social pain to opiate addiction and overdose. It's worth noting that substance use and abuse—one of the major health risks of adolescents and young adults—often begin when adolescents first enter the arena of high-stakes social sorting. With their Social Brain Networks at peak sensitivity to status descent and social pain, adolescents may reach for intoxicants because they can blunt the feelings of social pain.

In a related study, Eisenberger has shown that acetaminophen, best known in the United States by the brand name Tylenol, can mitigate not only physical pain, but social pain as well. As seen on MRIs, the pain of social exclusion activates essentially the same brain regions and pathways involved with physical pain. One of the ways acetaminophen may reduce pain is through activation of mu-opioid receptors, which also respond to THC, the active molecule in cannabis.

Beyond self-medicating to relieve social pain, using drugs, alcohol, and tobacco may play another role in making adolescent users feel as though they've raised their status. It may signal to a group that the user is older. As we've seen, hierarchies tend to defer to older group members.

Given the power of social descent to create social pain, adults who care for adolescents could consider talking openly about status. Popularity and status are deeply rooted in our evolutionary pasts and are the central obsession of many adolescent young adults. Questions about popularity and friendships may be more likely to yield information about social pain than direct questioning about mood.

THE TARGET ANIMAL

After about eight months in the communal den, Shrink, his twin sister, Prince Meregesh, and their peers entered another developmental phase, one with more independence. They began finding their own food and forming relationships with other adults in the clan. You might think that as older adolescents, hyenas would gain some freedom from the adult interventions into their hierarchies. But this is actually a moment when maternal intervention becomes even more intense.

High-ranking mothers continue to intercede in offspring conflicts—long after their young are big enough to fight their own battles. Dominant mother hyenas push subordinates out of the way so that their daughter or son gets first crack at a carcass. They rush to their offspring's sides during fights with older adults and help them win.

Not only did Queen Mafuta's maternal intervention ensure that Meregesh got everything he wanted, from the best food and sleeping areas to the most popular friends, it also kept him from experiencing the dreaded loser effect. The loser effect is something mother hyenas know instinctively. Once a winner wins, she tends to keep winning. Once a loser loses, a similar pattern is seen, and the losing often won't stop. And so avoiding the loser effect while promoting the winner effect becomes a way of training a developing adolescent (and his or her developing brain) to grow accustomed to the feeling of higher rank.

Practicing on easy marks accomplishes this. An adolescent can become what's called a target animal, an individual chosen by dominants to receive abuse. Low-ranking individuals may initially be selected as target animals because of a physical or behavioral difference. But they become prime targets if they have no allies or coalition to come to their aid. Target animals experience frequent, and sometimes unrelenting, social defeat.

Studies of social defeat in mice have shown that losing fights makes them much less aggressive in subsequent bouts, and more prone to losing. Over time, the loser effect leads low-ranking animals to simply give up trying. They often won't even engage with other animals—in rank battles or other socializing. Studies of lobster dominance have shown similar trends.

Being the target animal makes low-status adolescents like Shrink skittish and vigilant, constantly on edge. Without status, they can't make friends. But without friends, it's hard to get and maintain status. If subordinate hyenas were thirteen-year-olds, they might say they were depressed.

Seriously depressed adolescents and young adult humans often describe feelings of worthlessness, helplessness, and hopelessness. They say they feel as if nothing they do will help or change their state of mind. This is practically a definition of the loser effect at work in groups of fish, birds, mammals, and crustaceans.

If defeated lobsters and hyenas could express themselves with words, they might describe a sense of worthlessness (low status compared with their dominant aggressor); helplessness (no peers to come to their aid); and hopelessness (a sense that they can't win, so why bother trying?).

Worthlessness is among the symptoms listed in the *Diagnostic and Statistical Manual*'s Criteria for Major Depression, and other sources on depression cite hopelessness. In yet another connection to birds, Schjelderup-Ebbe in 1935 described subordinate birds as seeming "dulled by a premonition of hopelessness," compared with superior birds' "joy of satisfied pecking-lust."

Unlike adults who are better able to remove themselves from toxic hierarchies, adolescents and young adults are often stuck. Legally required to stay in schools where they may be mocked or bullied, and socially or financially tied to neighborhoods or families in which they may be disregarded, many adolescents just can't escape. Or at least that's how it feels to them.

Adolescents and adults who seem to have it all still get sad, sometimes even truly depressed. A human being's internal self-perception can be very different from how others see them. Social experiences during adolescence shape individuals' views of their status in ways that sometimes continue into adult life. The happiness that might come from success in adult life may be blunted by the enduring effects of social defeats during adolescence.

It should interest parents, teachers, mental health professionals, and teenagers themselves that some actions do seem to cause animal hierarchies to shift. Experiments on hierarchy stability show that removing individual fish or monkeys from a group and putting them back in at a later time can sometimes lead to a re-sorting, a new shuffling of the social deck. A human version of this might be a student leaving for summer vacation and coming back to find himself in a different place in the hierarchy. This can be good for an adolescent struggling at the bottom of a group who finds him- or herself in a better position upon return. (On the other hand, as almost anyone who has experienced it can attest, missing out on group activities during adolescence can sometimes result in being shuffled to the bottom.)

Increasing physical spaces can also sometimes loosen up fixed hierarchies. In late summer 2014, we traveled to Saskatchewan, where we had an opportunity to observe a herd of Canadian bison that had been brought into a large corral after having spent the summer on open rangeland in Prince Albert National Park. We trudged across a muddy paddock among the huge, beautiful creatures, hearing their low grunting. Suddenly they all started moving toward a water trough, where they lined up, peacefully and obediently.

Their order at the trough wasn't random. The dominants drank first, and the rank order progressed down the line. Visiting veterinary schools and dairy farms, we'd seen similar peaceful bovine linear hierarchies with groups of cows heading off to the milking parlor, the bossy dominants going first.

The veterinarian who cares for these Saskatchewan bison told us the drinking hierarchy emerges in late summer and winter, when the animals are boarded in barns. In the spring, however, on the vast acreage of the national park, the hierarchies become less fixed. Dominants and subordinates drink at the same time around an open lake. Relaxing a rigid hierarchy may be as simple as going outside. The key is that hierarchies become more rigid when resources are scarce. And having enough personal space is a valuable resource.

But even if physical spaces can be improved or adolescents can escape from a toxic group, the self-perception of low status can linger. While self-assessment of rank is usually accurate among schoolchildren, studies of depressed adolescents show that they perceive their status as substantially lower than how their peers view it. Many have an internalized sense of themselves as being on the lower end of the hierarchy. The loser effect may begin with an actual contest with another individual, but it lives on in the mind of the defeated, who subsequently feels beaten before even trying. It can create an identity, a lasting imprint. This effect may be especially strong during wildhood, a period of intense hierarchy-building, social experimentation, and brain reorganization.

FEELING LIKE A LOSER: BULLYING

One of the most predictable accelerants of depression in adolescents is bullying. Multiple studies have shown a close association between having been bullied and developing depression or anxiety. A 2005 study that compared bullying across eleven-, thirteen-, and fifteen-year-olds in twenty-eight countries reported that numbers varied widely, with the highest prevalence of bullying being among boys in Lithuania and the lowest among girls in Sweden. Some 20 percent of school students, grades nine through twelve in the United States report they've been victims, according to the National Institutes of Health (NIH), which has a special anti-bullying unit as part of its adolescent health task force. Defined by the NIH, bullying is "unwanted aggressive behavior by another person or group." It can be physical, like punching, kicking, and pushing, or behavioral, like hiding, stealing, and damaging belongings. Bullying can be verbal: name-calling, teasing, spreading rumors or untruths. Bullying can also be coercive and indirect: "refusing to talk to someone or making them feel left out; encouraging other individuals to bully someone."

Although much has been learned about bullying over the past decade, it is impossible to fully understand the complexities of human bullying without examining its function and form in animals. By doing this, we discovered that the current thinking on bullying—and possibly even clues for treating it—could be profoundly enhanced by applying to human behavior what animal behaviorists have long known about hierarchies in other species. Working across disciplines, we identified three different types of animal bullies that have human correlates. We call them "dominators," "conformers," and "redirectors."

Bully Type #1: Dominators

The main thing to know about bullying in animals is that it's almost always done in order to gain and retain status. High-ranking animals eager to maintain their ranks bully others as a dominance display, a public show performed for the community whose purpose is to reaffirm the high status of the bully. Remember, status is about perception, and so to get and keep status, a dominance bully needs an audience. If onlookers accept the dominance display of individuals or groups—and they typically do—the bully continues as dominant.

Dominators choose their victims carefully. They don't take on a social peer or near rival. They choose lower-status individuals to pick on. One difference between animal and human dominance bullying: in humans the aggression doesn't have to be physical. The threat of emotional injury through humiliation can be the weapon of domination.

As we've seen in hyenas and other animals, including primates, dominance bullies—who can be male and female—are sometimes nurtured by their bully parents and trained from a young age to grab power and to threaten, bluster, or overreact if anyone resists. Learning how to be a bully begins early and is self-reinforcing: the more dominantly an animal behaves, the more he or she will be perceived as high-ranking. Aggressions against target animals provide practice for the young dominants as well as educational theater for the rest of the group, who can witness their own falling positions in relation to the rising young elite.

Dominance bullies can be scary and unpredictable, because they need to constantly demonstrate their power. If the group stops paying attention, these bullies will not hesitate to make an example of someone weaker, in order to profit from the audience effect.

Without community support, the transgenerational reign of dominance bullies in a troop or school can be hard to dismantle. And remarkably, communities sometimes behave in ways that enable these parent-to-child bully legacies to persevere. Coalitions of older adults—often the parents of lower-ranking animals eager to stay in the good graces of the dominant family—sometimes do the dirty work, picking on adolescent members of their own rank in the hopes of raising it. The reluctance of onlookers to stand up to a bully comes from their fear of becoming targets themselves. But it may also be that the targeted animal is different in a way that lowers the status of the group or creates danger for it. The onlookers' reluctance to intervene may be part of an ancient tendency of groups to shun individuals who are different, and the reverberations of the oddity effect.

Bully Type #2: Conformers

We wondered if Shrink's low status might have been affected by his "special" ear—the one with the bend in it that made him look a little different from the other young hyenas. We asked Oliver Höner, who said it surely had no influence on Shrink's place in the group, although it may have had other effects, for example on Shrink's personality or even hearing. Höner emphasized that both of those possibilities were highly unlikely, but unstudied, so he couldn't say definitively. However, to our surprise, Höner did say that he has found a correlation between hyenas' rank and the condition of their ears. "Alpha females," he told us, "have much nicer ears than low-ranking hyenas." He explained that when hyenas fight, they go for the ears, and it's not unusual for the vulnerable appendages to be shredded or bitten entirely off. Hyenas that can't manipulate their ears to show submission when necessary are disadvantaged socially. Höner noted that he has found only correlation, not causation, between number of ear notches and rank.

We've seen that many dominator bullies choose target animals based on their differences. Similarly, appearance-based bullying—which involves exclusion, shaming, and shunning of individuals who are different in some physical or behavioral way—is prevalent in younger human adolescents. According to a 2018 report released by the nonprofit YouthTruth, 40 percent of middle school students report having been bullied, with appearance-based bullying being the most common type. Often this kind of bullying is the work of an individual dominator trying to retain power and status.

But there is another type of bully who also targets differences. Conformer bullies use the threat of shunning—social exclusion—as their weapon. But the underlying purpose is different from that of dominance bullying. Conformer bullies aren't motivated by the need to demonstrate and elevate their status by acting aggressively toward others. Consciously or not, they may be trying to protect themselves and their group by removing non-conformers. Being in a group with "odd" members can attract unwanted attention.

Like dominance behavior, conformity has a strong and ancient evolutionary foundation. As we saw in Part I ("Safety"), schools of fish, flocks of birds, and herds of mammals are in greater danger of predation if members of the group look or behave differently. The oddity effect—which you may recall has roots in antipredation behavior—is the shunning that groups show toward a member that is different in some way. Animals standing, swimming, or flying near the oddly colored or oddly behaving individual are at particular risk. They may feel it's a matter of life and death to get away from the odd animal, out of fear that they will also become easy targets, simply by proximity.

While humans don't live in flocks, herds, or schools, we do share some behaviors typically seen in other animals living in groups. The oddity effect may lead to conformer-type bullying, as individuals attempt to avoid those who might put them at risk of another danger: the dreaded social descent.

Middle and high school bullies may leverage their group's innate preference for conformity by pointing out differences (real, exaggerated, or fabricated) in their targets. Spreading sexual rumors or homophobic slurs is a common tactic. The differences of others are emphasized to reduce their status and create distance. This is a process sociologists call "othering." Once an individual has been "othered," the larger group is less likely to support them. They may even join in on the bullying. Fear of being "othered" induces conformity in group members—this is true among adolescents and in adult society.

Like adolescent bullies, some political leaders cannily leverage their seemingly instinctive understanding of othering and its connection to the oddity effect. Nazi Germany's portrayal of Jews as typhus-spreading vermin and Rwandan Hutu propaganda depicting Tutsis as diseased cockroaches are classic examples. The targeted groups were othered as threats to the safety of the group.

Bully Type #3: Redirectors

Some notions about bullies hold that these intimidating individuals are actually victims themselves. Perhaps they have poor self-esteem and take their frustrations out on others. However, because the majority of animal bullying is performed as dominance displays by higher-ranking animals against lower-ranking ones (it's very rare for a lower-ranking animal to aggress against a higher-ranking one), we believe the bully-as-victim may be a third type, which we call the redirector.

Unlike dominance bullying, which springs from confidence, redirection bullying has its basis in anxiety and fear. To gain perspective on how it may function in people, it helps to see how it works in dogs.

James Ha, an author and animal behaviorist from the University of Washington, has more than four decades of experience decoding animal actions and helping clients with troubling behaviors in their pets. He described for us a form of dog aggression that seems to come out of nowhere. It often shows up in otherwise well-behaved dogs that are anxious and have been severely punished or dominated, sometimes by a member of their human family. These dogs are fearful, especially in the presence of the terrifying human, and sometimes react by barking, lunging, and biting. But never do these dogs attack the thing they're actually scared of. They go after innocent bystanders, often the youngest member of the family or a smaller animal.

This behavior can be made worse by what Ha calls "trigger-stacking"—when the dog's usual anxiety triggers begin to build up and the dog feels it has no option but to attack. Anxiety triggers for dogs can be as common and obvious as loud fireworks and thunder, or as subtle and hard to read as time of day or a strange scent. But as the triggers stack, the dog may get more and more anxious until it directs its aggression outward.

Redirectors respond poorly to force, and overly rough discipline makes their fear and anxiety worse, which in turn can make their aggression worse. "We don't punish fear" is a refrain of equine behavior expert Robin Foster, because not only can't fearful animals process punishment in the moment but doing so also reinforces the link between fear and aggression. Redirection bullying, especially if it crops up during the sensitive developmental window of adolescence, becomes an animal's default way of coping with the normal anxieties of life. These dogs have been conditioned to associate fear with aggression. They've learned, explains Ha, that "if I'm aggressive when I'm scared, the scary thing goes away."

For dogs, the major contributing factor to developing this behavior, says Ha, is lack of socialization—not being around puppies and people during critical periods of development. The dogs most at risk of developing anxiety aggression are shelter dogs. And of that population, the ones at highest risk are dogs placed into shelters during adolescence. Particularly if they're attacked by another dog while they're there, shelter animals can develop what Ha calls "kennel dog syndrome." These are dogs whose fear-based aggression is so ingrained they're hard to adopt. Dogs that are isolated, attacked, or severely punished during adolescence can spend a lifetime struggling with behavior issues that make them hard to live with and have around other dogs. Many of these animals never do manage to settle into a happy or calm life, even with medication and optimistic, patient owners who want to help them recover.

And this is key. The effects of anxiety may be made even worse—

they may last longer, grow deeper, possibly even cause brain or genetic changes—if the anxiety begins during critical periods of development, like adolescence.

THIS IS YOUR BRAIN ON LOW STATUS

Linking status and mood allows us to see a range of other applications. Being bullied pushes an animal's status down and can be damaging. A study of mice demonstrated how status descent impairs learning. Eighteen mice were tested on their maze-learning ability. Then they were forced to spend three days cooped up with another mouse, causing one mouse to become dominant and the other subordinate. When retested, the performance of dominant mice improved, but the opposite was seen among subordinates. It is possible that dominant mice were performance-enhanced by higher testosterone levels (the winner effect). Or that subordinate mice were impaired by higher stress-hormone levels. Whatever the mechanism, the lowered learning and test performance by subordinates might have major relevance for adolescents who are trying to learn amid the status battles of classrooms and schools.

A study of rhesus macaques showed a different way that status interferes with the demonstration of talent and learning. Two groups of monkeys—one made up exclusively of the members of high-ranking matrilines and the other of members of lower-status groups—were given a series of tests. First they were evaluated on their ability to approach unfamiliar boxes and learn to dig for peanuts. Then they were timed on how quickly they could properly choose which colored boxes contained the peanuts and which contained rocks. The total peanut retrieval was measured.

The monkeys were given these tests under two conditions: in front of only their cohorts, low or high status, and also in front of the combined group. While monkeys from the high-status families excelled under both conditions, the low-status monkeys performed well only if no high-status monkeys were present.

The investigators concluded that lower-status monkeys were intentionally inhibiting their performance, effectively "dumbing down." It is probably an extension of classic subordination behavior, which helps to minimize conflict and the risk of aggression from dominants. However, the response is likely embedded in the social brains of our species too. If you've ever tried to concentrate on a conversation when a celebrity or bully is in the room—or on a mental task when a rival is staring you down—you understand how strong this effect can be.

Understanding that differences in status can impair learning and academic performance is essential for educators and students at all levels. It may be useful for an elementary school teacher struggling to understand why a bright child can't grasp a concept. Or why a middle or high school student who gets the material just can't demonstrate it on the test. And it should be part of campus conversations about clubs and societies that, through the exclusion of others by race, gender, and socioeconomic level, create the kinds of status hierarchies known to impair learning ability, academic performance, and possibly future opportunities for members of excluded groups.

Chapter 10

The Power of an Ally

Not every adolescent who is bullied will become depressed. Some are better able to cope with the stresses of it than others, and for humans one important mitigating factor is the presence of allies and friends. Jaana Juvonen, an adolescent bullying expert at UCLA, puts it this way: "The power of a friend is incredible. A kid with just one friend [has] a lower risk of getting victimized, getting bullied in the first place. Moreover, the distress of the victim is alleviated when they have that friend."

Höner confirmed the same is true for hyenas. "The number of friends you have will ensure you keep your social status," he told us.

Shrink's early life was defined by his lack of advantage. But after observing Shrink interacting with other hyenas in the communal den, Höner and his team started seeing something interesting. Shrink was especially good at something called "social coalition walking," the hyena version of inviting a friend for coffee or a pickup game of basketball. More colloquially, this winning hyena behavior is known as "friendship walking."

As Höner described it to us, "two males meet and somehow decide, 'Let's go on a little excursion together.'" Höner's voice had a smiling tone, almost affectionate, as he explained how Shrink would approach other males, and then the two would trot along together, their bodies touching, tails up in the confident position. Every few meters Shrink and his friend would stop and sniff a grass stalk attentively, even if nothing interesting was in fact to be seen or smelled. Sniffing during a friendship walk is hyena small talk, like commenting about the weather or sports or politics just to be sociable. Hyena friendship walks can last several hours, with the two animals stopping every so often for these sniffing conversations. The behavior is seen in adult hyenas as well—in fact, it's one of the main ways that adults preserve their social bonds. Doing a lot of it in adolescence, as Shrink did, sets a young hyena up for an easier social time later.

His skill at hyena friendship, his ability and willingness to invite other hyenas out for some bonding, served Shrink well. Höner hasn't looked into why it's easier for some animals to initiate this behavior than others, whether because of personality, temperament, or opportunity. But one thing is clear: learning to attract friends and maintain friendships during wildhood is vital, and it isn't automatic. Adolescents must practice having friends, repeating the give-and-take that underlies attachments. Particularly important are peer-peer relationships that don't have kin bonds to cement them. And they get this exposure by practicing on one another—through play.

STATUS IN PLAY

Nature is a giant playground, with young animals from fish and reptiles to birds and mammals romping and cavorting in rivers, meadows, oceans and skies. Karl Groos, a German philosopher and psychologist, proposed in an 1898 book that "Animals cannot be said to play because they are young and frolicsome, but rather . . . in order to . . . supplement . . . individual experience in view of the coming tasks of life."

While Groos's description takes some of the fun out of play, the "coming tasks of life" are indeed embedded in the play behaviors of many animals and humans. Young predators mimic hunting, practicing the stalking, pouncing, and clawing they'll need to feed themselves one day. Typically, these behaviors are encouraged by parents who bring "toys" to their offspring: injured penguins for young leopard seals and incapacitated scorpions for young meerkats, for example.

A wild African cat called a serval engages in "angling play," according to ethologist Gordon Burghardt. The serval will allow "captured mice and rats to escape under a tree stump or hole and then try to retrieve them with a forepaw . . . the serval picks up the prey cautiously by the back fur, carries it to the vicinity of a crevice, lets it run if the prey animal still doesn't slip into the hole, and the serval often pushes it in with its forepaws in order to be able to fish it out again."

Adolescent killer whales play at beaching themselves, mimicking adults who ride waves up onto the sand to nab prey and then slip back into the sea. Orcas who receive this training when they're adolescents appear to become better adult hunters, and their skills develop faster.

Similarly, courtship behaviors, which we'll discuss more in Part III, are critical for adolescents to learn early in order to interact appropriately and successfully with mates later in life. For bald eagles, whose pre-mating rituals involve a harrowing, sometimes fatal whirling sky-dance called the death spiral, play behavior involves adolescent eagles flying toward each other and touching toes—real-time, in-flight, aim-and-grab practice in preparation for when they will clasp their partners' feet and fling each other around.

One of the easiest play behaviors to identify is play-fighting, such as stylized boxing in kangaroos or head-butting in young rams. Australian wombats and tiger quolls chase, stalk, and wrestle one another. Red-necked wallabies have twenty-one different play-fighting actions, including skipping, grabbing, pawing, sparring, and kicking.

To a human viewer, play-fighting may look like practice for future self-defense against predators. It may seem to be preparation for staying safe. But actually, self-defense and play-fighting aren't really the same thing. Play-fighting prepares young animals for a different kind of combat: the battle for rank within their own groups. It is noteworthy that young animals from guinea pigs to capuchin monkeys who engage in lots of rough-and-tumble play with their peers when they're young don't become belligerent fighters. They become better friends. They fare much better in social hierarchies as adults. Play allows younger animals to try negotiating conflicts without harm. Additionally, subordinate animals can learn to communicate when they don't like what the dominant is doing.

As University of Massachusetts Amherst biologist Judith Goodenough puts it, "Without experience in a dominant role, young monkeys may grow up to be overly submissive, and without experience in the submissive role, they may grow up to be bullies. Play-fighting also may help a juvenile learn to read the intentions of others. Is the opponent bluffing? How motivated is this opponent? These social and cognitive skills may in fact prove to be more important than physical skills."

We asked Höner if Shrink learned this lesson, and he said that all hyenas have to learn the two sides of a play-fight—even alpha females. Queens, he pointed out, sometimes intrude into other clans' territories. There they are the interlopers and must be submissive to other residents of the territory. "Any hyena knows how to show submissive signs. That's a key to survive. If you don't do that, you get beaten up very badly," he said. And many of those submissive signs are learned in early wildhood during play-fighting with peers.

Young male white-tailed deer spend summers together in mixed-age groups essentially "playing"—learning and relearning the rules of group living. At the start of summer males of all ages shed their antlers. Coming together as a group gives them more eyes and ears for protection from predators during this vulnerable, antler-less time. It's also the deer equivalent of leaving their guns at the door. Play-fighting is less injurious if no one is wearing their weapons.

The purpose of play-fighting in these deer—and in many other animals—isn't just to prepare to fight off predators or compete with one another for resources or mates. Its hidden purpose is to teach them how *not* to fight one another. Because that's what stable animal groups do: they don't fight. Play-fighting trains young animals to understand different positions within a social hierarchy. And that can make them more flexible and effective leaders later in life, as well as more productive, secure members of their chosen groups.

For social animals, there's no substitute for this vital training. Human adolescents and young adults have a lot of options to practice the constant sorting and re-sorting of hierarchies. Organized sports, theater, and music can allow status-shifting in supportive environments based on specific skills—rather than on dominance contests of looks, size, strength, or family networks. They enable adolescents to move around a hierarchy with a modicum of control over what happens. Smart coaches and choreographers and conductors will give their players chances to star as well as opportunities to support.

Creating smaller hierarchies within a bigger hierarchy is an excellent strategy for surviving the sorting wars of human adolescence. Being a lower-status individual in a group can be an important and maturing experience. Internships and apprenticeships are examples of this, as are peer leadership relationships between, say, a ninth-grader and a nurturing twelfth-grader. But enjoying higher status in a different group is also educational. Becoming a member of multiple groups in schools, in local communities, and even online serves two equally important purposes: It builds adolescent social skills. And it makes the challenges of this phase more tolerable.

Given how physical animal play is, it's worth taking a moment to consider whether the virtual play of modern human teens is a proper substitute—or analogue—for the head-sniffing, antler-rubbing, tag-and-reverse, and coalition walking shown by deer, monkeys, elephants, and hyenas. For animals, physical contact seems to be key for the kind of socializing that prepares them to have friends as adults. Multiplayer video games aren't as physical, but they do encourage adolescents to spend time getting to know other people, sometimes around the world. Often gamers occupy a variety of positions within the virtual-world hierarchy of a game. The video games played with others—distinctly different from ones played alone—are fundamentally social experiences, and many gamers say they provide the same collective benefits as in-person play. New research confirms that video gaming isn't necessarily socially isolating. However, other studies suggest that, as one would suspect, chronic gaming does have a negative impact on the development of social skills. In adolescents addicted to video games, social skills including cooperation, accountability, altruism, and the ability to express feelings are reduced.

ASSESSMENT OVERLOAD

Wild adolescent animals have two advantages over their modern human peers. While they too face high-stakes tests and the stress that goes along with them, these periods of assessment start and stop. There are seasons to play and seasons to breed and seasons to migrate. But today's human adolescents on the other hand are never given a break. With internet-fueled social media, there is no off-season.

In addition, wild adolescents have their competition right in front of them. Shrink didn't have to prove himself to his clan members by day and then at night contemplate his relative rank to the eight other clans in the Ngorongoro, the dozens of clans nearby in the Serengeti, and all the other hyena clans in parks and zoos in Europe, the Americas, Australia, and Asia. He didn't have to ponder what life would be like as a jackal instead, or a hunting dog, wolf, or leopard seal.

A young human on social media never has the full picture of their competition—and yet they never fully escape it. Modern social networks are too vast to know everyone in them. That can make distant celebrities or politicians, whether menacing or friendly, seem to be nearby, when in fact their dominance or prestige has nothing to do with one's own daily real life. This is not a completely new problem, of course. It has grown with cities, better communication, radio, and TV. The internet hasn't created competition among hypersocial, self-aware humans. But social media has expanded the number of peers that adolescents may stack themselves up against to unprecedented levels.

Comparing oneself with others isn't a uniquely human habit. Remember that the Social Brain Network (SBN) helps animals process, decode, and act on social information. A crucial function of the SBN is to help an animal assess itself relative to others. But this system evolved in animals experiencing not constant, but periodic, assessments. In modern human life, the assessment has become perpetual, and it often starts well before wildhood is even under way.

For many, adolescence has become a tyrannical, nonstop sorting, rating, and ranking tournament. Middle school—the initiation arena—has become a relentless assessment zone where ratings are given for every aspect of adolescents' physical and emotional lives: body type and fitness, sports ability, eating choices, sexual expression and experience, social adeptness, academics, outgoingness, material possessions, and general appearance. These have eternally been the concerns of younger adolescents, but never before has a culture provided them with such continuous and public measuring tools.

After a full day of assessment from peers, teachers, parents, professors, bosses, and mates, students go home. Home used to exist as a sort of status sanctuary. But now, through laptops and phones, a direct sorting pipeline digitally spews into many adolescents' bedrooms, dinner tables, and car rides, while they're studying, TV-watching, game-playing, reading, or having downtime. And the assessment continues through the night as the metrics of status pulsate through the glowing screens of their devices.

What we have in the twenty-first century with the introduction of social media is an unmanageable and unhealthy amount of assessment. Our Social Brain Networks simply can't process it all. We propose a new term for the anxiety and suffering caused by adolescent SBNs saturated with near-constant evaluation: "assessment overload."

Assessment overload is what evolutionary biologists call a mismatch disorder. Mismatch disorders arise from discrepancies between modern human environments and the ancient ones in which our bodies and minds evolved. The modern obesity epidemic is a mismatch disorder, arising from the differences between food-scarce environments in which human and animal metabolic systems evolved and today's world of hyperabundant calories.

The rising levels of stress and anxiety in adolescence can perhaps also be understood as a mismatch disorder. SBNs evolved in mammalian groups with intermittent contests. The perpetual assessment in the lives of modern adolescents—the test-taking, sports performance, and now social media rankings, and so on—overwhelms SBNs, which evolved in environments with less constant evaluation. Like the hyperabundance of calories, unrelenting assessment is presumably unique to the modern human world. It is more intense and more constant than what young social animals have ever had to deal with.

STATUS SANCTUARIES

Problems emerging out of evolutionary mismatch may be best solved by re-creating a better alignment between physiology and environments. This can be done by restoring the conditions of previous environments in which the physiology (or behavior) originally evolved. In the case of obesity mismatch, it would mean paring back hyperabundant food items and restoring seasonality to modern diets. In the case of assessment overload, it means introducing assessment-free periods into an adolescent's life. Instead of letting adolescents drown in a flood of scores and rankings, adults could be throwing them lifelines in the form of assessment-free zones, times, and places where judgments are held at bay. Instead of relaxation spots or unplugging areas, or chill zones, they could be called what all those places actually are: status sanctuaries.

Status sanctuaries—where adolescents might engage in noncompetitive sports, reading for pleasure, private moments of rest without the intrusion of social media—would allow adolescents (and their developing SBNs) to take a break from the hierarchies, the real-life ones in front of them and the virtual ones on their screens. Assessment is a normal and important part of adolescent life. Assessment overload is causing illness and distress.

Although human friends are a must, thanks to the species-spanning nature of the Social Brain Network, adolescents can also benefit from relationships with animals. Horses, dogs, cats, and other pets have sophisticated SBNs that respond to people in surprising ways. Pet therapies have grown in popularity as their benefits for human mental health become better known. As Kathy Krupa, a certified equine assisted therapist, told the *New York Times*:

"A horse couldn't care less if someone has been in jail or has a learning disability. They only judge you by how you are at the moment. You're even allowed to be afraid around a horse as long as you admit that you're afraid. I've seen a horse walk right up to a terrified kid and put their heads in their chests."

In Philadelphia, a nonprofit organization called Hand2Paw matches at-risk youth with shelter animals that are similarly vulnerable. The founder was a nineteen-year-old University of Pennsylvania student who herself was looking for social connection during her second year away at college. The social relief of human/animal interaction goes both ways. Not only do the Hand2Paw human youth volunteers gain from the program, but the homeless dogs and cats they care for receive hundreds of hours of bonding, grooming, and socialization—staving off kennel dog syndrome and making them more adoptable.

YOU'RE NOT WHERE YOU'RE SORTED

For most hyenas, Shrink's inauspicious start would have spelled a lifetime of misery or early death from starvation or predation. But Shrink wasn't most hyenas. He didn't just plod along the path mapped out for him by genetics, environment, and the social order he was born into. His story took a twist with a single extraordinary decision.

One day, probably emboldened by hunger, he marched straight up to the most powerful hyena in the clan—Queen Mafuta—and asked for her help. She rebuffed him. He asked again. She turned him away again. Over and over he asked for her help until a few days later, for reasons that Oliver Höner still doesn't completely understand, Mafuta relented. Shrink's determination combined with Mafuta's unusual acquiescence changed the hyena's life.

What Shrink asked for was milk. And Mafuta gave it to him.

For weeks, Shrink and Meregesh, the prince and the pauper, nursed side by side at Mafuta's breast. With Mafuta's nutrient-rich milk and the plentiful quantities of it, Shrink quickly grew into a strapping and, to use Höner's word, "handsome" hyena.

It's extremely unusual for hyenas to adopt other youngsters. We'll never know why Queen Mafuta finally said yes, but Höner said that she and Shrink had a special relationship driven by Shrink's combination of charisma, social intelligence, and drive. While Shrink's hyena charm certainly helped, his social mobility was ultimately the result of the strong coalitions he built, his grit and gumption, and, of course, luck. Perhaps most important, Höner said, Shrink knew how to spot and take opportunities when they presented themselves. His friendship walks with other males, his experiences in rough-and-tumble play and role reversals, taught him valuable social lessons for getting along with others.

From the moment he allied with Mafuta, things changed for Shrink. He got better nutrition. He associated with high-status animals. Mafuta began defending him in encounters with other clan members. Along with his social skills, these factors came together and when it was time to migrate, Shrink got into a good position in another clan. He became a popular mate and fathered many offspring. Later, Shrink dispersed again and joined a third clan. Once more, his social skills served him in the transition to a new group.

Before you think this tale has an uncomplicated happy ending, not everyone was in favor of Shrink's rise. Changes in status sometimes require sacrifice. In this case, Shrink sacrificed his relationship with his mother to save himself. The entire time Shrink was begging, cajoling, and demanding that Mafuta adopt him, Shrink's own mother was desperately trying to keep him from giving up on her. Beba, Höner told us, "was not super happy that Shrink went to suckle from Mafuta. She even tried to physically carry him away from Mafuta's side. But he just kept insisting."

Like humans throughout history who put their painful or embarrassing—or sometimes loving but unworkable—pasts behind them, Shrink sensed that his ticket to survival was tangled up in his relationship with a mother who couldn't care well enough for him. So, he took the chance and went elsewhere.

Every day, all over the world, humans struggle with the situations they're born into and the ones they aspire to. Just one example comes from the historian and author Tara Westover, whose 2018 memoir, *Educated*, describes her childhood in rural Idaho growing up as the daughter of survivalists. Because her father didn't believe in public education, Westover attended no school until the age of seventeen, instead spending those years of her wildhood "working in her father's junkyard or stewing herbs for her mother, a self-taught herbalist and midwife." Determined to go to college, Westover began to teach herself, eventually graduating magna cum laude from Brigham Young University and earning a PhD in history from Cambridge University. A hyena's motivations would of course be vastly different from a human's, but, like Shrink, some of Westover's remarkable fortitude might be attributable to temperament and outlook, including what she described to the *Times of London* as "[t]hings that I now recognize as just part of my personality: willfulness and assertiveness, maybe even a bit of aggressiveness." In the same interview, she shared this opinion on family estrangement: "You can love someone and still choose to say goodbye to them, and you could miss someone every day and still be glad they're not in your life."

For Thorleif Schjelderup-Ebbe, adolescence and adult life didn't unfold so positively. Lacking social skills, he was unable to promote his work and ascend academic ladders. He struggled to gain recognition. And although he had pioneered the study of hierarchies in animals, he was never particularly successful at navigating the pecking orders in his own life.

Shrink's story has many lessons about how privilege, environments, and individual agency shape the fates of human adolescents and young adults. Shrink's life could have proceeded much in the way his mother's did—living at the bottom of the social group, derided, harassed, deprived, endangered. But maternity isn't always destiny. Young hyenas, and the young of many other species, sometimes take their futures into their own paws, talons, or fins.

Happy living on planet Earth requires first acknowledging a few grim realities. Level playing fields don't exist in nature. Parental rank inheritance is real, and parents intervene to help their offspring all the time. Perceived status drives mood—and, depending on where that perception lands individuals in their groups, can make them feel anxious and depressed or connected and happy. Animals do bully one another, and adults bully adolescents, although it all can get much, much better as one grows up.

An animal's best chance for increasing the good and blunting the harmful aspects of hierarchies is to develop social skills. While no animal can control the circumstances into which it is born, social skills are the most important tool for creating a bearable life. Strengthening them is important.

Shrink himself might say that even if you are not born a privileged creature and life at the low-status end of a den, pack, or classroom may feel terrible, status reversals are possible with practice, tenacity, and just enough of the good fortune he found in the Ngorongoro Crater.

PART III

SEX

During wildhood, humans and other animals must learn to balance desire and inhibition by interpreting the language of courtship. Such signals form the basis of sexual consent and coercion.

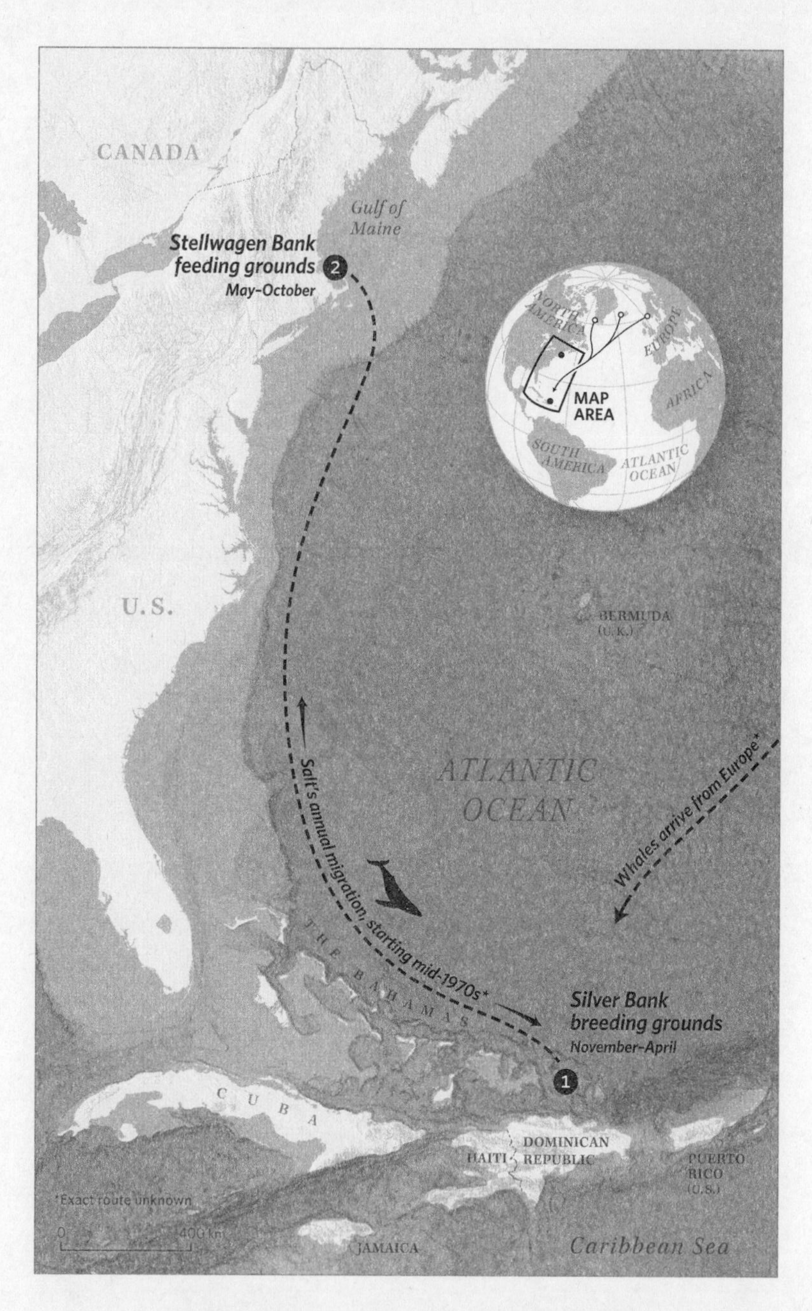

SALT LEARNS THE LANGUAGE OF LOVE

Chapter 11

Animal Romance

Every year since the 1970s, the grande dame of Stellwagen Bank has summered off Cape Cod. The season runs from May through October, and this prominent matriarch always arrives early, preferring to be in residence by the first or second week. She sets up in her familiar neighborhood near the best places to eat. She catches up with old friends and acquaintances. Often, she brings along a female companion. For fourteen summers the grande dame also arrived with a newborn at her side—a different youngster each time—and no male escort in sight.

The grande dame of Stellwagen Bank is a humpback whale. A large white scar on her gray dorsal fin makes it appear to be brine-encrusted, and so for decades she's been affectionately known to marine researchers and the public as "Salt." Salt belongs to a mega-clan of several thousand humpbacks living in the North Atlantic off New England, Norway, Greenland, and Canada. In winter, they all migrate to the Caribbean, thousands of miles south, to breeding grounds called the Silver Bank.

Since at least 1976, when she was a young adolescent, Salt has been swimming back and forth between Silver and Stellwagen Banks. Her first trip north to the Gulf of Maine would have been as a newborn calf, and she'd have been escorted there by her own mother after being born in the Caribbean. Now a great-grandmother several times over, Salt serves as the materfamilias of a well-documented humpback family tree. She's clearly mastered staying safe during the yearly trek to and from the Caribbean. And her social skills are solid, evidenced by her long female friendships and her stable rank in the overall pod hierarchy.

But there's something else to Salt's notable fifty-or-so years on planet Earth, and her dozens of offspring, grand-offspring, and great-grand-offspring point it out. As with Rose, the elegant matriarch in the movie *Titanic*, the more her story unfolds, the more we understand that Salt has lived a life of romantic confidence and adventure. Her past holds an intense history of attraction, rejection, longing, and passion.

Like staying safe and navigating hierarchies, finding and choosing mates is a skill that animals must learn. You may think that wild animals rush to mate the moment they're physically able, that they simply let instinct take over. But the reality is surprising and much more like the human experience than most people realize. To begin with, throughout nature, there's a noticeable lag, a waiting period between the completion of puberty and the onset of breeding. Becoming a sexually mature adult, behaviorally and emotionally, takes time. And so, even at the earliest stages of their lives, wild animals are receiving sex education. However, their curriculum focuses not on copulation but rather on communication. Maturing animals must learn how to express and read desire in themselves and others.

During wildhood these lessons become more urgent. Salt's romantic coming-of-age intensified sometime in late 1978. ABBA and the Bee Gees were on the Billboard Hot 100, *Grease* and *Close Encounters* were playing in movie theaters, and in the warm Caribbean waters off the Dominican Republic, a young whale found her first love.

LOVE IS IN THE AIR

Pitting passion against inhibition, longing against the sting of rejection, romance is, at essence, desire mixed with uncertainty. If you know where to look, our planet is alive with it.

Above a public park north of New York City, two bald eagles speed directly toward each other, seize talons in midair, and plunge to earth, spinning like figure skaters. Just before crashing, they release each other's claws and zoom skyward. Then, flinging themselves together again, they perfectly execute another death spiral.

In tropical Australia two flying foxes—the largest bat species in the world—scream at one another. The male brays like a donkey and the female yells back and either swats the male away with her wing or invites more contact by grabbing his ankles and pulling him close.

Along a creek in Grayson County, Virginia, two salamanders perform the "tail straddling walk," a slow-motion, amphibious tango with the female in the lead. Tail straddling often follows a round of stylized air kisses called "joint head swinging." The two salamanders face each other and bob their heads from side to side, cheek to opposite cheek. Until recently, scientists assumed that only males initiated these salamander courtship moves. But a few years ago, a veterinary student took a closer look, and she observed that the courtship is two-way—performed by both males and females.

And right there, in the banana bowl on a kitchen counter, two fruit flies meet for the first time, excited and curious. The male taps the female with his leg. She responds with a combination of chemical and behavioral signals. If the male senses she's not into him, he flies away to find another mate. But if she indicates yes, the pair launch into a fruit fly mating performance that includes singing, chasing, and wing fluttering you might call theatrical if you were feeling poetic.

Scientifically described by zoologists, the reproductive repertoire of animals can sound decidedly unpoetic. One journal described it as "complex, ritualistic behavior that involves many visual, olfactory, gustatory, tactile, and acoustic cues, as well as intricate motor output directed toward attracting a suitable mate."

Whether undertaken by a fruit fly or a human, what all these behaviors and "intricate motor outputs" add up to is courtship. Courtship behaviors help animals select mates. They signal interest and disinterest. Courtship rituals may follow common patterns, but the internal feelings—the drama, rejection, sparkle, torment, heartbreak, and pleasure—behind every act of intimacy are as unique as the individuals conveying them. Courtship is a complex expression and assessment of mutual desire.

Most important, although they're rooted in instinct, courtship behaviors are shaped by learning and experience. They develop as an animal matures socially. And they can take time to figure out. Not only do post-pubescent animals often wait to partner up until they've gotten the hang of courtship, some seem to be almost repelled by potential mates until they've reached a certain level of social growth.

It can be similar for young human adolescents finding themselves in sexually mature bodies without the full social and emotional knowledge of what to do with them. Much of the formal sex education offered in today's middle and high schools in America focuses on the consequences of the physical act, in particular on pregnancy and disease. This is wise, of course; young adult humans need to protect their health. Sex can be risky for animals too. With multiple partners and no ability to practice safe sex, wild animals can and do get STIs—which can sometimes be fatal.

But in nature, a successful wildhood involves not tutelage in the physical risks of sex, but rather a sophisticated and nuanced education in courtship, in reading a potential partner's signals.

How animals discover each other, express interest, assess the interest of the other, and decide what happens next—those are the complexities that must be learned. And practicing them begins well before an animal actually starts mating.

To be clear again, we aren't talking about the physical mechanics of sex itself. We mean practicing courtship—the thousands of special, fine-tuned glances and nods and tilts and shadings that unchain corresponding behaviors in potential mates. Practicing courtship is the process of learning to balance excitement and inhibition. It's learning the animal messages that signal "no" to partners as specifically and strongly as they signal "yes." In some species, including arguably our own, this part of becoming reproductively mature can take years.

In short, sex is easy. Romance is hard.

Let's say for a moment that a young humpback whale is interested in another humpback. What could she do next? Swim right up to that individual and start mating? Of course not. How would she know whether the other whale desired her back? How would they express their interest, their attraction, their consent?

The tender complexity of sexual communication between two young animals can be evoked by a word the Yaghan people of Tierra del Fuego may have had for this moment: *mamihlapinatapai.* The Yaghan language had many words to describe awkwardness while eating, bathing, paddling a canoe, making a spear shaft, or even climbing a tree. While the precise definition and history of the word *mamihlapinatapai* remain controversial, it roughly describes "a look shared by two people, each wishing that the other would initiate something that they both desire but which neither knows how to begin." *Mamihlapinatapai* perfectly describes the sometimes awkward and exciting moment when two inexperienced individuals mutually decide to take the next step.

But even just getting to this moment poses an epic challenge for whales looking for love across thousands of miles of open ocean. It turns out that humpbacks address this challenge in one of the same ways romantic young humans have since at least the invention of the mix tape: with music.

HOW TO ASK OUT A WHALE

Sometime in late 1978, in the warm waters of the Caribbean's Silver Bank marine reserve, a low, bassoon-like rumble reverberated under the waves. The bleating ribbons of sound rolled through the water, swelling and falling, for more than twenty minutes without once repeating a tonal phrase.

In humpback breeding territories around the world, mature males—each weighing as much as a loaded tractor trailer at forty or fifty tons—gather to sing in choruses, blasting out complex songs with unique melodies, dynamics, and flourishes. Humpback songs last around twenty to thirty minutes each, but concerts can run for hours. One musical marathon studied by whale biologists lasted nearly a day and a half. Humpback refrains are passed down from generation to generation, like sonorous family heirlooms. They remain constant in their essence but change over the years. In addition to ancestral ballads, individual whales also compose their own original melodies, which must be rehearsed. They vary a section and pause to go over it, small chunks at a time. With so much to learn, it can take a young humpback years to become a great singer.

In 1978, the study of humpback chorusing was just getting under way, and even fifty years later, some mystery still surrounds this incredible and people-pleasing animal talent. It's believed that male humpbacks sing to tell other whales to stay out of their territories or, more genially, where to find food. And the males sing during breeding seasons to locate potential sex partners.

With so much praise heaped on humpbacks' extraordinary singing, you don't often hear about the clinkers. But every once in a while, eavesdropping marine biologists can discern some singers who aren't quite getting it. They're hitting wrong notes; their tone is thin; they forget the order of the phrases or lose their place. The scientists call these musical renditions "aberrant." The whales creating them are labeled "aberrant singers."

But a University of Hawaii humpback expert thought these singers might not be aberrant at all. Louis M. Herman had a notion that perhaps the musical misfits were just inexperienced. Could it be that those whales were simply adolescents still learning the repertoire?

Herman and a team went into the waters off the Big Island and measured eighty-seven male humpback singers, expecting to find they were all mature males. Until this study, it was assumed that mature males, perhaps for competitive reasons, blocked younger males from joining their singing groups. In reality, Herman found that although most of the singers were mature, 15 percent were adolescents. Younger males, his research suggested, were being invited to croon love songs along with the adults in the chorus. But why would the older group accept these unrelated, up-and-coming interlopers?

The benefits were obvious for the younger males. Getting to sing with the adult males gave them valuable practice. Like junior violinists getting to spend a month in the back row of the philharmonic, adolescents could listen to the experienced musicians. They could observe their techniques, learn the playlist, and try out their own phrasings. How else would they learn all the ancestral songs? During this practice, the younger whales built lung capacity and the strength to hold their breath longer, a sign of stamina that adds to the beauty of the song. These singing internships weren't just for fun—they could make or break a whale's future singing career.

As for why the older males tolerated the mistakes and potential competition of the youngsters, Herman hypothesized that what the adolescents lacked in experience they made up for by adding volume. A louder chorus created bigger sound waves that carried farther underwater and could reach a greater number of the interested female whales listening for the songs.

For whales, becoming a great singer takes years. Time spent practicing their mating chants with older adults is a form of whale courtship education. There's no evidence that whales need specific instructions in having sex itself, but it's clear that they do need to learn from other whales how to sing the songs of love.

Singing is a central part of courtship for a wide range of animals. And learning from older singers is a common way for adolescents to get the necessary vocal training. Among singing bat species, the young are assigned tutors who teach them the vocal stylings they will use as romantically mature adults. For songbirds, receiving musical instruction from older birds refines the melodies they use to attract opposite-sex partners. (Both male and female birds are taught to sing. Bird courtship has been studied for centuries in males but overlooked in females and assumed erroneously, until recently, to be only heterosexual in focus.)

Humpbacks aren't the only troubadour-seducers of the high seas. Male and female leopard seals vocalize over vast ocean distances to find mates. During breeding season, lone leopard seals may spend many hours a day vocalizing. But they don't just wake up one morning knowing all the chords and lyrics. When they're about a year old, adolescent male leopard seals begin practicing their music even though they won't start breeding for another four years or so. During those practice years, as their stamina increases, their voices become fuller and richer, and their songs get fine-tuned. As they practice, both male and female adolescents also learn important social rules, such as the appropriate context in which to emit the moans, clicks, vibrato, and falsetto that later will signal erotic intent to potential partners.

That the singing voice ignites erotic desire has not been lost on human lovers throughout the ages—or on the music industry. Beyoncé knows it. Prince knew it. So did Elvis and Sinatra. The voice is a powerful aphrodisiac.

In many birds and a few mammals, erotically stimulating songs work by entering through the outer ear, activating the auditory cortex, ricocheting through the brain, and triggering a cascade of hormones leading to arousal. French researchers have even identified the specific "sexy syllables" that, when sung well by male canaries, encourage responsive females to return their desire.

Male birdsong can be sufficiently stimulating to actually induce ovulation in canaries, doves, and parakeets. "[S]inging may function in part to synchronize ovulation," reported one article in the *Canadian Journal of Zoology*, especially in animals with limited breeding seasons or separated by long distances. Humpback scientists have hypothesized that when resonant sound waves from booming male choruses reach distant female humpbacks, they may not only help the partners locate one another, but also spark ovulation. But stimulating female reproduction isn't the only function of these songs. Humpback songs also attract lone males whose visits to the singers include displays of tail lobbing and breaching which may be an example of male-male affiliation, social status displays, friendship, or all three.

Think about your favorite love song and imagine your fantasy vocalist singing it over and over, just for you, at a private concert. Multiply that sound by eighty-seven tractor-trailers loaded with stereo amps, and that may have been what Salt felt in those Caribbean waters during her first season of love. Her body receptive and excitable, she'd been out swimming, listening for songs, when she heard the call. Something stirred; she turned and sped toward the music.

Chapter 12

Desire & Restraint

About halfway through Disney's 1994 animated movie *The Lion King* comes a scene of teenage love. It's remarkably suggestive yet left unconsummated—on-screen, at least.

Simba and Nala, two lions who have been friends since cubhood, have reconnected after several years apart. Both are now firmly in post-pubertal, pre-adult wildhood. Simba sports a handsome mane; Nala has developed shapelier flanks and doe eyes. As Elton John's lush ballad "Can You Feel the Love Tonight?" swells across the soundscape, the two adolescent cats spend a day together frolicking in a waterfall, play-wrestling in a clearing, leaping through a field at sunset, and rolling down a hill in a tangle. Landing at the bottom, the pair find themselves on top of one another, and suddenly the playfulness is charged with an entirely different energy. Nala licks Simba on the cheek and then reclines on a bed of green grass, narrowing her eyes and lowering her chin as she peers up at her friend, and the two characters nuzzle.

The moment holds just long enough for any budding adolescent to start wondering what will happen next, but it's interrupted by the meerkat Timon and his flatulent warthog friend, Pumbaa, singing a coda about how their friend trio is doomed now that Simba has a girlfriend. While the abrupt break is in keeping with the film's "G" rating, adolescent sexual behavior that stops short of intercourse is not far off from what often happens in nature. In the wild, male and female adolescent lions may play-fight and court like Simba and Nala do in the movie, but they don't always start reproducing right away. In other words, although they're physically mature, they aren't yet having sex.

The fact that wild animal adolescents who are physically developed aren't necessarily sexually active is an important and surprising point. It may seem uniquely human, but it is observed across the animal kingdom. In some situations, as soon as an animal reaches puberty and has the biological capacity to reproduce, it does. But at other times, in species from fish to birds, reptiles to mammals, an animal's first time may happen months, years, or even decades after it has completed puberty.

However, watch most nature films and you'll be told a different story. The nature documentary, a genre invented and largely produced by mature men, has for decades been a primary source of information about wild animals for the general public. These entertaining, if misleading, portraits have reflected at least as much about the culture and characteristics of the humans behind the cameras as they have about the wild animals in front of them.

The films could be about giraffes, foxes, sloths, or sage hens, but they have typically featured a perspective in which eager-to-mate males pursue coy females who succumb to their charms. Or, in another version, all the males desperately compete with each another before a judgmental panel of females. In this case, the females usually all want the same thing—the strongest mate who is also the best protector and provider. No matter how many times they're repeated, these classic tropes are simply incomplete and often inaccurate. So, too, is the common human insult that people who jump immediately into bed are "behaving like animals."

Besides skewing toward the largely male perspectives of their creators, most nature films have also been adultocentric. Wild animals may be adult-sized and appear to be sexually active, but that doesn't mean they are. More tentative, inexperienced, cringe-y adolescent sexuality has rarely been the focus of these programs. The celibate reality of many adolescent animals perhaps doesn't make for the best TV.

The truth about adolescent sex in the wild is much more nuanced. Rarely in documentaries do you see adolescents being kept from mating by older, dominant group members, or young adults staying in the nest for a few seasons, passing up the chance to breed. You don't see young adult female gorillas, already sexually mature and cycling normally, continuing to play as if they were juveniles, ignored by mature males as possible mates. You don't see all the social practicing. The trying-on of roles. The testing of behaviors. You don't see a lot of younger animals being unreceptive to—leery of—sex.

To be clear, wild animals are not making humanlike decisions about the timing of their first sexual activities. As far as we currently know, our species is unique in having the ability to consciously balance the pros and cons of sexual encounters navigating the webs of ambiguity around morality, religion, and cultural norms.

But while religion, ethics, and popular culture don't influence the timing of first intercourse in animals, the environment surely does. Daylight and breeding season exert a strong effect on hormone production, leading to sexual readiness and desire. Availability of food and numbers of nearby predators contribute to when an adolescent animal becomes sexually active. During a season with little food or lots of predators, it may not make sense to use energy and incur the risks of having offspring who will likely not survive. A better strategy might be to wait. For example, by the age of about four, Antarctic fur seals are sexually mature, physically. However, if the fish and squid they eat are scarce or predatory orcas are especially abundant, the fur seals might not start breeding until they're seven.

Even members of an animal's community can exert a powerful influence on an individual's emerging sexual life and opportunities. In many species, older dominants essentially force adolescents to be celibate. Through intimidating behaviors, dominant males and females can literally shut down the reproductive systems of subordinate adolescents and young adults—which delays sexual behavior (intercourse) for these otherwise reproductively mature animals. Being a subordinate is stressful, and higher levels of stress hormones seem to suppress fertility. Dominant female mammals from yellow baboons and meerkats to hamsters and mole rats intimidate subordinates, causing temporary infertility, implantation failures, and pregnancy loss. In wolf packs where only the dominant pair reproduces, pheromones in urine may block the sexual drive of subordinate females and males. The premier pair's monopoly on breeding ensures more resources for their pregnancies and pups. In wild Bornean orangutans, only one male at a time has mating privileges, and only he develops the big cheek pouches characteristic of the dominant. The cheeks of adolescent males remain small until their status rises. After a ten-year puberty, male sperm whales finally become sexually mature around fifteen, but often aren't sexually active until their twenties, their reproductive opportunities held in check by dominant males. For the same reason, and with similarly long life spans, male elephants too develop physically over a long puberty but don't begin breeding until their late twenties, sometimes even thirties. Adolescent males with reproductive ability may want to mate—they sometimes try to. But their subordinate positions and social inexperience limit their opportunities.

READY OR NOT

The timing of an animal's first sexual experience has tremendous consequences for its future. When the first time is delayed, individuals will be older, physically healthier, and socially smarter, and that in turn can make them better partners and better parents. Animals who have offspring too early often lack the know-how and resources to properly feed or forage. Born to parents not ready to care for them, these animal offspring frequently suffer or die. Fish who incubate eggs in their mouths, called mouthbrooders, for example, routinely swallow their first clutches. First-time sheep mothers take longer to accept their lambs. The offspring of inexperienced brown bear and gorilla mothers are at increased risk of dying. In one study, first-time snow monkey mothers abandoned 40 percent of their infants.

An early-breeding, inexperienced female may find both her offspring and herself struggling to survive. Pregnancy in a still-small adolescent can be physically dangerous for both her and her baby. For young female mammals the physical burden of lactation can be additionally challenging. Young mandrill mothers have infants that develop more slowly;

infants born to young mother marmosets and rhesus monkeys are smaller, and their mothers make less milk; and the firstborns of young mother savannah baboons weigh less than offspring of older mothers.

This tracks with the human experience: "[A]cross human societies, adverse pregnancy outcomes are greater in mothers under 15," reported Margaret Stanton in *Current Anthropology*. The World Health Organization reports that children of teen mothers are more likely to be born prematurely and at low birth weight, which is associated with compromised health later in life. They're more likely to die in infancy and face challenges including blindness and deafness, cerebral palsy, and intellectual disability. They are far more likely to live in poverty and repeat the pattern of early pregnancy than the children of older mothers. For the mothers themselves, early pregnancy can be devastating. According to WHO, the leading cause of death worldwide for girls ages fifteen to nineteen is pregnancy-related complications.

The low social status of most human and animal adolescents and young adults means that a reproducing adolescent couple may find themselves forced to live on the less desirable edges of more dangerous territories. Low-ranking adolescent and young adult bird parents may be forced to build their nests and raise their young in low-quality outlying areas with harder-to-obtain food and more predators. The first clutch of eggs a bird parent produces, particularly if the parents are very young, are at higher risk of predation from the moment they're laid, because the new parents have less experience warding off predators.

Psychosocial dangers of too-early sexual experiences are also seen in many animals. Horse breeders know that stallions and mares who are mated when they're not socially ready can have lifelong trauma around sexual function. Premature sexual experiences for yearling or younger colts can irrevocably alter the temperament of the stallion that colt becomes, especially if those early experiences are with a "mean mare." With their tail-swishing and squealing, physically aggressive mean mares can damage the emerging sexual function of an inexperienced young male. As young stallions get older and more experienced around mares, this danger decreases.

In the world of horse breeding, humans are making decisions about when an animal is ready for intercourse. In our species, every individual should have complete control over their own sexuality and expression. But, tragically, this is often not the case. Among adolescents around the world, coercive first sexual experiences are alarmingly common. The negative consequences for both female and male adolescent victims are significant and enduring. Depression, self-injury, and substance abuse are common. And, compounding the damage, academic challenges often emerge in victimized adolescents, which place them at risk for further hardship.

The risks of early pregnancy and the benefits of waiting have become increasingly well known. Over the past twenty years, in many parts of the world, human teen pregnancies and births have been steadily declining. In wealthier modern human societies, delaying childbirth reflects efforts to maximize future safety and opportunities for offspring. Greater parental resources—material, social, educational—accrued over time may provide offspring with greater access to jobs, housing, and health care.

Today, many human adolescents and young adults are waiting to have sex for a reason having nothing to do with forestalling pregnancy. According to a Harvard study that surveyed some three thousand adolescents around the United States, more students are graduating from high school as virgins than at any time over the past twenty-five to thirty years. While other factors may be impacting teen sexual activity, such as economics and novel social interactions emerging out of digital technologies, this particular research indicated that students may be delaying relationships in order to protect themselves emotionally. They're afraid of getting hurt.

Waiting periods serve another crucial function. In the weeks, months, or years between developing their reproductive capability and actually mating, young animals are getting educated. Socially and romantically educated—in the courtship cultures and traditions of their particular species.

PRACTICE, PRACTICE, PRACTICE

The field truck took a sharp right turn and we bumped up a rutted incline before stopping atop a small hill. Spread below us was an expanse of rolling green, hazed with yellow wildflowers. The small valley held a lake, and around the edges of it milled a herd of Milu deer, a species we'd never heard of until observing them in person, there at the Wilds, a ten-thousand-acre conservation and nature preserve a half hour from Columbus, Ohio.

Milu deer are native to China but extinct in the wild. The world's largest known group of them resides at The Wilds, and has grown from fifteen individuals in 1995 to around sixty today. Pointing out across the lake, our veterinarian guide for the day explained how this herd had arranged itself for the afternoon.

Farthest away were three or four older adolescent males, slouching together on the shore. A little closer to where we were parked was a group of younger adolescents, their soft antler nubs just starting to grow. They raced in and out of the water, like middle schoolers at a pool party. Closer still was a mixed-age group of females, some lounging on the banks, flicking their ears, some dabbling in the water, some pregnant and some with babies nearby. Surveying the whole scene like a king, alone at the end of the lake nearest to us, was the herd's dominant male.

Like all fully mature Milu males, he was crowned with an enormous antler rack with multiple points and thick branches, forming a giant basket over his head. His crown was so huge it was as if small but sturdy trees had erupted from the top of his skull. But even more marvelous than the antlers themselves was what the male had done to them. By dragging his head through the water and surrounding shore, he had festooned his rack with weeds, grass, and other vegetation. Dripping in festive, botanical ribbons from every point, tucked like birds' nests in the notches, clumps of muddy leaves and merry wads of stems and stalks decorated his antlers. "Antler adorning," as it's called, is a common behavior of many South Asian deer.

No one really knows what, if any, biological benefit these remarkable decorations provide, but the more lavish the display, the more popular that Milu individual will be as a breeding partner. Like a peacock's tail-flare, the spectacle of antler adorning seems to have no direct function, beyond signaling sexual availability—and, more important, social maturity—to potential mates.

We observed that antlers weren't the only thing this male was employing to advertise his mature status. His body was a different color from those of the rest of the group. Where the females and younger males were a toasted-biscuit brown, this dominant was a dark chocolate. However, that wasn't because his actual coat color was different. It was because he had covered himself in mud and his own urine—another aspect of Milu sexual signaling. We also learned that mature males make a characteristic vocalization called a bugle, and they do a distinct, swaggering head-waving movement to show off their decorated antlers.

The adolescents at the far end of the lake were on their way to this mature look, but they weren't quite there yet. Their racks bore a few mossy fringes, but nothing like the baroque drapery of the single dominant male. They were still the biscuit color of their mothers and young siblings. They had gone through the physical changes of puberty—in body and antler-size, they rivaled the male. But still in wildhood, these adolescents didn't yet have the cultural experience that makes an animal fully mature. As the summer wore on, they would get better and better at rack decorating, bugling, and mud/urine rolling. Eventually they would get strong and confident enough to challenge the dominant male for his position. But not yet. For now, they were hanging out in this gang at the far end of the lake, observing, waiting, practicing.

Bowerbirds, native to New Guinea and northern Australia, are legendary in ornithology circles for the elaborate nests that mature males build to attract mates. But, like whales learning to sing underwater and Milu deer adorning their antlers, bowerbirds don't figure out how to build sturdy and attractive nests overnight. Adolescent and young adult bowerbirds observe master nest builders for a year or more. Hours spent watching, and more hours spent honing their own skills on practice bowers, plus, when the coast is clear, hours of practice on their mentors' nest sites make for well-trained and desirable bowerbird mates. Like strategic apprentices, these males are smart enough to work hard, but savvy enough not to threaten any of the adults until they're ready to take their jobs. While these male bowerbirds have the physical ability to reproduce by five or six, they often wait until they're seven to develop their mature plumage. Delayed plumage maturation—like a form of avian redshirting—gives these adolescents and young adults time to grow stronger and more experienced while staying safe from attack by potentially jealous adult males. To be clear, the delay isn't conscious; rather it happens automatically in response to certain environmental and social cues. Male bowerbirds who have extra years to learn and practice their unique courtship behavior are ultimately more successful at attracting females to their nests and mating.

LOVE ISN'T WHAT IT USED TO BE

For many thousands of years, Milu deer and bowerbirds have been signaling interest with the same basic behaviors. Human courtship behavior, by contrast, seems to change with every generation, adapting to economic pressures, cultural shifts, and new understandings of sexual wants and expectations. The dating landscape of modern adolescents and young adults in many ways looks nothing like that of their ancestors—even their most recent ones.

Jennifer Hirsch, a Columbia public health professor and author of *Modern Loves*, a cultural anthropology of marriage, writes that "around the world, young people are . . . deliberately positioning themselves in contrast to their parents and grandparents." In her fieldwork, she found that from Mexico to Nigeria to Papua New Guinea, dating norms and courtship communication have shifted.

"In rural western Mexico, young couples walk hand in hand in the plaza or even dance together in the dark corners of the town disco, rather than courting as their parents had, in secret whispers through a chink in a stone wall," writes Hirsch. She continues:

Among the Huli of Papua New Guinea, young spouses often live together, rather than in the separate men's and women's houses of the past, claiming that "family houses," as they are called, are the "modern" and "Christian" way for loving couples to live. In Nigeria, although marriage is still very much regarded as a relationship that creates obligations between kin groups as well as between individuals, courtship at least has been transformed into a moment for young men and women to demonstrate their modern individuality.

An artwork created by a Plains warrior that was part of an exhibition at the Peabody in 2012 depicts a nineteenth-century Lakota custom for young men and women who wanted to communicate mutual interest. A standing couple, encircled by a large red blanket, hold their faces close. Only their heads peek over the top edge. Dashed lines run between their mouths to indicate conversation. The exhibit label explained that "when a young man wanted to court a woman, he tried to talk to her when she went to fetch water in the evening. If she were receptive, he would envelop her in a double-sized wool courting blanket." With the huge blanket wrapped around them, creating a temporary structure, the couple could talk and assess one another's interest, out of sight of their protective parents and the rest of the community.

The digitally enhanced twenty-first-century dating landscape might have some young adults longing for a courting blanket to give them some privacy, or at least clearer rules about how to start a romantic relationship. And while it's true that elders have been hand-wringing about youth fads since at least ancient Greek times, and probably longer, it does feel as if we're in a special period of sexual behavioral upheaval that has left adolescents uniquely ill-equipped to understand their own sexuality. It may feel that way, but there is a simpler way of understanding what has happened. What's changed isn't adolescent uncertainty about sex and adult concerns about adolescents' having it. It's that modern adults no longer know or teach their young the complex and honest communication that contributes to bringing animals together when they're ready.

Teen life coach Cyndy Etler opined on CNN that sexual education ought to expand into the socio-behavioral in order to make adolescent young adults safer: "Teens say they want information about social, emotional, and behavioral topics, including what predatory behavior looks like. How to handle unwanted advances from people you know. How to broach these taboo subjects—as in, exactly what words to use."

The psychologist Richard Weissbourd offers a different interpretation of the same message. He believes that many twenty-first-century adolescents and young adults are yearning for lessons about love. They want more guidance on how to start a relationship and how to end one, how to deal with breakups, and how to avoid getting hurt.

Weissbourd believes that simply talking more about healthy relationships could go a long way toward blunting the misogyny and gender-based degradation he calls "rampant" in current attitudes about sex. In his opinion, "relationships are something we all need to practice—hearts get broken, it might end badly, but we can learn from that. We can learn how to do it honestly and with kindness. And that prepares us for mature relationships as adults."

One of the best ways for adolescents to do that, believes Weissbourd, is to hear stories about the give-and-take of relationships, including experiences with romantic kindness and heartbreak. Romantic role models can come from television and film, and from classic and contemporary novels. As the Australian writer Germaine Greer noted, "A library is a place where you can lose your innocence without losing your virginity."

In our species, books, films, and other media are an important part of social learning about sexuality. Observing Elizabeth Bennet and William Darcy balance their needs and desires with the expectations of their parents and community is a central thrill of Jane Austen's *Pride and Prejudice*. The story instructs whether you're reading Austen's words on the page or watching Keira Knightley and Matthew MacFadyen in the 2005 film adaptation. Its appeal even transcends its historical setting, as proven by the blockbuster success of Helen Fielding's *Bridget Jones's Diary*, an updated spin on the story that Renée Zellweger and Colin Firth (playing Mark Darcy) memorialized on-screen. The basics of courtship—balancing desire and uncertainty—continue through human culture even as clothing and hairstyles change.

Around the time of Salt's sexual coming-of-age, some members of her horizontal tribe—that is to say, human adolescents—were devouring a book that explored many of the same experiences. *Forever . . .*, published by Judy Blume in 1975, tells the story of Katherine Danziger, an eighteen-year-old college-bound high school senior, and her first sexual experiences with Michael, a boy from her hometown. *Forever . . .* has become a YA classic, but like other stories that contain scenes of adolescent sexuality, from Stephenie Meyer's vampire-themed *Twilight* series to Stephen Chbosky's *The Perks of Being a Wallflower*, to John Green's *Looking for Alaska*, *Forever . . .* is on the list of "Most Challenged Books" compiled by the American Library Association (ALA) since 1990. The ALA, a professional organization for librarians founded in 1876 that advocates for inclusion and intellectual freedom, reports that adolescent sexual content is by far the most common complaint they receive, outstripping violence, gambling, suicide, and Satanism as reasons people call for a book to be challenged.

What adolescents read and watch informs their understanding of many aspects of life, including sexuality. But it doesn't necessarily determine their behavior, and that is a good thing. Every millennial who grew up watching *Sex and the City* isn't consigned to choose a sexual future only as an uncertain Carrie, voracious Samantha, cold Miranda, or uptight Charlotte. Learning to make choices in a sexual realm can be enhanced by considering how sexuality is handled or fumbled by the characters met on pages and screens, and exposure to the larger world of human sexuality may help adolescents better understand their attractions and impulses.

But it's important to note that in many animals, particularly in primates, social learning is extremely powerful, and what animal adolescents watch their peers doing influences them. Animal adolescents can't read, but they do watch. They observe and learn when things go right and when they don't. Moreover, early sexual experiences—those that happen in wildhood—echo throughout adult life, for humans and for other animals.

The sexual education of wild animals isn't censored, but neither is graphic imagery as instantly and continuously available as it is for modern humans. Marketers of popular substances both legal and illegal have learned that it pays to make their products more potent and addictive. Just as marijuana farmers have succeeded in making cannabis two to five times stronger than it was in the 1980s, today's pornography is more explicit, more available, and often more a part of adolescents' lives than anyone could have imagined a generation ago. Exposure to sexuality is a part of most wild animal adolescence, but constant, streaming access to intense and exaggerated sexual imagery is not.

While wild adolescent animals obviously don't watch movies or TV, they do have plenty of opportunity to see mating behavior. And it turns out that observing the best practices of in-the-know elders is a common way that younger animals learn not only about sexuality but about how to communicate desire and understand responses. Again, this is not just about observing the sex act. It's about learning to understand how desire is expressed and returned.

THE MATING TREE

In the rain forests of Madagascar lives a carnivore called a fossa (pronounced "FOO-suh"). Put a round-eared teddy-bear face on a leopard's lean body and imagine it twisting up a tree trunk with the agility of a snake, while chasing prey with the single-minded ferocity of a wolverine. That's a fossa. And fossa courtship behaviors are as exotic and unusual as they are.

As described to us by Mia-Lana Lührs, a German evolutionary biologist, fossa communities designate certain tall trees as mating trees. Females looking for mating partners climb out onto branches of these special trees and start chanting courtship calls that carry throughout the forest. From far and wide come male fossas who climb up, indicating their willingness to mate. This Rapunzel-like scenario lacks a ladder made of magical hair, but it's got a fairy-tale quality to it, with an eligible and interested female attracting nimble and enthusiastic males.

But the suitors who respond to the female's courtship calls are not just males in their sexual prime. Notably, younger males arrive at the mating tree, too, and watch the older competitors clamoring to climb up. Lührs recalls a time she was sitting under a mating tree recording her observations during a particularly raucous session when two adolescent males raced up. They were very interested in the action of the older animals and zipped around the base of the tree, up and over Lührs and the chair in which she sat. They ran loops away from and then back toward the action. However, not once did either young male try to climb the tree. For fossa males, climbing the mating tree signals desire, and these two weren't ready. Like sixth-graders at a dance or high schoolers at a nightclub, they watched but remained on the sidelines.

The mating tree isn't just a place to have sex; it's a place to learn about courtship. Fossa females also learn courtship from older members of their community. Lührs described a time she observed a mother and daughter arrive at the mating tree together. The daughter climbed up and started calling. Meanwhile, the mother waited on the ground and even took a nap. After a while, when no males responded, the daughter descended the tree and mother and daughter left together.

When we asked Lührs to hypothesize about a possible evolutionary or social reason why a fossa mother would escort her young adult daughter to and from the mating tree, she commented that the action seemed guided by "tradition, some social learning" transmitted from mother to daughter. From an evolutionary perspective, Lührs observed, fossa daughters could benefit from being introduced to the mating system by their mothers before an actual high-stakes mating moment. A first-time fossa might be physically safer and reproductively more successful if she had practiced her species' ritual of call and response before having to do it for real.

Lührs added that if the mother is still breeding during her daughter's last years of dependence, the young female may have an opportunity to watch how her mother navigates consent—communicating desire, assessing it in mates, deciding what to do next—so the young female may move into her own early sexual experiences better informed.

Adult competence with courtship strongly influences their offspring's future behavior around sex. Adults can show adolescents what healthy and mature relationships look like.

FAR FROM THE MATING TREE

For fossas, wildhood can be a unique time of sexual fluidity. As females enter early adolescence around twelve months, their bodies and behaviors may transform to become more masculine. They sprout spiny appendages that resemble adult male genitalia. Female adolescent masculinization peaks between two and three years. Once they are mature adults, they generally resume looking and behaving as females. Male fossas can display the opposite: especially if they live and hunt alone, they sometimes physically present as female. Transient shifts between feminization and masculinization have been noted in other mammals, including spotted hyenas, moles, certain primates, as well as some birds and fish.

A special challenge arises for humans when adolescents' sexual identities differ from those of their parents. Even the most supportive and empathetic adults may be ignorant about communicating desire within spectrums of sexuality unfamiliar to them. Andrew Solomon explores this particular parent-child disconnect in the introduction to *Far from the Tree*. Solomon writes,

> Because of the transmission of identity from one generation to the next, most children share at least some traits with their parents. These are vertical identities. Attributes and values are passed down from parent to child across the generations not only through strands of DNA but also through shared cultural norms. Ethnicity, for example, is a vertical identity. . . . Often, however, someone has an inherent or acquired trait that is foreign to his or her parents and must therefore acquire identity from a peer group. This is a horizontal identity. . . . Being gay is a horizontal identity; most gay kids are born to straight parents, and while their sexuality is not determined by their peers,

> they learn gay identity by observing and participating in a culture outside the family.

Adolescents whose sexual identities differ from those of their parents may need to look outside the family to learn more about sexual expression. For gender-nonconforming and LGBTQIA adolescents, peers sharing similar sexualities and horizontal identities become an important source of information for one another.

Salt likely arrived at her first sexual encounter with some understanding of whale courtship behaviors, learned socially from her mother and other whales in their pod. As a calf swimming with her mom, she'd heard the yearly chorus of male humpbacks and seen how her mother responded. Salt may have felt the raucous exhilaration of a humpback courtship group (more on that in a moment) and had a preview of being the intensely desired female at the center of it.

From about age two to around ten, Salt would have gone through puberty and spent much of her time alone or with other adolescents. From albatrosses and penguins to elephants and otters, adolescent animals band together. They're typically focused on perfecting foraging and hunting skills, and trying to stay safe from predators, like the killer whales that specialize in adolescent humpbacks or the leopard seals that go after king penguins. Any sexual activity that occurs in these groups of adolescents and young adults is usually not procreative.

For ten years or so, Salt likely heard male chorusing during the winter breeding season and felt no desire, and no need to respond. But one winter that changed.

Chapter 13

The First Time

When you're watching a thundering plain of ten thousand migrating caribou or an oil spill–sized mega-school of five million anchovies, it's easy to forget that each and every mammal and fish within those groups is an individual, singular in age, sex, and size. The diversity within herds, flocks, and shoals includes differences in levels of attraction and desire. Not every female caribou is interested in mating with every male. Male starlings are not attracted to every female in the murmuration. Even fish have what veterinarians would call "partner preference."

We humans call it chemistry.

A short truck ride away from the Ohio lake at the Wilds where Milu deer practice adorning their antlers lives a population of cheetahs. The Wilds is one of nine centers in the world entrusted with selectively breeding these sleek big cats, whose numbers are declining in their native ranges in Africa. It's part of a worldwide conservation effort called the Species Survival Plan (SSP), a consortium of zoos, sanctuaries, and other expert advisors that manage animal pairings in order to maximize genetic diversity in populations of animals whose numbers are dwindling.

The day we visited, the cheetah keepers were frustrated. Two of the big cats who were perfect for each other—on paper—wanted nothing to do with romance. The male didn't want to approach the female. The female didn't want to approach the male. The chemistry wasn't there. We've heard the same story from selective breeders of giant pandas, bustards, ferrets, hyenas, and a range of hoofstock: zero attraction between two potential mates, although both were of appropriate breeding age and experience. As one zoo biologist put it: "Things don't always go well when a male and female first meet."

Veterinarians have found that in agricultural animals, even what we might call "the mood" needs to be right. Severely inclement weather can inhibit sexual interest. For breeding stallions, turnoffs include a slippery floor or too many human spectators. For cows, who show more signs of arousal at night, time of day can be important.

Chemistry—the special attraction that individuals have for one another that contributes to the desire to mate—is frustrating to try to manufacture if your job is to increase the planet's population of baby pandas or cheetahs. But as a creature with desire to give, searching for another creature with desire to give back, chemistry can be one of the most exciting parts of being alive.

The complex feelings of attraction and desire that make up chemistry can be at their most overwhelming and mysterious during wildhood. Like the neurobiology that shapes the perception of status, and the fear profiles that create animal defense mechanisms, the physical and behavioral infrastructure of courtship is common across groups, yet uniquely personalized to an individual adolescent. And it develops through experience.

Salt, of course, wouldn't have the words to describe her feelings, but physically, biologically, the inputs would be similar to those of other animals. We have no idea who Salt's first mate was, but one thing was likely: he and Salt had chemistry. He was possibly a Norwegian, Canadian, or Greenlander who had migrated every summer, like she had, a young calf swimming at his mother's side back and forth between this Caribbean breeding area and the northern feeding grounds. He too had probably witnessed courtship among older whales throughout his life and been part of an adolescent or young adult group that traveled the Atlantic together for a few years, learning to feed, avoid predators, and socialize. Perhaps one day he was invited to join the men's chorus, where he learned the old songs and composed a few of his own.

Salt heard the chorus and swam toward the males, listening. Once she was close enough, she bathed in the music, assessing the singers for accuracy and creativity, contemplating the attractiveness of these possible mates. Her presence at the breeding ground indicated an initial level of interest, but it didn't commit her.

Perhaps Salt's first mate sang with a particular lilt that caught her ear. Perhaps he held a note in a certain lovely way that gave her the impression he was a strong and deep diver. Maybe the krill he'd fed on all summer in Norway or Canada or Greenland gave him a quality she couldn't quite define but that told her he was healthy and a good forager.

Whatever the combination of factors, Salt approached and, with a signal that humpback researchers still don't completely understand, chose him as her primary "escort." However she did it—some say it's with a slap of her pectoral fin; others just call it a signal—she indicated her desire. Her chosen escort signaled back, and so the age-old process began.

What happens next in the humpback courtship ritual is thrilling for whale watchers. Judging by the excited reports of tourists and seasoned scientists alike, what's known as a "rowdy group" or more scientifically as a "competitive group" is one of the planet's most magnificent behavioral displays. Two whale-watch boat captains with dozens of years of experience running tours off the Silver Bank described it like this on their website:

> A receptive female will first acquire a single primary suitor known as an "escort." If another male whale believes he is a more suitable mate and imposes himself upon the pair with the intention of displacing the escort, he is known as a "challenger." When there is more than one challenger it is the beginning of a rowdy group, with each male whale vying for the coveted position at the female's side.
>
> A typical rowdy group consists of 3–6 males but groups as large as two dozen or more whales can be seen on the Silver Bank. The female sets the pace and the competition can cover many miles and last for many hours as the males jockey for position. The competition can be very physical, with males pushing or ramming with their rostrums [a beaky protuberance on their snouts]; striking with the bony "anvil" on the bottom of their chin; using flukes and pectoral fins to strike

> each other; snapping jaws or vocalizing, performing peduncle throws [powerful tail splashes], lunges and breaches to intimidate; and even trying to hold each other underwater to impair breathing in order to tire the opponent. Injuries can result, typically heavy scratches caused by barnacles on the chin and fins . . . rubbed raw and bloody, or even the cartilaginous dorsal fin can be snapped off.

Tony Wu, an underwater photographer who has captured shots of Pacific humpback competitive groups, calls the display "all-out pec-slapping, bubble-blowing, body-slamming, tail-whacking, guttural snorting, full-contact chaos." Another cameraman, Roger Munns, documented a Tongan humpback rowdy group for the BBC natural history series *Life* and described being in the center of it as "incredible . . . like standing in the middle of a motorway."

Despite the rowdiness, even seeming violence, experts on whale courtship note that the female controls the pace. Consent is hers to offer, to demand, and to solicit from the males. Males indicate their interest by participating in the run. Any male who doesn't consent presumably doesn't take part in the competition.

Competitive group behavior can last many exhausting hours. It ends abruptly with the mating of the female and her chosen male. Although rowdy groups have been observed in whales all over the world, surprisingly, whale scientists have rarely seen the moment of intercourse, perhaps because it appears to happen in a quick thirty seconds.

In 2010, a photographer reported having witnessed humpback intercourse in a Pacific pair. Near Tonga, the photographer said he caught a rowdy group that ended in an epic clash between two giant bulls, while off to the side the female mated quietly and quickly (a news report called it "brief and tender") with a smaller, younger male.

SEXUAL BEGINNERS

Wildlife biologists readily recognize sexual inexperience because it reveals itself in behaviors that are often exaggerated and poorly timed.

Flirtation and mounting are clumsy and sometimes miss the mark. We heard from a range of animal experts that despite this, or maybe because of it, animals from penguins to horses are more tolerant of sexual unsophistication when both partners are inexperienced.

Even a creature as unremarkable as a moth has a first time. You might not have had many opportunities to think about moth virginity, but you may be interested to know that even these insects are no more expert at first-time sex than whales or humans. We know this because of an intriguing study that an entomologist-turned-librarian undertook in a Minnesota cornfield.

Shannon Farrell was studying courtship behavior in the *Ostrinia nubilalis* moth. Because she needed sexual beginners, she started her research with 252 virgin moths—individuals she had bred in her lab, separated by sex and kept apart to guarantee their inexperience. She wanted to see how they acted during their first sexual encounters. In particular, she wanted to observe whether courtship behaviors were patterned—exactly the same, and innate—or whether beginners had individual variations, personal moth embellishments. *Ostrinia nubilalis* moths, like many members of the moth and butterfly family, assess and return desire with surprisingly complex behaviors described rather lyrically by entomologists as "fanning," "circling," "bowing," "kneeling," and even "embracing."

Farrell's research was funded by the U.S. Department of Agriculture, which was less interested in the whimsy of moth courtship dances and more invested in learning how to stop them. *Ostrinia nubilalis* by another name is the dreaded European corn borer, responsible for devastating millions of dollars in crops every year. The USDA was searching for behavioral alternatives to chemical pest controls. They wanted Farrell to learn whether courtship could be interrupted, limiting the spread of moths by preventing them from reproducing.

Farrell found that adult moths with breeding experience are very regular in the way they express sexual receptivity and desire. They do follow specific patterns. The moth beginners, on the other hand, are all over the place. Their courtship behaviors vary hugely. With time and experience their movements became more streamlined. We can't know whether these first-time moths experienced *mamihlapinatapai* (the desire to connect without the knowledge of how to initiate). But in Farrell's experiments, the moths' first attempts were full of fumblings and missteps. Signals were misunderstood or ignored, by both males and females. Mastering the multilayered harmonics of courtship was challenging for the sexually inexperienced moths. The same is true for human sexual beginners.

Whether mating partners are rainbow trout, anole lizards, bald eagles, or human beings, sexual intercourse follows essentially the same patterned behavior. What sets each species and each individual apart, what confers uniqueness and beauty on cultures and individuals, is not the sex act, but rather the special behaviors that allow two individuals to express desire and connect.

For every species on the planet, first sex can be blundering or sweet, exciting or embarrassing, intimate or intimidating. Of course, just because the sex act is a patterned behavior doesn't make it any less scary—or exciting, or pleasurable—at first. For some, the emotions involved in going from sexual inexperience to experience are a line of demarcation between childhood and adult life, which can be a profound moment. Gaining sexual experience can create a separation between children and their parents. Something feels different—even when nothing else in the parent/child relationship has changed.

It's impossible to say exactly what Salt's first underwater mating was like, or, for that matter, her fourteenth, some thirty-five years later. Whether her first rowdy group ran long or short, whether it involved lots of males or just one or two, and which male she chose will all remain mysteries of the deep.

IT'S COMPLICATED

Once adolescent animals have embarked on the journey, the landscape that lies ahead is as unique to each individual as every other aspect of their growth. Experiences are diverse, and those experiences create personalized sexuality profiles the same way that fear experiences create customized internal armors. Some animals remain together for a period of time after mating, resting, nuzzling, and touching. Titi monkeys, found throughout South America, bond by intertwining their tails after mating. Whether animals do it just for a few moments, a season, or a lifetime, the behavior is called "pair bonding."

"Maintenance of monogamous pair bonding," as scientists call it when it's repeated by the same two animals, struck us as an amusingly clinical phrase for the emotional labor that couples invest in the sometimes exciting, sometimes tedious task of remaining relevant to one another in long-term relationships. One of the most common questions about animal romance concerns monogamy: Do any animals besides humans mate for life?

The committed, decades-long marriage-type pair bonds seen in many human cultures and couples are indeed rare in the animal world. Few animals celebrate their species' equivalents of golden wedding anniversaries. Some birds, like swans and some hawks, do appear to maintain lifetime monogamous pair bonds. Some animals stick to one partner for a single breeding season and find a new one the next year. But most animals are not monogamous, and many are what biologists call promiscuous.

One animal, a seahorse relative called a pipefish, is notable for the way it maintains pair bonds over a lifetime. Pipefish couples have a remarkable daily ritual called a greeting ceremony. Every morning, the two fish meet at the same place and go through a brief swimming routine that includes back-arching, horizontal parallel-swimming, and vertical up-and-down bobbing. After several minutes of this, the couple separates, and they have no more contact until the next morning's greeting. They do the ceremony only with each other. And they do it even during the nonreproductive season, when they're not breeding. Judith Goodenough, an animal behaviorist, writes: "[I]t is thought that its function is solely to maintain the bond with the partner in preparation for the breeding season."

DNA paternity tests show that, unlike the monogamous pipefish, female humpbacks have many different sexual partners over the course of their lives. At the end of the 1979 season, Salt and her mate likely parted ways. Salt headed for the Stellwagen Bank with her kin. Her mate probably swam away with his group, back to Norway, Canada, Greenland, or wherever his summer feeding grounds were.

If they reencountered one another in some future season, they might mate again, or perhaps the spark would be gone. But the chemistry that attracted them that first time, and the experience they gained from courting one another, would inform their future pairings. Courtship, the ancient and planetwide guide through desire and uncertainty, had brought two individuals together across thousands of miles of open ocean.

And this behavioral marvel is shared by many creatures. Mastering courtship behaviors requires understanding one's own sexual interest, expressing it, accurately assessing the interest of another, and, crucially, mutually learning how to coordinate and synchronize behaviors. These steps, practiced over and over in a young animal's life, lead to what is essentially an agreement between two animals to come together sexually. This animal agreement correlates closely with our human behaviors around consent.

Chapter 14

Coercion & Consent

In a Boston neurobiology lab, grad students in sneakers and hoodies sat keyboarding on computers and peering into microscopes. Above their workstations hung video monitors, each showing an array of white circles on a black screen. The discs flickered with tiny movements, and closer inspection showed they were crawling with flies. We were visiting Michael Crickmore, who, with his colleague Dragana Rogulja, studies ancient motivational systems in the brain. The scientists (who also happen to be married to each other) foster a collegial energy in their lab, and we were soon watching the fruit flies chasing one another, tumbling, circling, and grooming, with Rogulja and Crickmore giving the play-by-play.

With just over a hundred thousand neurons (compared to humans' hundred billion or so), fruit fly brains are much smaller, simpler, and easier-to-study versions of the neural systems that drive equivalent impulses in humans and other mammals. Among the many circuits Rogulja and Crickmore study are systems regulating sleep, feeding, and aggression.

But another big aspect of their research has important possibilities for understanding how males and females negotiate sexual conversations. Rogulja and Crickmore have identified what they call the male fruit fly's "courtship control center." It's a special cluster of about twenty neurons dedicated to regulating mating behaviors. Intriguingly, those twenty brain cells don't just spark fruit flies to act on their desires. This courtship control center receives and regulates both "stop" and "go" signals.

How males behave is governed by a push/pull between excitement and inhibition centered in this brain area.

Say a male fruit fly wants to mate. He has desire but is uncertain whether a given female is interested. The need to resolve this tension between instigation and inhibition—to navigate desire and uncertainty—shapes what happens next. Remember that a male fruit fly initiates courtship by tapping a female with his leg. Fruit fly legs are covered in tiny receptors that can sense pheromones—airborne scent molecules that chemically transmit information among members of the same species. With that tap, the male is sampling chemical clues about a partner's level of interest. (Female systems haven't been studied as extensively as the male circuits, but researchers from Case Western Reserve University have determined that a female's decision about whether she is sexually interested in a male is made by a similarly small number—just nineteen—of neurons located in three brain areas.)

When a female senses male interest, her courtship control center weighs and assesses her levels of excitation and inhibition. The communication of courtship is two-way, and it sometimes but not always progresses to mating. If a fly senses the potential partner is not yet mature, not in the mood, too young, or too old, stronger inhibition signals are sent and received. In fact, Rogulja and Crickmore have found that more than half the time (56 percent), courtship stops right there. After the leg tap, the two individuals mutually express "thank you; next." However, 44 percent of leg taps do lead into the courtship phase—including the characteristic chase, song, and wing-vibration dance.

Moderating the whole complicated process is a brain chemical best known for its relationship with reward-seeking: dopamine. Dopamine, found in animals ranging from fruit flies to humpback whales to human beings, sparks motivation and guides desire. If the potential mate is also motivated, then the excitement provided by dopamine takes over—and the more dopamine, the less sensitive the flies become to inhibition.

Rogulja and Crickmore also found that dopamine was the key neurotransmitter propelling courtship behavior to progress into actual mating. Some flies would start courting but give up along the way. Those were found to have the lowest levels of dopamine. The flies most likely to continue were the ones with the highest dopamine levels.

To explain dopamine's power and how it works, Crickmore recounted a story from the neurobiologist Oliver Sacks's book *Awakenings*. "Mrs. B," a grandmother and émigré from Vienna, appeared to be in a catatonic state—conscious but unresponsive after decades of increasing emotional apathy. Sacks thought it might have something to do with dopamine; perhaps Mrs. B had lost her ability to produce or process the neurochemical in her body. Sacks administered a dopamine-building amino acid called L-DOPA, and a week after starting treatment Mrs. B began to respond. She became talkative and as Sacks writes, "showed an intelligence, a charm, and a humor which had been almost totally concealed by the disease." Mrs B was able to tell Sacks how she felt before receiving L-DOPA: like an "unperson." Sacks recalls her saying she had "ceased to care about anything. Nothing moved me—not even the death of my parents. I forgot what it felt like to be happy or unhappy. Was it good or bad? It was neither. It was nothing."

Dopamine, Crickmore explained to us, fuels motivation. It doesn't cause behaviors, but it shapes whether outside stimulus is met with a yes or a no. More than that, dopamine keeps behavior going once it's under way. Crickmore likened it to the fuel in a gas lawn mower with a cord-pull starter. If there's not a lot of gas in the tank, you need a lot of pulls to start the engine and it's not going to last very long. It's going to cut out. But if it's filled with gas, it'll start right away and keep going. Dopamine is like that—it can spark a behavior and fuel the motor that keeps it going.

But a key aspect of dopamine's role in motivation (and in turn motivation's role in courtship) is that none of it is deterministic. It's never automatic. Behavior can be dialed up and dialed down with the push of dopamine, but it's not an autopilot system. Fruit flies are categorically not, Crickmore stated, "little robots."

Rogulja and Crickmore are ultimately less interested in courtship and consent than in misplaced motivation—how the drive of addiction takes over desire, for example, or how inhibition tilts into depression. But two intriguing parts of their findings are relevant to understanding the connection between courtship and consent. First, the flies' decision to initiate courtship takes into consideration both personal libido and the libido of the potential partner.

There is literally a two-way fly conversation about sex.

This is so important it's worth pausing to say another way. Two-way conversations about sex are the building blocks of human sexual consent, but as the flies' behavior reveals, elaborate brain machinery isn't necessary for the sexual dialogue of yes and no.

Second, as Crickmore put it, the flies are not "reflex machines." They show, he says, an impressive amount of behavioral flexibility. Once courtship is underway it can be stopped or modified by either partner.

Fruit fly sexual chemistry is of course vastly different from human sexual chemistry, whose unique complexities and subtleties require special focus, attention, and respect.

But we share with other animals the ancient legacy of courtship, centered in a brain region and shaped by culture. This means we are primed to be attuned to the responsiveness of partners at every moment of a sexual encounter. And developing the back-and-forth of sexual communication starts in early wildhood.

SEXUAL COERCION IN WILD ANIMALS

Biological communication systems signaling interest and registering response are nearly ubiquitous in sexual animals. This challenges us to consider whether the messages are ever ignored. Do animals sometimes bypass courtship altogether? Bluntly, do they ever coerce unwilling partners into sex? The answer, simply stated, is yes.

One of the first known accounts of animal sexual coercion came from George Murray Levick, a scientist with the 1910–13 Scott Antarctic Expedition. His horrified description of "hooligan" male penguins forcefully copulating with female penguins and even chicks was considered too controversial for inclusion in a UK scientific publication at the time.

Since then, instances of sexual coercion have been documented in a wide range of animals, including insects, fish, reptiles, birds, marine mammals, and primates. As part of our original research for this book, we conducted a systematic review of the scientific literature and compiled a comprehensive list of species in which coercive sex has been described. Male sheep, turkeys, fur seals, mosquitofish, guppies, sea otters, and more sometimes use coercion to get sex. When we placed these forty-three species in a phylogeny (a model that shows the evolutionary connectedness among selected animals), what emerged was an uncomfortable but important truth: coercive copulation (male on female and male on male) is widespread across the animal kingdom.

There has been reluctance among some biologists to look to animals for insights into human sexuality. Initial efforts to understand human sexual behavior through an evolutionary and comparative lens have sometimes been scientifically flawed and marred by sexist assumptions and findings. Some have feared that identifying coercive sexuality in the wild could be misinterpreted, that its occurrence in nature—or calling it "natural"—could justify or excuse human sexual aggression. But recognizing sexual coercion in nature doesn't justify or excuse it among humans. Animals of the same species kill each other in many ways and for many reasons, but this doesn't make murder acceptable. In fact, the revelation in our research is how commonly sexual relations in the animal kingdom are governed by dialogue between the parties, the two-way communication of courtship.

Studies of animal sexual behavior show that much sexual activity in animals is also non-coercive. Signals of "yes" and "no" and "I'm not sure" are not only recognized: they are usually understood and, based on observations of animal behavior, "respected." When a stallion approaches a mare who isn't interested, her ears flatten, and she may shift restlessly and strike out, bite, or kick the approaching male. Most of the time, males encountering these clear signals of disinterest back off. Similar male response to female expressions of disinterest are seen in cats, dogs, and other mammals. Even reptile males take their cue from female receptivity. Male Amazonian red-necked turtles signal intent with a ritual that includes placing their nostrils on potential mates and biting. Females respond by swimming away if they're not interested. If, on the other hand, they are, they allow the males to rest their bodies on theirs. One study showed that females rejected a full 86 percent of males who approached them for sex. Remarkably, only 4 percent of the rejected males continued to try to have sex. The researchers concluded that in the majority of cases, female "no" signals were respected.

Some animals appear to have more coercive sexual activity than others. Human observers have noted that courtship seems to be completely absent in Indo-Pacific sand-bubbler crabs, and females appear to resist all copulations. In red-spotted newts, a female *can* signal willingness to mate by snout-nudging an approaching male. However, it appears she has no effective way to signal "no." If he approaches and she walks away, he restrains her for sex anyway. There could be forms of sexual communication among these crabs and newts that human investigators are missing. But seen through our limited human lens, coercion seems to be common in some species.

FORCE, HARASSMENT, AND FEAR: HOW COERCION HAPPENS

In humans until recently, sex was deemed to be coercive only if physical restraint or force was used. But just as we humans are becoming more nuanced in our ability to recognize the presence of coercion without physical violence, so too are animal experts trying to understand the range of coercive sex throughout nature. University of Cambridge professor Tim Clutton-Brock described in a 1995 paper three distinct types of sexual coercion in animals. The first type involves the use of physical force. The second type describes sex that is offered to terminate nonstop, disruptive harassment. The third occurs when the threat of violence, but no actual force, is used to intimidate another into sexual submission.

While animals can't tell us how they feel about their sexual encounters, in the cases where physical force is used, coercion seems clear. For example, when a male Antarctic fur seal sits on and copulates with a restrained king penguin pinned beneath him or a male sea otter assaults a juvenile harbor seal (often perforating internal organs and causing death), coercion is hard to dispute. The fact that these cases occurred between different species supports the bullying nature of the encounters. But even within the same species, human experts can distinguish between forced versus non-forced sex. Female white-cheeked pintails—a Southern Hemisphere duck—are at times receptive and other times non-receptive to sex. They signal willingness to mate by dropping into a crouching position, pushing the fronts of their bodies toward the ground. But sometimes sex is forced on unreceptive females by males who hide in vegetation, grab them, or chase them in flight. According to a Canadian study from 2005, "forced copulation is easily distinguished from unforced copulation in white-cheeked pintails by the absence of a courtship display, the grasping and mounting of a resistant female, and the absence of the typical prone posture of an unresisting female." The universality of these descriptions of sexual coercion through physical force across species is striking. But not all coercion is so blatant.

Sex between two animals that appears to be non-coerced, because physical force is not observed, may actually be coercive in a less visible way if the female has been harassed into submitting. Some males may persistently badger unreceptive females for sex, preventing them from foraging and feeding—a phenomenon documented in dolphins, sheep, quail, and coho salmon. Sexually harassed elephant seals, fallow bucks, and female tortoiseshell butterflies ultimately relent, submitting to sex simply so they can go about their lives. An uninformed observer, who sees no resistance or physical restraint, might not recognize these encounters as the coercive events they are. This is especially true because the intimidation and threats of violence may take place hours or even days before the sexual encounter.

Primatologist Richard Wrangham and anthropologist Martin Muller studied sexual coercion in chimpanzee males in Kibale National Park in Uganda. They observed that females who were fertile sometimes approached and initiated sex with males—but not just any males. They approached the ones who had been previously aggressive with them.

Conventional wisdom of the time held that fertile female chimpanzees were choosing the mates they preferred. But Wrangham and Muller realized these females weren't choosing; they were complying. They were approaching these males because they were scared not to. The males had intimidated them through aggression and violence in previous days and weeks so that the females would not resist, and would perhaps even initiate, sex during their fertile period. This was a clear demonstration of sexual coercion without physical restraint.

The use of intimidation to secure sex has been demonstrated in captive female gorillas, who reduced the likelihood of being attacked by males by presenting themselves sexually to the more aggressive gorillas in the group. Finding sexual coercion in the absence of physical force among nonhuman animals opens up a powerful and overlooked dimension for understanding coercion and consent in our species.

Among human beings, too, not all sex that appears to be consensual actually is. The 2017 #MeToo movement exposed the widespread sexual intimidation and abuses of power among men in many industries. Like the male chimpanzees and gorillas, these bosses used their power to punish and sexually coerce females who dared not refuse. The use of physical, financial, reputational, and other forms of intimidation to secure sex is common. From the threatening climate in a home with domestic violence to a workplace in which one's ability to feed a family is at stake, abuses of power and fear contribute to sexual coercion in our species.

Intimidation and fear work as weapons of coercion because victims have no alternatives. They are stuck. An alarmingly common form of sexual coercion against adolescents relies on a similar strategy: choosing victims who also can't move or escape. Being intoxicated can incapacitate, and predators of all kinds, including sexual predators, seek out targets who are physically compromised. The use of date rape drugs is one example of this principle. Adolescents who get drunk or high can become easy prey. Alcohol and drug education is crucial for the sexual safety training of predator-naive adolescents.

WITH A LITTLE HELP FROM MY FRIENDS

When individual animals are raised in isolation, the lack of adolescent training and practice in courtship leaves them ignorant of how to pursue sex. For example, young and adolescent guinea pigs that aren't brought up with other guinea pigs are more forceful and less successful sexually than guinea pigs who have social practice earlier. Rats raised without play partners often don't mature into sexually capable adults. And studies have shown that young American mink—both males and females—need rough-and-tumble play to prepare them for adult sexual behavior. As Jamie Ahloy Dallaire, one of the scientists studying the mink behavior, noted, "Rough-and-tumble play is commonly known as 'play-fighting,' but at least in some species it might be more appropriate to think of it as 'play-mating.'"

Growing up without role models and playmates might also diminish libido. "Having absolutely no social interactions—that is, being raised in isolation from weaning to adulthood—suppresses sexual behavior," wrote Katherine Houpt, a behaviorist from Cornell, in her textbook on animal behavior. Boars raised alone have a lowered libido with little interest in sex. Similarly, dogs who grow up without contact with other dogs may have a normal libido but lack sexual mounting ability, having missed out on play-mounting as puppies.

In keeping with human medicine in which research on female health, including sexuality, has lagged behind similar studies on men, socialization's effect on females "hasn't been studied as thoroughly," Houpt observes. She does note, however, that female cats not adequately socialized may reject Toms.

A range of social experiences is necessary to help adolescents learn the signals of receptivity and non-receptivity, because even though they can sometimes be plain and bold with clear meaning, courtship behaviors can also be subtle and impossible to fully understand without participating in a social community.

Sexual communication in animals contains insights for human sexuality. The back-and-forth of expressing, assessing, and responding is at the heart of courtship. And courtship learning intensifies during wildhood. At no other time is a sexual creature—human or nonhuman—more plastic, open-minded, and able to learn about the communication of desire. Courtship in humans and other animals is essentially a conversation about sex—specifically whether or not it will happen. In our species and others, sexual conversations often don't end with the desires of both individuals satisfied. But if there is no conversation—no sexual dialogue of courtship—it means that one participant is not there willingly. In other words, sex without the two-way conversations of courtship is coercive.

The same three types of sexual coercion seen in animals also occur in humans. Men and women are coerced through physical force, harassment, and intimidation. The solution to the devastating issue of sexual coercion in humans may not immediately emerge from the study of animal sexuality and courtship. But it exposes the singular importance of learning about sexual communication during a young animal's wildhood.

HOOKING UP IN THE WILD

An adolescent arriving at a moment of intensifying sexuality may feel there's nothing new about sex: it's something everyone knows about (except him). At the same time, he or she may also get the message that sex is instinctive and innate—"animal"—and somehow just happens. But, in fact, none of Earth's sexual beginners knows what they're doing at first, as we know from watching moths go through it. For our species, add to the mix alcohol, immaturity, pressure to perform, and a shifting understanding of consent, and things quickly become even more complicated.

One of those complications in the twenty-first century is hookup culture, which the American Psychological Association defines as "brief uncommitted sexual encounters between individuals who are not romantic partners or dating each other." A 2013 academic review of hookup culture in the United States by researchers at the Kinsey Institute and Binghamton University found that dating culture has shifted toward more openness and acceptance of uncommitted sex among young adults.

"We're dealing with a culture among emerging adults that views sex in a noncommittal way, emphasizing experience over committed relationships," they said in an interview about their research. Acknowledging that the culture has changed, the authors advised continuing discussions between potential partners. "We neither condemn nor condone any consensual sexual activity," they wrote, "but we do endorse the need for emerging adults to be aware of, and honestly communicate, their own intentions, desires, and the comfort levels of themselves and of their partner(s) during engagement in sexual activity." Instead of speculating on possible effects this shift might have on the mental and emotional health of adolescents and young adults, the authors cautioned: "Hookups, although increasingly socially acceptable, may leave more 'strings' than public discourse would suggest."

Sociologist Lisa Wade says that on college campuses, "hookup culture is everywhere." It's "an occupying force, coercive and omnipresent. It's more than just a behavior; it's the climate," she writes in her 2017 book *American Hookup*.

However, while hookup culture is real, Wade notes that hookups themselves are less frequent than the hype would suggest. It's a myth, she argues, that college students are having a lot of sex. "Students guessed that their peers were doing it fifty times a year. That's twenty-five times what the numbers actually show." Students can opt out of hooking up, she writes, but they can't opt out of hookup culture.

Psychologist Richard Weissbourd agrees. Hooking up is far from the norm, he says, although the myth persists that it happens all the time. This contradiction is backed by research from the APA and reporting by students on college campuses. What adolescents most crave, believes Weissbourd, is romance. They want and need to connect.

The communication between two people hooking up may not be as intricate as a humpback rowdy group, or for that matter a fruit fly dance. But the point of hooking up is supposed to be that both partners want to do it. Even if it boils down to quick and mercenary steps, both partners must signal, interpret, and accept sexual willingness. If that's not the case, then it's not really a hookup—it's probably something more coercive. As Wade writes, a hookup has "oddly strict parameters. It's spontaneous, but scripted. . . . It is, in short, a feat of social engineering."

There's widespread discussion these days about confusion over how to discern what a potential partner wants. This confusion comes from immaturity, pressure to perform, and a dearth of information passed down from previous generations. But, notably, a single factor reliably blurs the communication of consent: intoxication. Alcohol and drug use, researchers say, drastically increase the physical and emotional risks associated with hooking up. It's perhaps instructive that, for wild animals, intoxication is not an issue.

To what extent hookup behavior is a myth may be a question, but there is no doubt that in addition to creating an anxiety of expectations for many young people, hookup culture presents a false picture of adolescent young adults as sex-crazed, indiscriminate copulators. Sensationalistic if not totally inaccurate, perhaps even ephebiphobic in some way, hookup mythology ignores the subtler yet powerful social-behavioral forces in the sex lives of our species and others: chemistry, courtship, and romance.

LESSONS OF WILD LOVE

By 2018, Salt had been the object of humpback desire for at least fourteen seasons over thirty-five years, and possibly more. It's reasonable to assume that all her couplings followed the same essential pattern of finding a desiring and desirable mate, diving into a rowdy group, and, away from human eyes, consummating the relationship, somewhere in the Caribbean.

Salt is now around fifty years old, and her distinctive white markings are easy to identify when she surfaces to breathe. She knows what she's listening for when the chorus of humpback tones reaches her body. She knows how both to express and to respond to desire when she's with a male. We'll never know how Salt's sexual life as a mature female differs from what it was like when she was coming of age in the 1970s. We'll never know about the misfires and fumblings of her first experiences, why she chose the males she did, if she had a favorite, or, even though all the research points to humpback reproduction being a receptive process on both sides, whether or not Salt entered each encounter willingly.

As we've seen, courtship across the animal kingdom is a form of communication. It's a language, but it's not confined to words. It's about understanding and experience. When you're effectively communicating, you're each folding in information given by the other, not engaging in two separate monologues.

Great love stories are generally not about sex. Many don't even include it at all. Great love stories are about the excitement, the motivation, the missed signals, and the pulling in and sometimes pulling away embedded in the moments leading up to actual coupling. The problems recounted in love stories are problems of communication.

The lessons of Salt's romantic adventures in the Atlantic Ocean, and of Earth's other romantic creatures, from moths and fruit flies to mink and fossas, are these:

First, waiting to have sex sometimes makes a lot of sense. Many animals around the world experience a lag between being physically ready and being socially ready for mating.

Second, while you're waiting, learn and practice courtship: how to send and receive signals of desire and mutually decide what to do next. Chemistry is shaped by communication that is willing, honest, and reciprocated.

Third, remember that among animals, partner preference varies and behavioral flexibility is part of sex and sexuality. Even fruit flies aren't little robots.

And last, know that at any given moment, all over planet Earth, pairs of adolescents and young adults are regarding each other with the feeling understood by ancient peoples on a faraway frozen beach: *mamihlapinatapai*, the mutual desire to connect, and the excitement of figuring out how to start.

PART IV

SELF-RELIANCE

Leaving home signals the start of adult life for some animals in wildhood. Others stay in the territories where they were born but take on new roles and responsibilities. Either way, adolescents and young adults gain assurance as they begin to provide for themselves and others.

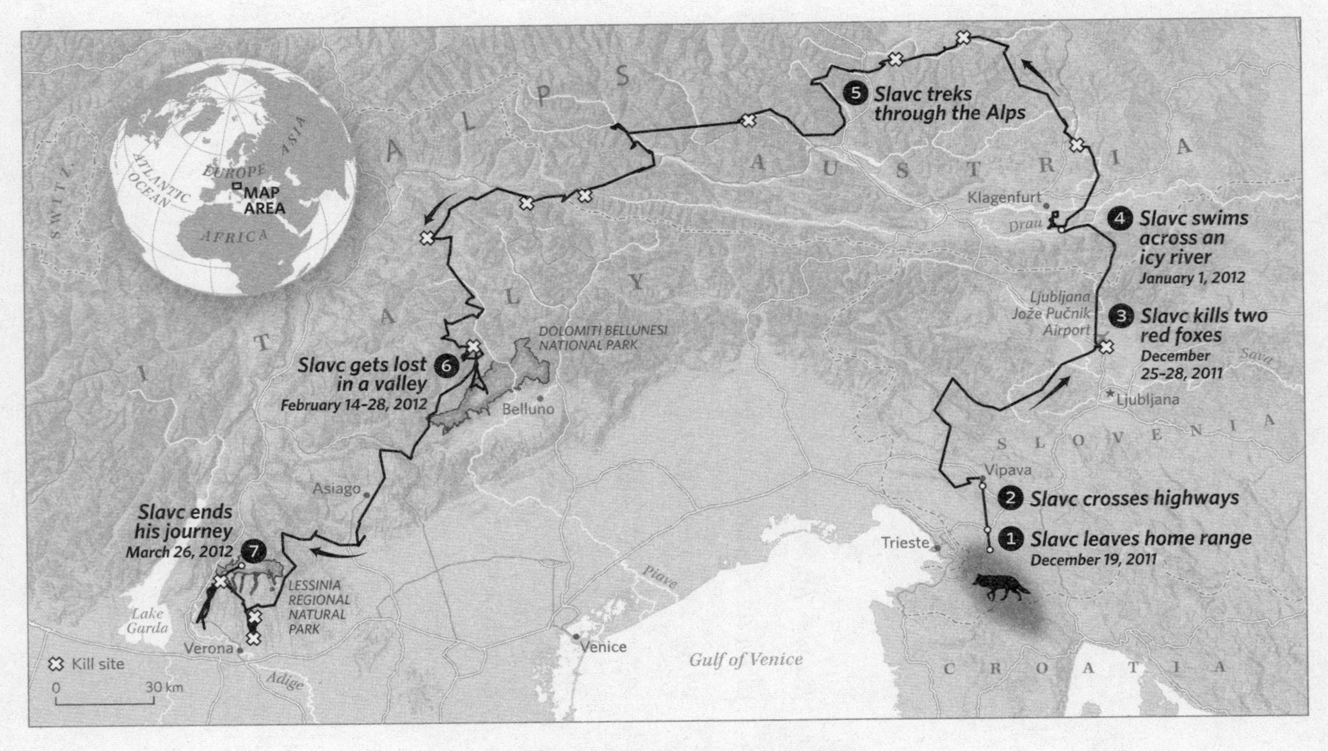

SLAVC SETS OUT ON HIS OWN

Chapter 15

Learning to Launch

All was dark when Slavc (pronounced slahv-tz), a young wolf, awoke in a Slovenian forest outside Trieste, Italy, on December 19, 2011. The nights had been colder than usual that winter, and the sun wouldn't be up for another couple of hours. That morning, Slavc decided to depart. He turned north, toward the Italian Alps, and left the only home he'd ever known.

Some months later, six thousand miles away in a Los Angeles ravine, a mountain lion also woke before sunrise. Down a dry creek bed he crept, unheard and unseen by the people slumbering in the mansions that backed up on the arroyo.

The wolf and the lion, both adolescents, had recently been living with their families. No longer juveniles, in the weeks leading up to their departures they had started separating from the places where they had lived as pup and kitten. Fully grown in body but still green in experience, each would make his way into the world alone.

Their very different fates offer dramatic examples of the defining power of this moment of life. Of course, neither realized as he set out that morning that driving him forth was a powerful legacy hundreds of millions of years in the making. The two animals were part of an overwhelmingly dangerous, ancient, and planetwide phenomenon that consumes young adults on the verge of independence. It's both a moment and a behavior called "dispersal."

A WILD BILDUNGSROMAN

If you were going to write a coming-of-age story about an adolescent animal, you would want to use dispersal as the plot structure. Dispersal is what screenwriters call the "inciting incident"; it sparks the action. It shapes the protagonist's quest. Dispersal—often referred to as "leaving the nest"—forces characters to face fear, form friendships, find love. Leaving home, they follow their dreams, seek their fortunes, and discover themselves. As a plot motor, dispersal is hard to beat. It tests adolescents and young adults by confronting them with isolation and conflict, and then it eventually delivers them into another state: adulthood.

Dispersal behavior is marvelously complex, but simply defined it's the process of separation, the moment when a young adult begins to live independently. If you're a dispersing adolescent, you begin taking over some or all of the responsibility for your own safety, socializing, and food-gathering. You likely start roaming farther from the place you were born, and your absences from home are typically longer and eventually often permanent. Not all dispersing animals leave home forever, though. And some never leave at all.

Modern human adolescents and young adults around the world disperse for the first time in a number of ways, whether to work or apprentice, attend school, or join the military or other service. For some, marriage marks a moment of dispersal. Financial independence, or even just stability, contributes for many to the sense of becoming a "real" adult. For others, dispersal means life on the streets.

The range of adolescent dispersal patterns that we see in human societies is mirrored by the many different ways wildlife leave home for the first time. At one extreme are Australian possums that one evening suddenly get up and march away from their birth nests in dogged, straight lines. At the other are *Parus varius* birds, and their exaggerated begging. These songbirds stay in the nest so long their parents finally have to cut off their food supply to get them to move out.

In world literary traditions, males tend to be the ones to strike out on heroes' journeys. Countering this are many examples from nature that show it's a lot more equal. For wild horses and zebras, females leave their families and join new clans. The same is true for our close primate relatives, bonobos, and also for baboons and tropical bats: it's females who seek their fortunes away from their birth homes. For other animals, including penguins, whales, meerkats, sharks, and many humans, adolescents of both sexes take off on heroic quests. Some go alone, but others swim, fly, gallop, and scamper together in adolescent bands exploring the world, sometimes for several years, before settling down to start adult lives.

Biologists believe that the drive to disperse provides a number of biological benefits, including discouraging animal family members from interbreeding. But the advantages of branching out can also carry a downside: an animal's first dispersal is one of the most dangerous times of its life. Recall young Ursula, the penguin leaving South Georgia Island for the first time. She and her peers were biologically primed to leave home yet couldn't do it successfully without first running the deadly leopard seal gauntlet. Young dispersers frequently encounter danger, and many don't make it through alive.

Slavc, the wolf, left home abruptly. It was impossible to mistake it for anything besides dispersal, said Hubert Potočnik, the Slovenian scientist who radio-collared and nicknamed the sixteen-month-old adolescent. Potočnik told us that he'd been following Slavc's meanderings for about a year when the young wolf suddenly took off. The California mountain lion, whom scientists didn't name but we'll call "PJ," was likely driven by a similar urge to rove. Dispersing adolescent mountain lions can cover dozens of miles in a single day as they seek new territory.

Whether alone or in groups, dispersing young animals must avoid the things that can kill or harm them, including in our modern world, motor vehicles. But even if they stay safe from outside danger, all animals confront another unrelenting deadly threat, one that will stalk them for the rest of their lives: starvation. Hungry animals take risks their well-fed peers do not. Some of the seemingly reckless behaviors seen during wildhood may very well be efforts not to die of starvation. More to the point, whether in the wild or on the crowded streets of a modern city, an adolescent or young adult that doesn't figure out how to feed itself is in grave danger.

Learning how to eat regularly—we're talking about literally making a living—is one of the most complex of all a young animal's tasks.

LEARNING TO LEAVE: PRACTICE DISPERSALS

Ursula the king penguin had no ocean training ahead of her plunge into the Atlantic. She knew nothing of the leopard seals lurking offshore. Her parents hadn't taught her how to fish. She didn't even really know how to swim. She had what biologists call an "uninformed" dispersal.

Slavc the wolf and PJ the puma had the opposite. Instead of heading into the world totally without experience, they received life-skills training before they left home. Their "informed" dispersals came with pre-departure coaching from parents and other adults, which started when they were still young.

Many mammals, birds, and even fish species are lucky enough to undergo informed dispersals. Possums are a good example because they train to leave home in multiple, phased steps. First, as joeys, possum siblings take turns riding on Mom's back. From this safe perch they begin to learn how predators look and smell, how to defend themselves, and how to find food that is safe to eat. Once they've grown too big for piggybacking, the maturing possums transition to being "at foot." During this exploratory phase, they scurry around Mom and stray farther day by day but always circle back for parental protection and care. Next comes sleepaway training. Each adolescent chooses a tree near the home nest for a solo sleepover. The dispersers-in-training are on their own, but their moms are nearby in case they need backup.

"Possums are really quite good mums," Hannah Bannister, an Australian conservation biologist, told us. "They seem to be setting up their babies for the best life possible." She recalls one possum mother whose oldest offspring took much longer than his siblings to enter the exploratory phase. This possum mom allowed her son to remain with her for extra time, supporting him until he became ready to go it alone. Sleep-aways, school trips, summer camps, overnights with relatives: humans too practice dispersal during wildhood.

Wolves, with their exquisitely complex societies, have even longer, more involved dispersal training for male and female adolescents. During puphood, Slavc would have been given "toys"—bones, feathers, skins—to practice with. Before learning to hunt live prey, he and his siblings would've pounced on these playthings and carried them around like trophies. As Slavc's voice began to change around puberty, dropping from high yips to lower barks and howls, he would've practiced his intonations with other wolves in his pack—crucial communication skills for effective group hunting. Next, Potočnik told us, something momentous in the life of a growing wolf happened for Slavc. Although not yet a competent hunter, he was invited to join the pack on family hunting trips. This hunting school—or what wolf expert David Mech calls "finishing school"—is a time for young wolves to make mistakes and learn. If they flub an aspect of the hunt during this training period, they are protected by their puppy license, and avoid the punishing reprimands that might be doled out to older wolves.

During hunting school, Slavc wasn't only honing valuable physical skills and practicing the required give-and-take of life in a social community. He was also still living with his family, and so he was guaranteed to get something to eat, even as he was developing the proficiency to make his own kills. Informed dispersals offer a kind of insurance against starvation.

Unlike wolves, who generally live and hunt in family groups, pumas are solitary as adults. Before dispersing, however, they live with their mothers for a year or even two. Attacking and killing a three-hundred-pound mule deer would've been far beyond PJ's abilities until he was well through adolescence. Until then his mother would have shared her food when she made a kill and at the same time been teaching him to hunt. Like house cats bringing their kittens injured mice or crickets, mother pumas create teachable moments for their young. PJ's mom would have presented him with wounded prey to sharpen his innate stalk-and-pounce behavior. Before heading off on his own, PJ might have practiced on fawns, rodents, and other small animals.

Clearly, dispersal is physically and socially stressful for adolescents, and leaving home without enough training and preparation can be treacherous. For instance, when African elephants are orphaned, often due to illegal poaching of their parents, they're forced to disperse without sufficient experience or information. On their own, these orphans often starve. Without adult elephants to guide them, they lack social skills and tend to be conflict-prone around other elephants and occasionally violent to people and other animals, both of which can get them killed.

SHOULD I STAY OR SHOULD I GO? DELAYED DISPERSAL

Dispersal-aged, adolescent animals don't all leave home on the same schedule. Just like humans, some of whom might drag their feet a little before departing, animals may also delay the inevitable. Ursula couldn't leave home until she had shed the downy feathers of chick-hood—she literally couldn't swim until this change happened. But once she'd grown her mature—and waterproof—black-and-white plumage, she was a step closer to ready.

Nestling barn owls have distinctive soft white plumage that turns brown as they get older. However, occasionally you'll see coats of white feathers lingering on adult-sized owls. Retaining the costume of juvenile birds can help an older adolescent hang on to the protection and opportunities that come before adult responsibility.

Prolonged immaturity, seen throughout nature, postpones the dangers and challenges of independent life—finding enough to eat, avoiding predators, navigating new terrain, meeting new groups, exploring sexuality—until the young creature is better equipped to handle them.

Humans who revert to an earlier developmental stage by dressing young for their age or using baby talk are showing a defense mechanism called regression. What triggers a regression in a young human is not always clear. But delayed plumage maturation physiology and behavior in nonhuman animals suggests that something in the environment may be cuing that young person that it isn't safe to grow up yet.

Delaying dispersal can also lead animals to put off breeding, which in many kinds of birds sets them up to become so-called nest helpers when their parents have a new brood. Helpers babysit and protect their new siblings and bring them food. These older bird helpers who have tapped the brakes on their own breeding will often be rewarded with better success from their greater size, experience, and status when they do breed the following season. Babysitting experience also prepares them to become better parents. As a bonus, just by staying at home, they improve their chances of inheriting their parents' territory.

Whether they're wisely delaying their departures to gain important skills, or they just don't feel ready, some adolescents need a parental push. Wolves have been seen wrestling and pinning maturing males in dominance displays biologists interpret as pre-dispersal harassment. Adolescent animals don't always appreciate the reminder. Young adult rodents sometimes put up a fight when their moms encourage them to move along. They beat at their mothers with their paws and refuse to be picked up.

Mountain lion mothers have their own strategies for forcing dispersal on their maturing cubs. Sometimes they lead them to the edge of their birth territory, turn around, and walk away. If the cubs try to follow, they're met with snarls and perhaps the cuff of a maternal paw. Puma mothers may also fail to show up at the prearranged rendezvous sites set up in case mother and cub get separated. The adolescent puma waits until finally getting the message that Mom's not coming back. When siblings are ditched this way, they often stick together for months, hunting, sleeping, and roaming as a group until each is big and skilled enough to separate. A singleton, however, is on his own.

WHY CAN'T WE GET ALONG? PARENTAL MEANNESS AND PARENT-OFFSPRING CONFLICT

Since the 1970s, biologists studying how and when animals leave home have based their research on a central concept: the interests of an animal parent and its offspring aren't always exactly the same. Animal parents want to maximize the number of healthy offspring they launch into the world. Offspring, on the other hand, are trying to monopolize parents' resources for themselves. What emerges is "parent-offspring conflict," a battle royale for attention, protection, and care. Offspring want as much as they can get; parents need to prudently invest and distribute their limited resources across the offspring they already have and any they might have in the future. In this framework, dispersal is forced by an irreconcilable imbalance between the amount of care a parent is willing to give and the amount of care the offspring demands.

Parent-offspring conflict has a lot to do with how and when animals leave, from the Siberian jay parents who entice grown chicks to stay with food bribes to the mountain lion mother who snarls when her nearly grown son comes too close to her kill. As much as an adolescent's behavior changes during this sensitive period, parental behavior changes too. Whether or not conflict forms the basis of the parent-child relationship, across species, one thing is clear: conflict peaks around dispersal.

When parental behavior shifts from encouraging and supportive to indifferent or outright aggressive, it can come as a shock to the offspring. They may not feel ready for life on their own—in fact, many animals at this stage don't even know how to hunt for themselves. Other important skills like flying and running, self-defense, interacting with adults, and making friends may be present but still developing in the young adult—not yet polished enough for the competitive arena of the real world.

Many human parents can tell whether their kids seem ready to leave the house. But others wring their hands right up to the day of departure and beyond, worried their offspring just aren't prepared. This is exactly when conflict peaks. Parent-child clashes may feel like proof the young adult isn't sufficiently safe or savvy enough to go it alone. But, in fact, conflict may be a signal of readiness.

The Spanish imperial eagles of Spain's Doñana National Park offer one of the clearest examples of this pivotal, sometimes behaviorally explosive, moment. And the researchers who study these birds have a special term for what's going on. They call it "parental meanness."

THE TRUTH HURTS

Picture a big brown eagle soaring above the oak trees of coastal Spain, scanning the terrain below with tiny flicks of her golden eyes. Suddenly she spots her target. She folds her wings against her body and plummets, picking up speed like a skydiver in free fall. Faster and faster she drops, until, mere feet from crashing, she rears back, wings spread, and thrusts out her talons. That move—wings-out, feet-first attack—is called a "stoop." Stooping makes eagles excellent and terrifying hunters, because they're so fast and nearly silent during the dive-bomb and so deadly and accurate when they strike.

But this mother eagle is not hunting prey. Her target isn't a rabbit or a mole, but rather her own son, who is fully grown but hasn't yet left home. He still depends on her to feed him. This mother eagle is stooping her son just as she would stoop prey, with one difference—her talons are not extended for a kill. They're interlaced, forming a club. And she uses that club to swat her eagle offspring, knocking him off balance. She stoops him while he's perching, and she stoops him in midair, while he's flying, sending him spiraling until he can regain his stability.

You can see why the researchers called this "parental meanness," but they observed that it doesn't signal a lifetime pattern of abuse. Eagle parents generally show no aggression to their offspring until this moment. In fact, they're usually exemplary avian mothers and fathers, providing all the nurturing and guidance that give eagle nestlings such a strong start in life. Parental meanness only happens at a very specific moment in these eagles' lives: just before dispersal.

In Spanish imperial eagles, the longer grown offspring linger around the nest, the more parental meanness occurs. But parents work up to it. Before they start stooping, they deploy smaller unkindnesses. They grow indifferent to the demands of their adolescents. They decrease feeding and don't respond to begging. Little by little when those cues don't work, the eagle parents' irritability intensifies into aggression. This hostility starts out as near-miss flight harassment, where parents accelerate toward their offspring and fly away at the last moment. But then it amps up into full-on stooping with contact.

Eventually, after a few days of this, the eagle offspring gets the hint and begins to move away. The researchers interpreted this to mean that "in the Spanish Imperial Eagle it is the adults who finally impose independence on their offspring, by cutting their rations and behaving aggressively." Although pushing out a son or daughter, perhaps in order to make way for a new brood of chicks, would seem to be the point of this behavior, there may be an added benefit—one that helps the offspring.

The scientists noted that while the soon-to-disperse adolescents were being stooped by their parents, the young eagles were also getting better at flying. What looks like aggressive behavior may also be a form of coaching—an intensive, last lesson before the bird leaves home. While of course the birds don't plan or think about it this way, dispersing eagles pressed by their parents may become better at the most important physical skill they'll need as adults: taking flight.

Without the aggression, human parents sometimes impart life skills by allowing the natural consequences of a child's actions to play out, while ensuring at every moment that the child is safe. It's sometimes called "tough love." Learning a hard but important lesson this way—and later realizing that their parents had their interest at heart all along—can be eye-opening, as the comedian Trevor Noah shares in *Born a Crime*, his memoir about coming of age as a biracial man in 1990s South Africa during the dismantling of apartheid. One day when he was a teenager, Noah took his stepfather's car without permission and when the cops pulled him over, he was arrested for auto theft. He spent a distressing week in jail, trying to navigate the legal system on his own while keeping his predicament secret from his mother. Noah eventually got out on bail with the help of a lawyer who mysteriously showed up and took on his case. It turned out that Noah's mother had hired the lawyer and put up the bail money after hearing through friends and relatives what had happened. "I'd spent the whole week in jail thinking I was so slick," writes Noah. "But she'd known the whole time." He recalls his mother explaining to him, "Everything I have ever done I've done from a place of love. If I don't punish you, the world will punish you even worse. The world doesn't love you."

NOT READY

We've already looked at many of the physiological and emotional changes adolescents go through during the transition into maturity, all of which may provoke conflict with parents. Dispersal can intensify that parent-offspring conflict.

As their children get close to moving away, parents often take stock and assess their offspring's readiness. Imagining all the things their child will soon face on their own can make parents anxious, especially if the skills are ones the parents feel they should have already imparted.

Parents of dispersers-to-be may find themselves worried about their adolescents' ability to accomplish a range of seemingly simple adult tasks: waking up on time, cleaning up after themselves, managing money. As the time unwinds toward departure, a parent may even provoke some fights as he nags his daughter to remove the wet towel from the sofa, check the gas tank, or get a flu shot. But it's useful to consider that the parent isn't trying to drive her away. He's trying to teach her.

A study of free-ranging dogs in Bengal, India, that followed groups of mothers and their pups from birth through dispersal illuminates this shift in parental strategy. The scientists found that when the pups were younger, the canine moms cleaned their dens for them. Once the offspring got close to the age when they were supposed to leave home, the moms didn't do that anymore. Weeks before their pups left, mother dogs tapered off the cleaning, leaving their rambunctious adolescent offspring to figure out how to tidy their own dens.

These mother dogs also trained their pups how to find food in the environment they would soon be alone in: the streets of Bengal. Shortly before weaning, mother dogs brought home scraps of garbage to begin incorporating into their kids' meals. This adjusted the young dogs' palates and smell sensors to a variety of nutrition sources. As we know, dispersing adolescents are constantly hungry, and parental preparation can help them sniff out the meals they'll so desperately need when parents are no longer around to provide them.

DO-OR-DIE: THE LAUNCH

Ready or not, the emotions that come with separation from home can be explosive or subtle for humans. While our species is unique in our expression of the doubt, excitement, fear, and thrill of setting out for an adult life, it doesn't mean that other animals don't experience something at the moment of departure. In his textbook on animal societies, the Cambridge behaviorist Tim Clutton-Brock movingly describes a mare leaving her herd to join a new family:

> In the dry mountains of the Granite Range of Nevada, two groups of mustangs graze on dry brush. In one group, a young mare is restless and is repeatedly herded back into the group by the resident stallion. When the group moves off along a winding trail, she lags behind. While the stallion's attention is temporarily distracted, the stallion from the neighboring group gallops out and places himself between the erring mare and the rest of the group, turns and herds her back towards his own group. The first stallion notices the maneuver and wheels to attack but is too late and he soon returns to his other mares. In the space of a few minutes, the young mare has made a decision that will affect the rest of her life, leaving the protection of her mother and father for life in a group of unrelated strangers that she has previously encountered only as rivals for patches of good grazing. She treads cautiously as she

approaches the resident females and the stallion guards her closely to prevent her returning to her neighboring group.

While in horse culture females disperse to other clans, in vervet monkey societies females stay with the groups they're born into and it's the males who leave home. When males reach sexual maturity at around five years of age, they disperse, often with brothers, peers, or other allies, and in the months leading up to the big day, these young primates become agitated, withdrawn, and moody, or even, in the words of one vervet expert, "depressed." Although they may not know precisely what's coming, these pensive male vervets have a big task ahead of them. When they find a new community, the first thing they must do is go up to the mature male leader and challenge him to a fight. This is sometimes mere weeks or even days after they've left their birth families. But these young adult monkeys not only have to summon the nerve to challenge a mature male; they also have to be extremely diplomatic about showing aggression. The final decision about whether they can join the group belongs to the females, and vervet females don't tolerate brutes. The most successful vervet males have to have had lots of social practice to be able to pull this off, and often that social training enables them to move in without a fight. In other words, social practice during wildhood, even for a monkey, is crucial for a successful launch into independence.

ROADKILL

We don't know how Slavc and PJ felt, predawn, on the mornings they departed on their wild adventures. Biochemically, mammals like us experience a natural surge in cortisol, a stress hormone, early in the morning, which raises blood pressure and accelerates hearts. So it's possible puma and wolf felt a hint of what a human would call excitement, the kind that makes setting off on a road trip at the crack of dawn somehow feel more thrilling than pulling out of the driveway after lunch.

But no sooner had Slavc started out than he was stopped by traffic:

the A-1 motorway running between Trieste and Ljubljana, loaded with speeding cars and trucks.

Motor vehicles are the leading cause of adolescent and young adult death in humans all around the world. For our species, largely unbothered by natural predators, these machines pose the greatest mortal danger to our young, whether they're in front of the car or behind the wheel. Starting at age ten and into their early twenties, human adolescents are injured by vehicle strikes more than any other age group. Older pedestrians over sixty-five have the highest fatalities—they're more likely to die if they're hit—but adolescents and young adults have more hits and more injuries.

And, as we know, motor vehicles ravage animal populations too. On American highways, a million animals are killed each day. We flippantly call them "roadkill," but underneath the fur, feathers, and guts are animal individuals. As urban environments encroach on wild spaces, greater numbers of animals find themselves faced with the challenge of crossing a road without getting hit. And it turns out that a disproportionate number of those proverbial deer in the headlights are dispersing adolescents. Away from protective parents, they are navigating novel environments for the first time.

Inexperienced adolescent animals and cars are a dangerous combination. Vehicles take out naive pukeko birds in New Zealand, possums in Australia, elephant seals along Highway 1 near Hearst Castle in California, and meerkats in the Kalahari Desert forced by their elders to cross roads first like minesweepers. Even adolescent whales growing up near shipping lanes off North America are struck more often by tankers and barges than their more experienced elders.

Wildlife biologists observe that with experience, animals, including deer and squirrels, can learn road smarts that keep them safe. Some urban coyotes have even learned to cross with traffic lights. Learning how to cross the street without getting hit by a car is often one of the first safety lessons a modern human learns. Children nine and under have the lowest risk, and some health advocates suggest that until the age of fourteen, children and adolescents should have adult guidance to cross streets safely.

Inexperience behind the wheel is even more dangerous—in fact, driving a motor vehicle is the single deadliest activity in modern adolescents' lives. New drivers are four times more likely to be killed and three times more likely to be injured than any other group. Lacking a full appreciation of the risks, they also have the highest rates of drinking behind the wheel and the highest noncompliance with wearing seat belts. Texting while driving is a dangerous distraction for all drivers, including adolescents. According to the U.S. National Highway Traffic Safety Administration, texting while behind the wheel makes it four times more likely a driver will have an accident. And according to a 2012 report, nearly 50 percent of adolescent drivers studied had texted while behind the wheel in the preceding month. In fact, among all ages who died in car accidents, fifteen- to nineteen-year-olds were most likely to have been distracted.

Of course, teens infamously seek sensation, misjudge distances, get distracted by their peers and electronic devices, and act impulsively. These characteristics all contribute to the grim statistics. Drivers as young as sixteen who went through graduated driver-licensing programs—which slowly over time give new drivers more privileges as they gain experience on the road—have improved safety records.

On the topic of Slavc's experience with motorways, Potočnik provided some fascinating background. From years of studying Slavc's pack, Potočnik knew these wolves regularly navigated around busy roads. Slavc had been safely crossing practically since he could walk, and Potočnik told us he'd even seen some social learning between the older and younger members of the pack. Bettering Slavc's chances of surviving were the safety structures that many European countries build for wild animals. Big motorways often have under- or overpasses engineered to enable animals to cross roads. When his journey brought him right up to one of the biggest killers of adolescents, Slavc knew what to do. He found an overpass and trotted across. When later in the day Slavc encountered the A-3, another busy motorway, the young wolf located a viaduct, slipped under the road, and continued on his way.

PJ

In vast Los Angeles, freeways are as abundant as the sunshine, but wildlife crossings are not. It's not uncommon for mountain lions to be struck trying to cross. Fortunately for PJ, his route down the creek bed that morning didn't take him near the roaring 405 or 101. But he did have to cross Sunset Boulevard, a curving twenty-two-mile four-lane road that traces the northern edge of the Los Angeles Basin from downtown to the Pacific Ocean. Predawn would've found Sunset relatively quiet, and PJ may have loped easily across. However he did it, he made it safely. The sun was starting to rise, and soon morning drive-time would begin, with its cacophony of commuters to herald the start of the human day.

Shortly after crossing Sunset, PJ found himself in a new environment, no longer surrounded by chaparral and trees. The rocky terrain under the young lion's feet had turned to concrete and asphalt. The brushy vegetation offering hiding places and resting spots had turned into manicured lawns and flat, landscaped walls of leaves. Disoriented, PJ kept going. He began to run, cantering down a shady street that intersected a wide boulevard called Arizona Avenue. Perhaps a car playing loud music zoomed past, or a truck blew its horn. PJ was suddenly in the middle of a city, alone and far from the environment he'd grown up in. Startled, he bolted up Arizona, scanning for somewhere to hide.

News reports and eyewitnesses later described what happened next. He found an archway and crossed through. He didn't know the path he'd chosen wasn't an escape route. It was a U-shaped courtyard. A dead end. PJ was trapped. He ran toward what looked like a break in the wall, not knowing it was the reflection of a glass door. Desperately he pawed at the glass, when he heard a sound behind him. He turned and saw a human being. PJ approached. The man turned and ran. PJ didn't know the man was running to call the police.

PJ was circling the courtyard, pawing at the glass, when suddenly a small army of humans appeared. Holding sticks, they began moving slowly toward him. Terrified, PJ alternately cowered and lunged, trying to get away, but was penned in on one side by the building and the other by the Fish and Wildlife agents. Suddenly there was a bang. Not knowing it was a tranquilizing dart, PJ tried once again to make a run for it and had almost lunged past the humans when he was thrown backward by a force he'd never felt and probably didn't understand. It was a wall of water from a fire hose, deployed by Santa Monica firefighters helping to contain the lion while they waited for the drugs to take effect. As PJ struggled to regain his feet, more bullets hit his body. The Santa Monica police who fired the shots weren't aiming to kill—these were nonlethal bullets intended to keep PJ where he was for at least ten minutes until the tranquilizer kicked in. PJ's eyes started to burn. Some of the nonlethal bullets were pepper spray pellets.

Terrified, PJ made one last attempt to flee. He had no way of knowing that this courtyard was close to a preschool and the Third Street Promenade, a busy downtown shopping area of Santa Monica. The police, the firefighters, and the wildlife agents couldn't explain to the agitated lion that they wanted him safely tranquilized so they could take him back up to the mountains and release him. But the tranquilizers weren't working fast enough. Disoriented, frantic, PJ heaved himself forward one last time trying to escape. In this last lunge, PJ almost got past the humans and back out to the street. In the interest of public safety, the officers aimed their guns.

SLAVC

The sleepy town of Vipava in southern Slovenia advertises itself to tourists as an Eden of wine and prosciutto. In summer, the white buildings topped with terra-cotta-tiled roofs are draped with the leaves of huge trees; tourists and residents alike drink coffee and beer and eat gelato at sidewalk cafes.

On a cold December night in 2011, Vipava received an unusual visitor. With data from Slavc's radio collar, which beamed a location every three hours, Potočnik could see that the wolf was in the back garden of a Vipava farmhouse. But reading the GPS points as they came in, the scientist was in despair. Vipava was so far away from Slavc's birthplace it seemed unlikely the wolf could've traveled there by himself in a single day. Slavc had logged so much distance that Potočnik was sure this adolescent he'd known since puphood had been shot dead by hunters and then driven to this faraway house.

But, in fact, Slavc was very much alive. All day he'd been alert and on the run, crossing motorways, avoiding cars, and evading people. Undetected by the humans in the house, he'd slipped into their back garden to spend his first night all alone. Farther from home than he'd ever been, he huddled by himself and, without the familiar warmth and companionship of his parents and siblings, curled up for a rest.

PJ

When you hear a news report about a wild animal wandering into a mall, apartment lobby, or playground, you can be fairly certain that one of the next sentences will contain the phrase "adolescent male," or "dispersing adolescent," or sometimes "animal teenager." More often than not, wild animals that go astray and encounter humans are in their wildhoods, making them desperate, low in status, hungry, territory-less, and biologically impelled to roam. Dispersing adolescents can find themselves in trouble when their inexperience meets the real world.

Even though PJ had been staying safe and finding his own food possibly for months, his skills were no match against the wrong environment. In the Santa Monica courtyard, a gun fired. This one contained real bullets. PJ fell back, dead.

SLAVC

Potočnik had worried all night about Slavc. The next morning, when the GPS indicated that the wolf had started moving north again, Potočnik was relieved. But he knew the bullet the wolf had dodged was metaphorical. A real one, delivered by a rancher protecting his livestock or an officer defending people, could easily end Slavc's dispersal. So Potočnik began rallying others around the adolescent wolf. He created an extended community of people invested in Slavc's survival, a community that would keep an eye on the wolf as he made the most dangerous journey of his life.

Potočnik called all the biologists and scientists he knew along a number of Slavc's possible routes. He contacted local wildlife and law enforcement officials and put the word out to hikers, ranchers, and anyone else who might come in contact with a roaming wolf. And he alerted the media. Soon, Potočnik wasn't alone in following Slavc's progress. News reports and websites gave near-daily updates on his whereabouts.

Conservation scientists who monitor individual animals in the wild can use location information to look after the animals' health and safety and intervene if necessary. The scientists can leave out extra food, treat them for infections and parasites, fix broken legs or wings, and redirect them if they seem lost. Like a human parent checking on a recently dispersed nineteen-year-old using smartphone tracking, text messages, or an in-person visit, these activities, called "post-release monitoring," enable scientists to check in on the young animals they care about and help with provisions or protection.

But even after all this, Potočnik was worried. There was still a chance Slavc would fall prey to a hunter, a speeding truck, another wolf, or illness. And, to add to the challenge, Slavc was heading straight toward the Alps and it was late December during a cold winter. With temperatures plunging, Potočnik knew the lone adolescent wolf would soon face the biggest challenge of all: finding his own food.

Chapter 16

Making a Living

Think about your most recent meal. Try to remember how much of it you prepared yourself. Did you choose, sample, forage, pay for, hunt, kill, pick, or pluck the ingredients? How much processing did you do? Did you cut, grate, shell, chop, peel, or gnaw? Did your food satisfy your hunger? Did it nourish you?

For something that must happen consistently—or else you die—eating on planet Earth is surprisingly hard to do for animals. A lot of skill must go into hunting adeptly, and the same is true for non-carnivorous forms of food gathering, like grazing and foraging. Another challenge about eating is that it's costly—it takes energy and time. If you're a wild animal, you have to wonder if a predator will attack you while you're searching for your own meal. If you're a predator, you have to know the Predator's Sequence and hope it goes well each time you hunt.

Being able to procure your own food is one of the most important markers of animal adulthood, perhaps even a stronger sign of full maturity than departing a home range, having offspring, or developing the physical characteristics of adulthood, such as horns, antlers, manes, or deep voices. Just as animals need to learn how to avoid predators, make friends, and communicate sexual desire and consent, they're not born knowing how to find food perfectly in the wild. Starvation is a major risk adolescent animals face—often bigger even than predation.

Dispersing animals are hungry, because learning to feed themselves is hard. And not all animals do it willingly—or well. The world is full of adolescents and young adults newly out on their own who haven't yet learned the necessary skills to find their own food. Hunger and the fear of starvation are so deeply ingrained in us, they even haunt protagonists of coming-of-age literature.

Katniss Everdeen, the eighteen-year-old main character of Suzanne Collins's aptly named Hunger Games series, survives in large part because she is able to hunt her own food. The archery and foraging skills she learned from her father keep her and her starving family alive at the beginning of the story. Later, they're keys to her survival in the arena. At one point, when she can't find food, Katniss digs through a garbage bin and agonizes over whether to steal a loaf of bread.

Katniss's desperation in that moment recalls that of the also-eighteen Jane Eyre, who, like a "lost and starving dog," begs for food at a bakery, shamefully offers to trade her leather gloves for a morsel, and finally eats porridge from a pig trough. Luckily for the starving Eyre, a farmer finally gives her a slice of bread. That smidgeon of food sustains Jane until she reaches the home of her long-lost relatives, ensuring her survival.

Frankenstein's creature—as poignant a coming-of-age disperser as any in literature—fares less well in his attempt to charm an old man in a hut and some alpine villagers into sparing him a bite to scrounge. He is chased away and threatened with weapons and finally has to eat berries in the forest and hide in a hovel, stealing food from cottagers in order to stay alive.

For modern, affluent young adults heading into nonfictional worlds on their own for the first time, dying of starvation is no longer a central concern. But this is definitely not the case for everyone. According to the Urban Institute, a Washington, DC–based nonprofit research organization, nearly seven million young people between the ages of ten and seventeen face food insecurity on a daily basis in the United States.

Individuals who lack resources have less choice in what they eat and how they obtain it. This is a serious, overlooked fact when it comes to understanding adolescent behavior. Across all human societies and across all species, hungry individuals are forced to take greater risks. Starvation is not the only way hunger can be deadly.

Hungry adolescent animals run into exposed open meadows or hunt under a full moon even though it makes them more visible. They edge out onto thin branches and into rushing water, forced to take the most dangerous, least desirable routes because of their lower status and lack of experience. They're also forced to eat poor-quality food—less nutritious and less tasty. By contrast, dominant and older animals can wait in safer positions. They eat choicer food and more of it. Less driven by the feeling of hunger, with its underlying fear of starvation, sated animals are also safer ones.

Hunger impels already vulnerable human adolescents to take risks. The Urban Institute reported that food-insecure teenagers in America steal, sell drugs, and may become involved in sexual transactions simply in order to eat. In addition to short-term physical risks, these food-seeking strategies may result in incarceration or criminal records, which further compromise their future opportunities.

For Slavc, even with the hunting practice he'd had with his parents back in the Slovenian forest, finding food was a challenge. Growing hungrier with each passing day, Slavc would have gotten more and more desperate. His hunger would have pushed him to take risks he wouldn't otherwise consider.

PACKING UP

Humans prepare for dispersal and its dangers, including starvation, in many ways. They save money; they pack food; they stockpile information; they gather the supplies they'll need. Young wildlife don't make such complex arrangements. Nonetheless, animals may get a biologic assist from deep within their own bodies, one that human adolescents likely share.

One of the first things many parents notice when their kids move out of the house is that the grocery bill goes way down. Yes, it's because there's one fewer hungry person in the house, a hungry person whose growing body needed a lot of fuel. But teenagers' legendarily insatiable appetites may have another cause, one with its roots in ancient pre-dispersal biology.

Research shows that just before some young animals get ready to leave the nest, some of their genes are activated (or "upregulated"), causing their bodies to change in ways that prepare them for life in unknown and possibly dangerous worlds. The outside environment cues the genes, which in turn direct body growth, a two-way relationship called genetic/environmental interaction.

Before migration, many mammals begin to store fat (unconsciously, of course), which helps if food is scarce on the road ahead. A young adult marmot, heading into a dangerous world without her parents, benefits from a fat-grabbing metabolism that packs extra energy supplies right onto her body. With this genetic assist, she may be a little less likely to die of starvation. Marmots and other dispersing animals may be further protected by another body system that revs up in adolescents: the immune system. Activated immunity will help them fight off the new pathogens and infections they'll encounter on their expeditions away from home.

Shifts in appetite and resistance, occurring silently and unseen deep within the bodies of animals on the brink of leaving home, have useful parallels for humans. Researchers are only beginning to hypothesize how these findings might be applicable to our own species. Perhaps ancient animal dispersal physiology encourages the body to hoard calories against starvation, thus playing a role in the obesity rates of teens and young adults. Or take the timing of autoimmune disorders like lupus, multiple sclerosis, and ulcerative colitis. They often come on in adolescence and young adulthood. Perhaps the life cycles of these diseases can be understood by considering the immune system overhaul seen in adolescents prior to dispersal.

With changes in their DNA preparing their bodies for an uncertain journey ahead, ravenous hunger becomes a central preoccupation for dispersing young animals. The lucky ones have informed parents who teach them tricks of independent living and support them while they get the hang of it even after they leave home. Animals with no parental backing have to figure it all out on their own.

TEEN TASTES

A week into his journey, near the Ljubljana airport, Slavc finally found a meal. He killed and ate two red foxes. His choice of prey shows some desperation. Wolves prefer to eat deer, but taking down a deer is a difficult feat. And so it's common for young adult predators in the early stages of independence to eat food that's easier to catch.

According to wolf expert David Mech, North Carolina red wolves eat a range of game, from white-tailed deer, raccoons, and marsh rabbits to rodents. But what is interesting is who eats what. In one study, adolescents consumed mostly rats and mice, while adults feasted on deer. Raccoons and rabbits went to the younger adult wolves with developing experience, who were building the stamina and smarts to take down deer. By the time they were adults, the wolves almost never had to eat a rodent. Like lower-paid, entry-level employees working their way into better salary ranges, young wolves saw their food quality improve as they gained expertise.

Being able to hunt a deer is also the standard used by the Endangered Wolf Center, a wildlife sanctuary we visited outside St. Louis. The oldest wolf sanctuary in the United States, the EWC takes in orphaned pups, trains them in life skills, and then releases them back into the wild. While they're being trained, the younger wolves mostly go after raccoons, possums, and rodents that wander into their enclosures. But the rehabbers are carefully watching for those skills to get good enough to take down big prey, like a mule deer. They will supplement a wolf's feedings until they've witnessed him or her kill a deer. Ideally, they like to see more than one confirmed kill before deciding a wolf is ready to go back to the wild.

Ben Kilham, a black bear conservationist and wildlife consultant in New Hampshire, takes a similar approach to the orphaned bear cubs he rescues, rehabilitates, and releases back to the wild. "I don't exactly teach them how to find food on their own," he says. "But I keep them protected while they learn how to do it."

As we've seen, finding food isn't easy in nature. Inexperienced meerkats often fumble with and even lose a precious scorpion before they can consume it. Even the most iconic hunters on the planet have to make many attempts before they have a successful kill. On average, lions hunting on the Serengeti and wolves in North America have an 80 percent failure rate—chasing and attacking five different animals for every one they take down. For Indian tigers and Arctic polar bears, a 90 percent failure rate is the norm. The availability of prey in the landscape plays a major role in hunting success. But so does the experience of the hunter.

Ursula would have spent months learning how to fish; for king penguins, learning to hold their breath long enough underwater is a big step. Similarly, dispersing humans with little food know-how typically fill themselves up with cheap, easy-to-obtain, lower-quality items, aka junk food, until they have the skills and resources to do the human equivalent of taking down a deer: paying for and/or preparing a meal.

There's another fascinating reason adolescents and young adults may be attracted to foods their elders reject. Changes in sensory perception during wildhood can make animals literally see food differently. Capuchin monkeys, for example, are suddenly able to see a wider range of colors as adolescents than they can as babies and adults. The colors fade out as they age. Biologists think this may make them better at seeing certain fruits, giving the younger monkeys a head start when they have to start competing with adults for foraging patches.

Sockeye salmon, when they're juveniles and adults, can see in the ultraviolet range. But for a brief period of time when they're adolescents, they suddenly lose that ability. It may have something to do with avoiding a certain kind of prey, because much of what sockeyes eat is marked with ultraviolet designs. Biologists postulate that temporary blindness to this food "packaging" somehow gives these fish a survival advantage they need during wildhood.

Adolescents may actually see, smell, and taste differently to protect them from foods that might harm and attract them to foods that might benefit them during this period, like food cravings and aversions during pregnancy may support survival of the growing fetus. Adolescents eat poorly, because they're lowest-status, last in line, and have to forage in the most dangerous areas with the poorest range of choices and the least-nutritious options. As we'll see, the oddities of adolescent appetites also help dispersing young animals adjust to novelty as they learn what and how to eat in their new food landscapes.

THE ADVERSITY ADVANTAGE: THE SQUIRREL WHO HAD GRIT

Among applied animal behaviorists (those who use their expertise directly to train or motivate animals), it's common knowledge that not all food is created equal. They know that dogs will find a bit of cheese, a chunk of liver, or a dab of peanut butter much more exciting than a mundane piece of kibble. Applied behaviorists call these special morsels "high-value treats," and use them strategically to encourage animals that may be distracted or struggling to learn a new skill. High-value treats are out-of-the-ordinary and desirable, and they both elicit and reward motivation. Wild versions of high-value food can be found in nature too. For the northwestern crows of coastal British Columbia Japanese littleneck clams are that treat.

In terms of crow time and energy, feasting on these clams is very expensive. The mollusks are difficult to source and preparing to eat them is time-consuming. First the birds have to locate mudflats where the clams are likely to be. Then they have to dig them out of the sticky ooze with their beaks. Lugging the heavy shells, they fly high into the air and drop the clams on the nearby rocks to break them open. If the shell doesn't crack, they have to retrieve it, fly it up in the air again, and re-drop it. Some shells need four or five flights to finally open. That's a lot of time and energy per clam.

When two Simon Fraser University scientists decided to take a closer look at the crows' seemingly inefficient behavior, they noticed something strange: an extra step in the process. The crows spent a great deal of time finding the clam, digging it out of the mud, and lifting it off the ground. But then sometimes, a crow would drop the shell and leave it behind, without taking it up into the air and dropping it on the rocks. Why would the crows go to the trouble of digging up perfectly good clams only to abandon them halfway through the process?

It turned out that the rejected clams were too small. By hefting the clam after dragging it out of the mud, experienced crows could calculate the potential energy content of the meat against the output they would have to spend on the flying-dropping cycle to extract it.

And who was better at making that calculation? Older, more experienced crows. Inexperienced crows invested more in the process of locating, digging, hefting, calculating, flying, dropping, and eating. And even once they had done all that, they had to outwit the scrounger crows—freeloading birds that don't bother with the whole exhausting extracting ritual and simply lurk near the action to sweep in and steal the hard-won gains of other crows. With scroungers too, the younger and less experienced were at a disadvantage. Their meals got stolen more often than those of older birds. But the crows got better the longer they stuck with it.

Some of the birds may have had a special attribute fueling their success: persistence, stick-to-itiveness, tenacity—the special combination of passion and perseverance that the psychologist Angela Duckworth calls "grit." Grit is a lot of things, part temperament, part biology, part training, and part environment, with expectation and opportunity thrown in. According to Duckworth, for humans, achieving one's goals requires sustained effort, practice, and high motivation.

Shrink the hyena from Ngorongoro Crater would probably have measured very high on a grit scale. From hyenas to tropical blackbirds to meerkats, individual differences in levels of persistence have been studied in animals. To measure it, scientists use food, often hidden in a puzzle that forces the animal to put effort into getting it. Confronted with the same obstacles, some animals give up more readily than others.

Some hyenas continue struggling to extract raw meat from a puzzle box while others give up after a few tries. Similarly, some meerkats keep trying to extract a crunchy scorpion from a jar long after their peers have given up.

Just as it does for people, grit improves outcomes for animals. Persistent animals displaying sustained effort, repeated attempts (practice), and high motivation are the most likely to solve problems and innovate. Time spent working on a problem correlates with success, whether it's the meerkat who finally succeeds in prying the scorpion out of the jar or the hyena who finally figures out how to extract the meat from the puzzle box. Sustained effort improves an adolescent's ability to face every challenge of wildhood. Staying safe, building social skills, communicating sexuality, and ultimately survival itself are all improved by repeated attempts during wildhood.

Animal grit contains lessons for our species: necessity may be the mother of persistence. It turned out that the most persistent animals were not the dominant, mature adults. They were younger and subordinate. As we know, subordinates have something else that may dial up their heightened stick-to-it-iveness: hunger. They have fewer resources. To get what they need to survive, adolescents and young adults must be tenacious, sticking around longer after older, dominant animals—better fed and therefore perhaps less motivated—have given up.

Animal studies add another notable observation that Duckworth similarly champions in humans: grit isn't a fixed trait. It can be developed. Squirrels who kept at the challenging task of extracting a treat from a specially rigged container became more persistent—and more successful—the more they tried. Motivated by a hazelnut, the squirrels kept at it. And in not giving up, their persistence increased. In other words, grit grew more grit.

INFORMED MOTHERS

When white-tailed ptarmigan mother hens want their chicks to eat a healthy meal, they use special peeps to point out plants with higher protein loads. The lucky chicks whose mothers are better nutrition "guides" continue to seek out more nutrient-rich plants even when they grow older and Mom is no longer around to remind them. When lambs graze with their ewe mothers and calves graze with cows, the young eat a wider variety of more nutritious foods.

Young humans with nutritionally knowledgeable parents are also lucky. Studies of parental influence on eating behavior show that parents are strong nutrition influencers, with a time-sensitive window of opportunity, during young childhood, to shape enduring food preferences and habits. Human parents transmit information about food to their adolescent offspring in a number of ways, by cooking with them, teaching them to market or read labels, and passing down techniques for making, storing, or preparing food.

But learning what to eat is quite different from knowing how to hunt, forage, find, or steal your own food. With animals, wildhood is when food-handling skills are imparted by parents and the larger community. It begins as soon as the young animal's strength and focus are strong enough.

Orcas use a technique called "stranding" to surf up onto a beach, grab a seal or penguin, and then slide back into the water. Adult whales teach their offspring this behavior by pushing them up on shore with an incoming wave, directing them to a prey item, and then intervening if they get in trouble surfing back out with the wave. It's a really dangerous move; if they don't learn how to do it exactly right, they risk getting stranded for real. After learning it from their parents, adolescent whales will sometimes practice this behavior with their peers, without parents around, riskily play-stranding with friends.

For human parents, transmitting food safety knowledge to their children is not nearly as dramatic. Today in our species it generally comes in the form of information about foods that may cause illness and even death in the short term (poison, allergy, anaphylaxis) or long run (diabetes, heart disease, cancer). While the specifics are quite different, the parental intention—to help offspring eat well and stay safe—is the same.

Young animals living in cooperative groups who hunt and forage must learn to participate in and contribute to collective efforts. Again, mothers appear to play an important role. Humpback whales like Salt catch fish using an ingenious system called "bubble net feeding." Working as a group, four or five whales swim around a shoal of fish, creating a kind of water tornado that traps the fish in its vortex. The whales blow bubbles into the spiral, which confuses and blinds the fish, preventing them from escaping. Once the shoal is all rounded up, the whales can swoop in from below, their giant mouths open, and slurp down an easy cone of seafood. This behavior must be learned and practiced, and there's evidence that it's passed from mother to calf and within groups.

In the 1980s, in the Gulf of Maine, around the time Salt was pregnant with her first calf, humpback whales were inventing a new and improved flourish on the bubble net technique. Called "lobtail feeding," it involved an extra tail smack on the surface of the water before the bubble-blowing commenced. The start of this new behavior coincided with the whales' shift to a different kind of fish that has a tendency to jump out of the water when it's scared. The tail smacks likely created a sound-wave "lid" for the top of the bubble net, keeping the fish where the humpbacks wanted them. It's pleasing to think that one of the many skills Salt may have taught her fourteen offspring when they were adolescents was how to spin a bubble net and how to deploy the lobtail method. Whether she did or not isn't known, but her fellow humpbacks in the Gulf of Maine certainly did, and these informed mothers were likely teaching the behavior to their lucky calves and their peers.

HUNGER GAMES

Hunting other animals is hard to do. Making the Predator's Sequence look easy can take years of practice. Some animal predator parents create opportunities for their young to rehearse before needing to perform this life-sustaining dance by themselves. This often involves bringing live prey to them.

Mother leopard seals present their pups with wounded penguins and encourage them to practice the kill. Cheetah mothers bring their kittens injured gazelles. Similarly, puma mothers provide fawns, beaver pups, skunks, and porcupines for predatory "edu-tainment."

Carnivorous meerkats send their young to scorpion school. There, the offspring learn to safely kill and consume the venomous thick-tailed *Parabuthus* scorpion. Experienced adults remove the stingers from these arachnids before presenting them, still alive, to the young meerkats. Gradually, as the maturing meerkat becomes more scorpion-savvy, adults bring a live specimen—stinger and all—and supervise while the offspring disable, disarm, kill, and consume it.

Predator training is part of the dispersal process for Spanish imperial eagles as well. In fact, it's what kicks off the separation process that ends with stooping runs and parental meanness. Like other birds of prey, these Spanish eagles follow a reliable pattern in how they taper off feeding their emerging young adult birds.

At first, the parents (both Mom and Dad) find their young wherever they happen to be perched and bring food to them. The parents alight next to the young bird, tear off a piece of meat, and offer it beak-to-beak—just like they used to when the chicks were nestlings. But after a few days, that babyish kind of feeding starts to decrease. The adults still bring rabbits or rodents for their son or daughter, but instead of cutting up the meat and feeding it to them, the adults land at a distance. If the youngster wants to eat, he or she has to fly to get it. And once the youngster arrives, the parent stays there, but doesn't help feed it. The youngster has to figure out how to tear off its own meat while the parent watches.

That phase transitions to a third stage, in which the young bird must fly to the parent who has the food, but now the parent flies away immediately and the youngster feeds all alone. As the parents feed them less and less, the young start to get more and more stressed. And they don't suffer this change in food supply and schedule quietly. They all beg—but their parents ignore their cries (the start of parental meanness). Standing firm, they bring food less and less frequently until their kids finally leave for good.

What makes this eagle process even more interesting—and high-stakes—is that the eagle parents do it before the offspring even know how to hunt. The young eagles aren't competent, because they haven't done it before. That seems illogical. How are they supposed to know how to listen for a mouse or grab a rabbit? How to swoop and stoop and seize and rip? Or, said another way, to detect, assess, attack, and kill?

The Spanish researchers postulated that parental meanness combined with "indifference" to begging forced eagle offspring to practice a skill vital to success as an eagle hunter: flying. The parents seemed to be using their offsprings' hunger to motivate them to learn the skills that would keep them alive as independent adults.

Whether it's bubble-net feeding, scorpion school, or beach stranding lessons, adolescent animals rely on their elders for crucial knowledge about what and how to eat. The ability to feed oneself translates into better survival. Self-reliance of this kind—and the confidence that comes with it—also prepares human and other animals to begin taking care of others, whether offspring, kin, or members of the community.

THE SELFISHNESS OF SACRIFICE

Animal parents pour time and energy into making sure their offspring learn the skills they need to survive in the world. Sometimes animals invest more in offspring who look like they are going to make it into adulthood. A study of leopards found that as their offspring got older, nearing independence, the mother cats actually *increased* the amount of time they spent finding prey for their adolescent cubs to practice on. In fact, these mothers put more time into doing that than they did into getting food for themselves. In human terms, that's not unlike a parent taking an extra job or two to fund the education of a promising son or daughter.

But life as an animal parent-teacher is not all altruism and encouraging growls and purrs. Learning from parents has limitations, and as a parent, having your own offspring as students can be taxing. A young wolf or orca can mess up the group hunt; cluelessness and adolescent antics can cost the family a meal. Animal parents may tire of teaching naive young to hunt and forage, especially as the offspring get older and their puppy license nears expiration.

These parent-child teaching conflicts are bypassed by some species in which the mentoring of young animals is done by a non-parent adult from the community. When banded mongooses in Uganda are about one month old, they leave the den and choose a foraging advisor to guide them in the intricacies of gathering food from their preferred diet. These non-related adults show their mentees how to steal reptile eggs, hunt snakes and birds, and scavenge for fallen fruit. The adolescent mongooses become territorial over their mentors and won't let other peers near them. A few months later, when the mongooses have learned what they need to learn, the escort relationship ends. But those young adults will forever prefer the foraging areas and techniques their mentors taught them over the ones their own parents use.

I'LL HAVE WHAT SHE'S HAVING

Adolescent animals, with their less-than-stellar hunting and foraging skills, and their suboptimal diets, are culinarily vulnerable in another way. They often copy their friends' food choices.

For example, if young Norway rats are given a choice between a tasty food they prefer and an unpleasant food they don't, they'll choose the tasty one every time. But in one study, when rats reached puberty and were placed with peers, their food choices shifted. They became twice as likely to defy their own taste preferences and copy those of their peers. To test whether this peer pressure went beyond taste preference, researchers offered sodium-deficient rats food containing health-improving amounts of salt. Once again, the adolescent rats rejected the fortified, healthy-for-them food in favor of whatever their peers were having. Remarkably, adopting peers' food preferences extended to poison: despite having been sickened by tainted food in the past, when a rat saw peers eating it, that rodent ate the toxic food too.

Rats carry scent clues on their fur and whiskers that reveal to their friends what they've been eating. But an even stronger influence is the smell of food on peers' breath. Even when it's bitter (a taste rats don't like), adolescents will want to eat it, as long as they've smelled it on their friends' breath. Similarly, they will follow along and develop aversions to the same foods their peers avoid.

To a human parent, the influence of peer pressure on adolescent food choice may be frustrating, especially if it comes after nearly a decade and a half of careful food education. Or food issues can seem like a warning sign of adolescent rebellion. Throwing away a healthy lunch, refusing to participate in a family dinner—these common adolescent behaviors can upset parents and hurt their feelings. But there is an ecological reason that explains the behavior in rodents and may influence humans as well. Peer information about a local environment is often more up-to-date than parental information. Set in their ways, benefiting from resources, status, or tradition, older animal parents can be out of touch with changes in nutritional ecosystems that younger animals are much closer to—and much more affected by.

Peers not only offer up-to-date information, but during dispersal they can also be a valuable, sometimes lifesaving, support system. When male fossas, the lemur-eating carnivores native to Madagascar, transition into adulthood, they often team up with another male—either a brother or a peer-friend.

"By pairing up, male fossas can cooperate in their hunts, which might allow them to eat more and grow bigger than solitary fossas," explains fossa expert Mia-Lana Lührs. Finding a compatible companion may be easier for two brothers, and the benefits of pairing up go far beyond hunting help. Finding another fossa he can tolerate enough to turn into a hunting partner eases a fossa's future. If he doesn't, says Lührs, that fossa will "find himself on the path to becoming a loner."

Hanging with friends at the burger place or coffee shop eating waffle fries and drinking sweetened milk tea may not lead to the healthiest diet. But a parent concerned about their adolescent's eating may be able to look at this behavior differently knowing that their kids are sharing the wildhood instinct to indulge in—and enjoy—food with friends.

A final point: as long as the eating isn't terribly out of control, the most important part of the picture is not really the food, but rather those friends. In all social species, individuals define themselves within groups—independence comes from self-reliance, not isolation.

Chapter 17

The Great Alone

After more than a week on the road, Slavc's stomach was finally full, albeit with an unsatisfying meal of red fox instead of roe deer. Over the course of the wolf's journey, Hubert Potočnik and his team investigated Slavc's kill sites and recorded what he was eating. They observed his hunting skills improve. Soon he would be eating a steady diet of deer—about one a week. But for now, Slavc may have felt the pressure of hunger pushing him forward.

Continuing north, Slavc crossed into Austria. There, on New Year's Day 2012, he suddenly found his path blocked by a massive river. The Drava springs from a source high in the Italian Alps and travels east through Austria before flowing into the Danube near Osijek, Croatia. In winter, its surface crusts with ice that floats above the deep currents. If he was going to continue on this route, Slavc had no choice but to cross the Drava. Not knowing the terrain, unable to find a bridge, Slavc dove in at one of the river's widest points. For 280 meters—the length of three football fields—he powered through the icy current.

On the other side, Slavc emerged from the freezing water, shivering and dripping wet. But there was no stopping now, and so he pushed on. For all of January and into February, the young wolf traveled west through the Italian Alps. Trudging through six-meter-deep snow in temperatures below freezing, he climbed to an elevation of 2,600 meters.

On Valentine's Day, Slavc entered the Dolomites, a range of mountains in northeastern Italy, through a pass called the Col di Prà. And here his journey seemed to end. Winter still gripped the landscape, and for the first time Slavc lost momentum. For days he circled an area appropriately named the Eternal Plains, searching for a way out. Then Potočnik saw something even more unusual. For five days Slavc's GPS marker barely moved. The wolf wasn't hunting, and no longer seemed to be trying to find a path forward. Lost, alone, in the cold, and with gnawing hunger, Slavc took an uncharacteristically long rest.

Isolation is a ubiquitous experience for most humans at some point in their lives and learning to handle it is part of growing up. In fact, surviving on one's own is a feature of coming-of-age traditions around the world. For Inuit boys, coming-of-age traditionally involved isolation from the group while learning to hunt and build igloos. You can see a snow scoop made of antler horn and sinew on the first floor of the Peabody Museum. The nineteenth-century owner of this implement likely learned how to use it from his father and other men in the community before spending time on his own to prove his hunting skills and establish his adult competence.

Traditional Australian aboriginal walkabouts challenged young men to spend months alone, fending for themselves; one stage of the Lakota vision quest involved spending four days alone on a hill. Adolescents and young adults in modern militaries are given survival training that emphasizes the ability to survive in isolation. In the United States, programs like the National Outdoor Leadership School and Outward Bound teach adolescents the skills necessary to survive in the wilderness. Among the most powerful experiences offered by Outward Bound is "Solo"—a stint spent alone in the wild that lasts from several hours to several days. Participants are given help with food and with shelter; dealing with the solitude is the central challenge and point of the exercise. Surviving alone in the wilderness is about much more than finding something to eat and a place to sleep. It's also about handling the psychological and even physical pain of loneliness.

We have no way of knowing whether a lone wolf suffers from loneliness. But research gathered from human subjects confirms negative physiologic effects of isolation—from inflammation to immune suppression to changes in cardiovascular function. In our species, at least, isolation can cause physical harm.

Some adolescents have a naturally greater preference for solitude than others, and experts point to the developmental benefits of periods of alone time for all adolescents. But an ongoing sense of isolation, loneliness, or disconnection can be a dangerous harbinger of depression and other health effects. Social isolation is a risk factor for suicide in adolescence. Solitude can be fortifying in wildhood; isolation can sometimes be deadly.

THE WILD SAFETY NET: EXTENDED PARENTAL CARE

For the past decade in the United States and other countries, parents have been criticized not only for helicopter parenting—hovering over their kids' every activity and mood—but also for raising so-called boomerang kids: children who return home after an initial period of independence in the work world or at college.

In fact, in the United States, this has become the norm. As of 2016, eighteen- to thirty-four-year-olds were more likely to be living with their parents than with a romantic partner. More than 60 percent of Polish, Slovenian, Croatian, Hungarian, and Italian eighteen- to thirty-four-year-olds live at home with parents, as do two-thirds of twenty-two- to twenty-nine-year-olds in China, Hong Kong, India, Japan, and Australia. And in most countries in the Middle East, the custom is for young adults to live at home until marriage.

Humans are notable for the long time we spend in dependent childhood and adolescence compared with some other species. But this may not make us such outliers after all. Many wild animal parents do not cut off support the minute their offspring leave home. In fact, many dial up the help and the training. If a youngster is having trouble getting enough to eat, animal parents will often feed them. If a youngster isn't meeting peers, parents will sometimes provide introductions. Like prudent parents finally spending a college fund, they bequeath territory and offer access to food larders they've been stocking away for this exact moment. The ecological term for these boosts that come after dispersal is "extended parental care." Conditions that lead human and animal parents to offer their grown offspring extended parental care are remarkably similar across species. Dangerous environments, food shortages, competition for territory, and pressure to find mates keep young adults living at home.

If boomerang kids were birds being observed by ornithologists instead of humans being critiqued by social scientists, a supportive parent-offspring relationship might be called the more bird-specific "post-fledging care." And maybe instead of bemoaning it, critics would recognize, as biologists do, that it can improve the future success and survival of offspring.

It's useful, and perhaps reassuring, to put the human versions of extended parental care into a larger historical and cultural context. Steven Mintz, a historian of the human life cycle, writes that in America, "a protracted transition into adulthood is not a new phenomenon," and "the decade stretching from the late teens to the late twenties has long been a period of uncertainty, hesitation, and indecision." He tells the story of a young man who after graduating from Harvard in 1837 at the age of nineteen was "hired as a school teacher only to resign two weeks later. He then intermittently worked in his parents' pencil factory, served as a tutor, and shoveled manure." The young man also worked for a while as an editorial assistant. Eventually Henry David Thoreau found footing as a writer and land surveyor, although he continued to be involved with and supported by the family pencil business.

"Contrary to what many people assume," Mintz writes, in the early United States, "the overwhelming majority of young people in the past did not enter adulthood at a very young age . . . During the early nineteenth century young men in their teens and even twenties tended to swing between periods of relative independence and phases of dependence when they returned to the parental home."

It's been like that during most of the history of the United States. With the sole exception of a brief period at the end of the Second World War, marriage, what we often think of as a traditional dispersal demarcation line, didn't happen for most young people until their mid- or even late twenties or early thirties. Even in the pre–United States colonial era, "young men generally had to delay marriage until they received an inheritance, which usually took place after a father's death," according to Mintz. He describes the transition to the adult world as having been traumatic throughout American history. Parents died early, education was often intermittent, and living arrangements were uncertain. Young immigrants, often female, traveled alone to find work.

In many species of birds and mammals, young adults old enough to be "ready" to move out are sometimes allowed—even encouraged—to stay in the home territory and help out. Occasionally these maiden aunts, and more often uncles, stay in their birth home for life. The arrangement is win-win-win for parents, offspring, and any new younger siblings. The young adults care for siblings by bringing food and acting as babysitters and mentors. They help the group by adding vigilance and security and extra numbers for mobs. Rarely are they freeloaders.

Staying in the home nest for extra time before dispersing is also not a sign of failure to launch. The benefits for these lingering young adults are many. If the environment has too many predators, young adults may be physically safer staying on longer with parents. If it's a year with a lot of peer competition, waiting a season can boost a young bird's chances at finding food, territory, and mates. Another boon of hanging around is that they're on-site in the event a parent dies and succession is up for grabs. They might inherit territory. For a low-ranking female meerkat, for example, the best strategy for getting her own territory is to stay close to home and wait for Mom to go. This strategy is also seen in chimpanzees, although males tend to be the siblings to inherit the territory.

Western bluebird sons who stay home over winter with at least one parent not only are more likely to survive the season, but they also tend to inherit some of their parents' territory come spring. The territory often comes with what Cornell scientists call "mistletoe wealth," stocks of the plant that serve as shelter and food for these birds.

Humans are not the only animals to bequeath their worldly possessions to their young. Any given plot of land may already have many "owners." North American red squirrel mothers leave their territories to offspring, usually adolescents, and, if they can, parcel in as much nearby vacant territory as they can manage. The mothers don't only give the gift of real estate. They can also stock the land with extra food, hiding larders all over it before relinquishing the whole package to their offspring. For these squirrel mothers, death doesn't mark the transfer of the estate. Instead, it's during middle age that a mother bestows this gift, packs up, and goes off on her own new journey just as her adult children are ready to take over the property.

One of the most powerful ways for animal parents to help their offspring with dispersal is to point them in the right direction before they actually leave. Parental excursions, a behavior seen in some mammals and many birds, involves a parent traveling out into the world with an adolescent offspring, scouting out sources of food, securing territory, and introducing the offspring into society. Like the social-climbing mothers of Jane Austen novels, a songbird called the *Parus major* takes her eligible bird offspring on visits to other flocks to introduce them to the best and highest-status potential mates to produce her future grand- and great-grand-chicks.

It's clear from many studies of a range of species that extended parental care saves lives—preventing newly independent young animals lacking in life skills from dying in the first dangerous days and weeks after leaving the nest. But the benefit of extended parental care comes at a cost: a delay in learning to feed yourself. A study of white-winged choughs, an Australian bird, showed that youngsters who stayed at home with lots of adults got more food and emerged from winter in better physical condition. But the trade-off came once they were on their own. Lacking experience, they were poorer foragers than birds who'd gotten no help.

Birds who receive extended parental care also show delayed antipredation behaviors. Young Mexican jays who spend extended periods with mature adults don't learn crucial mobbing skills.

Young animals must ultimately strike a balance between receiving care to keep them safe and fed in dangerous environments and honing the life skills they will need when they are truly independent. In this context of extended parental care in animals, it's interesting to think about the criticism leveled at contemporary parents who remain involved in their kids' lives through adolescence and young adulthood.

A report from the Harvard Graduate School of Education noted that "especially in affluent communities, their parents are hyper-involved in their academic and social lives, so it's unusual for teenagers to study, arrange a meeting about a bad grade, or even resolve a disagreement with a friend without parental help."

The excesses of some parents are easy to mock, and robbing young adults of opportunities to practice resolving their own conflicts is clearly misguided. Yet amid the criticisms of millennials moving home with their parents, the clear importance of continued parental involvement gets muddied. Mintz puts it like this: parents "have good reason to be standing by with a rescue rope as their children try to make their way through the overgrown and traditional paths to adulthood that may no longer secure employment. The twenties have replaced the teens as the most risk-filled decade. Problematic behavior—binge drinking, illicit drug use, unprotected sex that leads to disease or unplanned pregnancies, and violent crime—peaks during this age, and missteps during these years can impose lifelong penalties."

In the rush to criticize parental over-involvement, a larger problem gets lost: the *lack* of sufficient parenting for many. For young adult humans without parents or parent-like mentors, dispersing into the adult world can be exceedingly dangerous. According to an analysis by social scientists at the University of Pennsylvania, young people aging out of foster care in the United States—the eighteen-year-olds without family to provide financial or emotional support—show increased rates of unemployment and reliance on public assistance. Their physical and behavioral health is worse than same-age peers; they've often reached only lower levels of education; and they've had more brushes with the criminal justice system.

Mentorship—what you might call a human version of "post-fledging" social care—vastly improves the lives of this vulnerable population. The report found that dispersing foster children who had a relationship with a competent, caring adult mentor fared much better during adolescence and the transition to adulthood. The best outcomes were for adolescents who found a "natural" mentor, which the researchers defined as "a very important, nonparental adult that exists in a youth's social network, like a teacher, extended family member, service provider, community member, or coach." These familiar adult mentors, chosen by the foster children themselves, as opposed to unfamiliar adults chosen for them by state or nonprofit groups, became "a protective factor" for foster kids in transition, providing "ongoing guidance, instruction, and encouragement aimed at developing the competence and character of a young person."

Extended parental care is seen all over human societies, at all levels of wealth. It may come in the form of a place to live, food, and direct financial support. But it can also come without a price tag, such as career advice, teaching a skill, moral support, social introductions, and companionship.

How much extended care is given depends on the parents' resources and the needs of the offspring. But it is widely, if not universally, seen throughout nature. And with good evolutionary reason: a parent's genetic legacy lives in their offspring but persists in their grand-offspring. So why wouldn't parents do everything they can to help? You can think of that behavior as motivated by personal selfishness or evolutionary fitness, or you can think of it as motivated by love. Either way, it is irrefutable that parents all over the planet are invested in the safety, health, and, yes, happiness, of their children.

Recognizing the advantages and disadvantages of extending parental care in the animal world can help humans develop a more realistic and perhaps even compassionate understanding of how and when to continue to support their older and adult children. It is true that white-winged choughs and scrub jays who stayed on with their parents might not have learned to forage or chase off predators as well as those who left. But if the world outside is dangerous and a young animal lacks the skills to protect itself, it may be safer to stay at home. Ecology may be at least as influential as psychology in determining behavior that favors continued dependence. Mintz puts it even more bluntly: "Parental support can play a crucial role in preventing their offspring's lives from going severely off track."

Examples in the animal world suggest that extended parental care is as much an evolutionary strategy as an indulgence.

POST-RELEASE MONITORING

Conservation biologists can keep track of their study animals in ways most parents of dispersing adolescent young adults might envy. Satellite telemetry, remote cameras, drone surveys, and binoculars keep eyes on the vulnerable young, like Potočnik keeping an eye on Slavc through a radio collar and networks of observers and other community members. Post-release monitoring in effect allows conservation biologists to offer high-tech extended parental care to young wild animals that hit rough patches. This usually has to do with food, and it's an easy fix: just as a parent might tuck some cash into their young adult's pocket, the biologists drop off some food where they know the struggling animal will find it.

Mark Elbroch, a wildlife biologist from UC Davis and a project leader for the nonprofit wild cat conservation group Panthera, blogged about a pair of orphaned mountain lions he tracked in the Grand Teton Mountains in Wyoming. The cats had lost their mother when they were seven months old and didn't have the benefit of learning how to hunt from her. Pumas in this range generally stay with their moms for about two years.

Elbroch wrote that a few weeks after their mother's death, the two young adolescent cats began to starve. "They became skeletal zombies moving about in the day, oblivious to their surroundings." For one of the youngsters, it didn't end well. At the base of a Douglas fir, Elbroch found her remains curled inside an abandoned bed she had slept in weeks earlier with her mother and sister. Elbroch noticed that her adult canine teeth were still erupting, indicating she was in the physical transformations of puberty when she died.

And so Elbroch got permission from the Wyoming Game and Fish Department to stage an intervention for the surviving sister, the other orphaned and hungry cub. Elbroch writes that he and his team located her and "dropped the hind leg of a road-killed moose in her path. Fifteen minutes later, she discovered the boon; she fed for four days straight, and just like that, became a different cat . . . Whereas she had been aimlessly wandering, seemingly unaware of dangers or anything in her surroundings, the moose meal provided her the sustenance to become a mountain lion."

Elbroch's team continued monitoring the adolescent lion, feeding her two more times, until they saw that she was successfully killing and eating small prey on her own.

Support keeps solitude from tipping into dangerous isolation. The ecological lesson from this is as paradoxical and yet as clear as the idea that taking risks makes you safe: sometimes a little help actually increases independence.

HOME SWEET HOME

Slavc spent ten days lost in the Dolomites. When he finally found a way out, his old energy returned. He shot straight south in the direction of Verona.

In early March, Slavc reached the outskirts of the fair city. Likely attracted to a nearby wolf sanctuary, Slavc stayed in the area for twelve days.

The cultivated vineyards and farms of Verona's suburbs lacked the dangers of the high Alps, but they also lacked the abundant wild deer he'd found in the mountains. As we've seen, novel environments are risky. And for wild animals, novel environments containing humans can be especially unpredictable. Like PJ finding himself surrounded by speeding cars and shouting people in the middle of Santa Monica, Slavc had to adjust to his new surroundings, and it forced him into some bad decisions. With fewer deer around, Slavc began to prey on domestic livestock. Over a few days he attacked a sheep and a goat, and fed on a horse. Watching via the GPS data points, Potočnik was tempted to intervene to get Slavc out of a dangerous dynamic that he knew would invite threats from angry farmers.

Luckily, resourceful Slavc found his own way out. The wolf traveled north out of Verona and to the Lessinia Regional Nature Park, a protected forest. And there, according to Potočnik, the "dispersal switch flicked off" and Slavc's journey abruptly ended. He was home.

Slavc hasn't left the Verona region since he arrived. But many animals, like the hyena Shrink in the Ngorongoro Crater, and lots of human adults continue to disperse throughout their lives, moving to seek new territory, opportunities, and loves, or to escape strife or impoverishment. Some humans relocate again and again because they're just curious about the world; simple adventurousness seems to motivate some adult animals to roam too.

And in certain important ways, with each new "dispersal" adults reenter wildhood. They make themselves small in experience all over again. In new territories, they are vulnerable to forms of predation and exploitation. With new social structures, they're in flux, with the anxiety and excitement that comes with finding a place in a group. Among new peers, they may need to learn different languages for expressing and responding to desire. And even though for most adults finding food is not as hard as it is for adolescents, making a living will always be one of the most important things they have to do. Each reentry into wildhood retraces the patterns first etched during the highly impressionable days of adolescence.

Chapter 18

Finding a Self

At what moment did Slavc become an adult? Was it when he survived his first day away from home and had his first solo sleep in the garden of a Slovenian farmhouse? When he caught his first meal of red fox or when he made his first deer kill? Was it after his isolation in the Dolomites when he found a way out? Or would he not truly become an adult until he had learned how to court another wolf and maybe even produce offspring of his own? Animals, like people, don't come of age in a single moment. Maturity is a collection of skills and experience, the competencies that come from realizing and facing the four fundamental tests of wildhood.

There are a few important lessons young adults setting out on their own can learn from the trillions of animal dispersals that have come before.

First: because dispersal is usually a process, learning to leave can really help. Having early practice dispersals—camp, a school trip, visits with relatives, maybe with new routines and responsibilities—makes a successful dispersal later on more likely. Excursions with parents in advance of leaving also help create an "informed dispersal." But even if an adolescent lacks parental support, mentors can teach resource-gathering skills, which can also be learned from peers and through everyday trial and error.

Second: no matter how well an adolescent is trained, no matter how talented he or she is relative to peers, an enormous factor that determines a dispersing animal's success is the nature of the world it is entering. A crucial lesson from nature is that the environment—whether resource-rich or resource-poor, competition-heavy or competition-light, predator-dense or predator-sparse—will strongly influence dispersers' outcomes. Personal agency, ability, and grit do contribute to "making it." But even the luckiest young adult animals see their successes and failures—sometimes their very fates—hinge on the worlds they're privileged or doomed to head into.

Third, physical dispersal away from home doesn't happen in all animals or humans. But whether an adolescent leaves or stays, for all maturing creatures becoming nutritionally independent is one hallmark of adulthood.

Fourth, novel environments are dangerous and sometimes lonely. Dispersing with trusted peers or developing relationships away from home can make a tremendous difference.

This challenges parents and societies to do two things: first, teach adolescents and young adults to provide for themselves, and second, give them opportunities, time, and motivation to practice what they have learned. Today, while modern high schools and colleges offer many kinds of important education, the practical instruction needed to become self-reliant is often not part of the curriculum.

Not to put too fine a point on it, but this means helping adolescents and young adults understand what it means, and what it takes, to literally *make their living*—to connect the sometimes abstract concept of job and career with the vital, species-spanning, and very human task of feeding oneself and contributing to one's community.

Soon after Slavc arrived in the forest north of Verona, he met a female wolf the biologists couldn't resist naming Juliet. Together Slavc and Juliet have parented offspring for several seasons—one litter contained seven pups. Others never return. Although farmers are keeping a wary eye on these new arrivals, who will each have a dispersal of his or her own, scientists were ecstatic about the pairing of Slavc and Juliet. For the past two centuries, Europe's wolf populations have been torn apart by deforestation and human encroachment as well as systematic culling—

killing of wolves to rid the landscape of these predators. Slavc's journey reconnected two populations, and two gene pools: his Dinaric-Balkan line with Juliet's Alpine heritage. Slavc's dispersal quest strengthened not only his local community and family, but his entire species as well.

For all living creatures, growing up means leaving the past behind for an unknown future. Once the tests have been taken, the skills practiced, and the experiences coalesced, there may be an ineffable, unmarked moment when a creature feels safe enough, social enough, sexually confident enough, and self-reliant enough to begin focusing attention outward—toward others. To recognize one's responsibilities beyond the self. Perhaps this is the moment wildhood begins giving way to adulthood.

Epilogue

Ursula, Shrink, Salt, and Slavc are no longer members of wildhood's planetwide tribe.

Ursula, if still alive, would be well past adolescence and into penguin middle age. King penguins in the wild can live as long as thirty years, but because Ursula's tracking signal went quiet, how long she survived will remain a mystery. We'll never know how experienced she became with the Predator's Sequence, the rules of groups, the conversations of courtship, and the challenges of hunting for fish. Perhaps she regurgitated meals for chicks of her own or offered extended parental care during a difficult season. Maybe she watched a son or daughter waddle to the beach and dive in at the beginning of its own journey to adulthood.

In February 2014, Shrink's body was found close to a river frequented by lions, and it's very likely he died the way many hyenas meet their end: at the fangs of these apex predators. Near Shrink's body was that of another hyena, also dead. Oliver Höner doesn't know whether the other was a male or a female but did know that Shrink had recently immigrated into a new clan and was working his way up the ranks. This companion might have been another male accompanying Shrink on a coalition walk. Or it might have been a female he was courting. Either way, it seems that Shrink was social until the end.

Salt has become one of the most beloved and well-studied humpbacks in the world. Around fifty years old, she still makes the yearly migration to and from the warm waters of the Caribbean, where choruses of males resound. Counting her grand-calves and great-grand-calves, Salt has at least thirty-one direct descendants. Sriracha, a recent calf spotted and named in 2016, joined siblings Salsa, Tabasco, and Wasabi on Salt's family tree.

Slavc's radio collar was programmed to drop off shortly after he arrived in Verona in 2012, and Hubert Potočnik no longer knows exactly where he is, although the wolf has been spotted several times, parenting his pups with Juliet. Slavc is believed to still live in Lessinia, Italy.

Creating offspring that survive and reproduce is how biologists measure the success of penguins, hyenas, whales, and wolves. But for humans, reproduction is not the measure of successful maturation. Staying safe, navigating hierarchies, communicating respectfully about sex, and learning the satisfactions of self-reliance are the true markers of adulthood. Acquiring the four life skills of wildhood prepares us for many kinds of adult success, from the professional and the public to the personal and private.

Not all wildhoods have happy endings, and sometimes we learn the most when things go wrong. But with multitudes of species facing the same four challenges over hundreds of millions of years, many solutions have emerged to improve the chances that things go right.

Turning to nature to find solutions to the challenges of human life is an emerging field called "bioinspiration." Bioinspiration (or as it's sometimes called, "biomimicry") is based on the knowledge that over evolutionary time, the Earth's animal species have faced essentially the same pressures. Over countless generations of life on Earth, organisms have evolved adaptations—or solutions—to the problems they confront. Bioinspiration leverages these solutions, producing what is essentially nature's ancient, massive R&D lab, to help human life.

Finding bioinspired solutions to benefit humans and other animals is the focus of the conferences connected with our previous book, *Zoobiquity*, which bring together physicians and veterinarians from universities all over the world. As we've shown in this book, the natural world also contains insights for growing up and becoming an adult. Understanding the wildhoods of other animals can bioinspire approaches for compassionately and skillfully guiding human adolescents toward adulthood.

The universality of wildhood extends beyond actual physical and mental development. When humans say something is in its adolescence, we don't always mean people and other living things. Between birth and maturity in any human endeavor comes an in-between stage, a time when the promise of the new arrival must give way to the realities and responsibilities of growing up. This is true of businesses, creative projects, relationships, careers, fields of academic study, even political movements, governments, and countries.

Although beginnings can be painful, difficult, and risky, they are also often the easy part. A birth, a launch, or a fresh start is filled with hope, with dreams of a better future and of new success. It's effortless to begin a marathon with energy and enthusiasm; it's during the miles that come after, as your body tells you how it's really feeling, as you start sizing up the competition and jockeying for position, when the outcome of the race is ordained.

As we've seen, an animal's wildhood can be an awkward and unflattering phase of development, and that's true of these non-animal enterprises as well. Consider all the celebrated tech start-ups of the past few decades, and take, for example, an exciting new app that has attracted millions of dollars in funding and promises to solve a problem we never knew we had. Without a track record, the public and the venture capitalists are often willing to extend the company a puppy license (the same goes for debut novels or albums, new hires, or first-term politicians promising change). But once the company is out in the marketplace—competing for its rank among every other app in the store, learning to protect itself from predators as it grows, and struggling to reach a sustainable and profitable maturity—that's when the promise of launch gives way to the thornier realities of growing up. As we know, many apps don't survive the transition, and they simply vanish.

Careers, too, mature through a wildhood. When medical students graduate from the classroom portion of their educations, they are technically physicians with "MD" after their names. But what comes next, called residency, is wildhood for doctors. As their puppy licenses expire, and lacking full experience, they journey through several do-or-die years during which they must learn to keep patients safe, navigate hospital hierarchies, develop professional partnerships, and become seasoned physicians.

In these cases, adolescence seems to serve as a metaphor—but is it really just symbolic? In any human undertaking, we replicate these same steps. Something is born, with all the promise of the unknown and without the weight of history. It then has to go through the challenging, awkward, and even dangerous phase of maturing. Only by surviving that phase (and not everything does) can real mastery and success follow. This pattern could apply to individuals learning a language or even starting a marriage as much as to group enterprises like launching a company, an administration, or a war. Any endeavor that begins with great fanfare can quickly go off the rails without a successful navigation of the early challenges.

Nearly ten years after we gazed over the Triangle of Death, we returned to Moss Landing in northern California. We found it changed. The commercial fishing economy has given way to sustainable aquaculture and ecotourism, which have grown under the stewardship of the Monterey Bay Aquarium Research Institute. New apartments, a couple of hotels, and restaurants are being built to accommodate visitors from all over the world who come here to whale-watch, walk the beach, and catch a glimpse of seabirds. Five hundred square feet of greenhouses have been planted with recreational and medical cannabis. In an old power plant nearby, Tesla is installing its new "Megapack" energy storage system—a colossal 1.2 gigawatt-hour network of linked batteries, each the size of a shipping container.

But in spite of the human changes, the area is still home to rafts of California sea otters. Observed by kayakers, they learn to crack open sea urchins, wrestle with their peers, socialize with elders. Ten years on, some but probably not all of the joyriding adolescents we first observed have become the discerning, gray-haired elders, relatively safe from sharks. As is the way of nature, adolescents grow up, and a new generation enters wildhood.

Acknowledgments

For their scientific contributions and for generously sharing their scholarship with us, we thank: Klemens Pütz, Phil Trathan (Ursula's story), Oliver Höner (Shrink's story), Jooke Robbins and the Center for Coastal Studies (Salt's story), and Hubert Potočnik (Slavc's story).

Other scientists and experts kindly guided us on paths both literal and intellectual: Athena Aktipis, Andy Alden, Hannah Bannister, Rachel Cohen, Pierre Comizzoli, Michael Crickmore, Luke Dollar, Bridget Donaldson, Penny Ellison, Kate Evans, Daniel M. T. Fessler, Bill Fraser, Douglas Freeman, Chris Golden, James Ha, Renee Robinette Ha, Joe Hamilton, Kay Holekamp, Andrea Katz, Ben Kilham, Annika Linde, Diana Loren, Mia-Lana Lührs, Tona Melgarejo, Katherine Moseby, Diana Xochitl Munn, Miguel Ordeñana, Benison Pang, Jane Pickering, David Pyrooz, Niamh Quinn, Dragana Rogulja, Matt Ross, Joshua Schiffman, Fraser Shilling, Todd Shury, Judy Stamps, Stephen Stearns, Tim Tinker, Richard Weissbourd, Charles Welch, Viola Willeto, Cathy Williams, Barbara Wolf, Anne Yoder, Sarah Zehr, and Joe Q. Zhou.

Special thanks to our colleagues at UCLA and Harvard: Daniel T. Blumstein, our teacher, guide, scientific collaborator, and great friend; Patty Gowaty; Kalyanam Shivkumar; Daniel Lieberman; Rachel Carmody; Carole Hooven; Peter Ellison; and Richard Wrangham; plus all our undergraduate students at both institutions, whose ideas helped shape our thinking.

Thank you to New America and to other supportive colleagues, readers, and friends, with special gratitude to: Annie Murphy Paul, Debbie Stier, Wendy Paris, Randi Hutter Epstein, Judith Matloff, Abby Ellin, Cyd Black, Deborah Landau, and Sidney Callahan; Karol Watson, Tamara Horwich, Holly Middlekauff, Gregg Fonarow, Corey Powell, Wiley O'Sullivan, Zach Rabiroff and Sonja Bolle.

Thank you to the Duke Lemur Center, The Wilds, the Endangered Wolf Center, the Peabody Museum at Harvard, Tozzer Anthropology Library, and the curators and staff of the Museum of Comparative Zoology: Mark Omura (Mammals), Jeremiah Trimble and Kate Eldridge (Ornithology), Jose Rosado (Herpetology), Jessica Cundiff (Invertebrate Paleontology).

With deep gratitude to Oliver Uberti for his remarkable images.

Thanks to Susan Kwan, who makes everything happen and does so with grace, insight, and brilliance.

To our superb editor and passionate advocate, Valerie Steiker; our visionary publisher, Nan Graham; and the entire extraordinary team at Scribner: Colin Harrison, Roz Lippel, Brian Belfiglio, Jaya Miceli, Kara Watson, Ashley Gilliam, Sally Howe, Kathleen Rizzo, and Kyle Kabel.

And to Tina Bennett, our incredible agent, whose guidance, sense of humor, and inspiration made this book possible.

Finally, thank you to our families—Idell and Joseph Natterson; Zach, Jennifer, and Charles Horowitz; Amy Kroll and Paul Natterson; Diane and Arthur Sylvester; Karin, Caroline, and Connor McCarty; Marge and Amanda Bowers; Porter, Emmett, and Owen Rees; Andy and Emma.

Glossary of Terms

adultocentrism: *A perspective that overvalues the adult and undervalues the pre-adult stages of life.*

alarm calling: *Defensive behavior of social animals in which individuals alert others to nearby predators.*

alarm duetting: *Alarm calling understood by and between two individuals, often breeding mates or pairs.*

association with high-status animals: *Tendency of social animals to prefer the company of higher-ranking individuals, sometimes used as a strategy to rise in a hierarchy.*

audience effect: *The influence on an animal's behavior arising from its knowledge that other group members are observing. Important in dominance displays and bullying.*

behaviors of last resort: *Responses by prey who have already been detected or captured by predators, to evade death at the last moment. Examples include death-feigning, release (auto-amputation) of tail, limbs, or claws, and defecation.*

bullying: *Repetitive and aggressive behaviors directed toward another individual. Three types of bullying in animals are dominance bullying, conformer bullying, and redirection bullying.*

- **dominance bullying:** *Repetitive and aggressive behaviors directed toward lower-status individuals in a group to demonstrate and reinforce the bully's high status and power. (See also "audience effect.")*

conformer bullying: *Repetitive and aggressive behaviors directed toward individuals whose appearance or behavior differs in ways that could endanger the group's status or attract unwanted and dangerous attention. (See also "oddity effect.")*

redirection bullying: *Aggressive behavior toward other group members that represents displaced aggression emerging from the bully being victimized by others.*

conformity effect: *Tendency for prey species to school and flock with similar-appearing and -behaving individuals to reduce their individual and group's risk of predation. (See also "confusion effect" and "oddity effect.")*

confusion effect: *Reduction in a predator's attack success rate due to difficulties targeting individuals when prey are moving in larger groups with others sharing very similar appearance and behavior.*

defense mechanisms: *Psychological responses (unconscious) to stress that arise to protect individuals from emotional pain.*

delayed dispersal: *A form of prolonged parent-offspring association in which an adolescent animal remains in its birth territory beyond the age at which it might leave, usually for at least one season longer than peers.*

delayed plumage maturation: *Suspension of the typical progression of an adolescent bird's feathers from sub-adult to adult for at least one breeding season.*

dispersal: *The departure of an adolescent or young adult animal from birth territory to a new area, usually to begin breeding and other aspects of adult life.*

domain of danger: *Location and position with the highest risk of predation within schools, herds, and flocks.*

dominance displays: *Behaviors and signals used by some individuals to demonstrate and reinforce their higher status over other members of the group. (See also "dominance bullying" and "bullying.")*

easy prey: *Individuals perceived by predators as weaker or less protected and therefore better targets for attack, given their reduced ability to escape.*

ephebiphobia: *Fear, hatred, or trivialization of adolescent-aged individuals.*

extended parental care: *Resources and protection given by mothers and fathers to offspring, post-dispersal.*

fight, flight, faint: *Three antipredation physiologic responses activated by the autonomic nervous systems of vertebrate animals. The sympathetic nervous system activates fight-or-flight responses. The parasympathetic nervous system activates the faint response.*

finishing school: *The inclusion of not yet fully grown wolves in adult hunts to help them develop their hunting skills.*

high-value treat: *A strongly preferred food item that elicits and rewards motivation.*

hookup culture: *An early-twenty-first-century approach to sexuality characterized by casual sexual relationships free of long-term or emotional commitments.*

hygiene hypothesis: *Theory that underexposure to pathogens in early life contributes to later risk of allergy and autoimmune diseases. (See also "mismatch disorder.")*

informed dispersal: *Departure of an adolescent animal from birth territory advantaged by advance knowledge of optimal territories, groups, and mates. (See also "parental excursions.")*

informed mothers: *Term used by whale biologists to describe mothers who have good knowledge of and ability with crucial life skills. This benefits the young whales—their own offspring and others—who learn from them.*

island tameness: *The loss of evolved fear responses due to absence of predators over time.*

loser effect: *The tendency for an animal who loses one contest to be more likely to lose a subsequent one. This is facilitated by specific brain changes associated with losing that reduce competitive ability. (See also "winner effect.")*

maintenance of pair bonds: *Activity and time spent between individuals, before, during, and post-copulation, as part of investment in a long-term relationship.*

maternal intervention: *Behaviors by mothers on behalf of offspring to help them win contests and attain higher status within a group.*

maternal rank inheritance: *The transgenerational transfer of status from mother to offspring, seen in several mammalian species and indirectly in some egg-laying animals.*

mechanisms of defense: *Animal behavior, anatomy, or physiology that provide protection from predation.*

mismatch disorder: *Disease or pathology arising from discrepancies between the ancient environments in which human bodies and minds evolved and the modern world in which we live.*

mobbing: *Defensive behavior of animals who come together to collectively intimidate, chase off, and/or inspect a predator.*

oddity effect: *The tendency for animals who differ in appearance or behavior to be disproportionately targeted by predators and suffer higher rates of predation. This is believed to underlie the tendency for fish and birds to sort themselves into schools and flocks with similar-appearing and -behaving individuals.*

othering: *A process in which an individual's differences are emphasized by other members of a group, leading to their shunning or exclusion.*

parental excursions: *Instructional scouting trips taken by an adolescent animal and parent prior to dispersal to survey potential territories, groups, and mates and to identify optimal opportunities for soon-to-depart offspring.*

parental meanness: *Parental behavior aimed at encouraging or precipitating dispersal of offspring who are reluctant to leave. Examples include ignoring the intensifying begging of offspring and aggressively stooping them. (See also "stooping.")*

parent-offspring conflict: *Clashes that arise when a young animal demands more resources from a parent than the parent—who must also consider the needs of other current and future offspring—is prepared to give. A secondary definition: clashes that arise between human parents and offspring who disagree about best behavior.*

pecking order: *Term coined by Thorleif Schjelderup-Ebbe from his observation of pecking behavior of chickens to describe the ranking of individuals within a hierarchy.*

popularity: *The state of being well liked by peers. Often used interchangeably with "perceived popularity," which indicates social dominance, prestige, and influence.*

post-release monitoring: *Techniques—including microchipping, satellite* *tagging, and radio transmitting—used by wildlife biologists to keep track of how animals released into the wild are doing on their own.*

practice dispersal: *Brief and frequent departure from and return to birth territory and parents in the period preceding true dispersal.*

predator deception: *Defensive strategies, including hiding and camouflage, used by prey to evade detection or discourage predation of them.*

predator inspection: *Safety behavior in which prey (individually or in groups) approach and observe predators to gain knowledge about them. Also used to signal to a predator that it has been detected and has therefore lost the element of surprise. (See also "signal of unprofitability.")*

predator-naive: *A state of high vulnerability in an animal due to its lack of knowledge and experience with potential dangers.*

Predator's Sequence: *The four stages used by predators to find, choose, restrain, and consume prey: detect, assess, attack, kill.*

prestige: *Freely given admiration for and deference to admired members of a group, which may elevate their status.*

privilege: *The advantages available or given to (not earned by) certain individuals and not others within a group. The advantages available or given to (not earned by) certain groups but not others.*

pronking: *A behavior in which an animal (typically an ungulate) jumps up and down repeatedly, with all four legs stiff and lifting off the ground at the same time. Believed to be a form of stotting. (See also "stotting.")*

puberty: *Physical changes leading to reproductive maturity.*

puppy license: *A period of time in which leeway is granted to young animals by older animals for immature behaviors that would not be tolerated in a more mature individual.*

quality advertisement: *Signals of an animal's strength and stamina used by prey species to discourage predators from selecting them for attack. (See also "signal of unprofitability.")*

receptivity: *Physical and behavioral signs displayed by female animals indicating fertility. In spite of the terminology, it indicates female fertility only, not necessarily a desire for sexual contact. There are many examples of fertile female animals rejecting sexual advances of certain males at specific times.*

redshirting: *Strategy in humans of keeping young athletes out of formal competition for a year to allow them to have a developmental advantage when they return to play. Also used strategically in decisions related to kindergarten entrance. (See also "delayed plumage maturation.")*

reminiscence bump: *Enhanced memory for events occurring during adolescence and young adulthood.*

rowdy group: *The high-intensity humpback whale mating display and contest in which a mature female leads a few to as many as two dozen competitive males on a chase through the water; also the group of male humpback whales pursuing and displaying for a fertile female. Also called a competitive group.*

serotonin: *A chemical messenger involved with many brain mechanisms, including those related to mood and status.*

sexual coercion by force: *The use of physical strength or restraint to secure copulation or other sexual contact with an otherwise inaccessible individual.*

sexual coercion by harassment: *The use of harassment to secure copulation or other sexual contact with an otherwise inaccessible individual.*

sexual coercion by intimidation and fear: *The use of fear, threat of harm, or intimidation to secure copulation or other sexual contact with an otherwise inaccessible individual.*

sexual consent: *The affirmative, conscious, and willingly offered agreement between human individuals who both wish to engage in any sexual contact.*

signal of unprofitability: *Defensive behaviors or appearances that advertise an individual's strength and stamina to discourage potential predators from targeting them. (Examples include lizard push-ups, kangaroo rat foot-drumming, and skylark flight songs.)*

Social Brain Network (SBN): *The network of brain regions involved in social perception, cognition, and decision-making.*

social descent: *A fall in rank and social status by a member of a group.*

social hierarchy: *A social structure in which individuals within a group are stratified by rank.*

social learning: *Acquisition of relevant information from other members of one's group, often peers.*

social pain: *The unpleasant sensations occurring following social exclusion or descent in status.*

social rank: *An individual's position in a social hierarchy.*

social status: *An individual's position in a hierarchy relative to other members, influenced by the group's perception of that individual.*

startle response: *Sudden physical movement in response to a perceived threat across invertebrate and vertebrate animals.*

status badges: *Physical signals of an animal's relative rank within a group. Some animals use deceptive signals simulating high rank, a form of cheating by lower-ranking individuals, to rise in a status hierarchy. These are "false status badges."*

status descent: *See "social descent."*

status mapping: *Mental representations of an individual's and other members' status within a group.*

status sanctuaries: *Protected time or physical space in which no assessments occur.*

stooping: *A hunting behavior of predatory birds involving pulling in wings, stretching out talons, and rapidly descending on prey. Can also be used aggressively by bird parents on offspring reluctant to disperse.*

stotting: *A signal of unprofitability behavior in which an animal advertises to potential watching predators that they will not be easy prey. (See also "signal of unprofitability" and "quality advertising.") May also be used as a social signal.*

sturm und drang: *German for "storm and stress," the expression introduced by G. Stanley Hall in 1904 to describe adolescence.*

target animal: *Individual animal, often already lower ranking or "odd," singled out for bullying.*

territory inheritance: *The passing of territory from animal parents to offspring, seen in a wide range of vertebral species.*

transitive rank inference: *The ability of an animal to determine its position relative to many others in a group, from its known relationship with one member.*

winner effect: *The tendency for an animal who wins one contest to be more likely to win a subsequent one. This is facilitated by specific brain changes associated with winning that increase competitive ability. (See also "loser effect.")*

zugunruhe: *From the German for "migration restlessness," the sleeplessness and hyperactivity in animals (birds, typically) prior to migration.*

Notes

PROLOGUE

3 *what primatologist and ethologist Frans de Waal calls "anthropodenial":* Frans B. M. Wall, "Anthropomorphism and Anthropodenial: Consistency in Our Thinking about Humans and Other Animals," *Philosophical Topics* 27 (1999): 255.

4 *a herd of wildebeest:* YouTube, "Amazing Footage of Wildebeest Crossing the Mara River," https://www.youtube.com/watch?v=5XBxE_AohVY.

6 *"horizontal identity":* Andrew Solomon, *Far from the Tree: Parents, Children, and the Search for Identity* (New York: Scribner, 2012), 2.

6 *exquisite evolutionary sense:* Paraphrasing Theodosius Dobzhansky: "Nothing in Biology Makes Sense Except in the Light of Evolution," *The American Biology Teacher* 35 (March 1973): 125–29. Presented at the 1972 NABT convention.

6 *Margaret Mead's* Coming of Age in Samoa*:* Margaret Mead, *Coming of Age in Samoa* (New York: William Morrow and Co., 1928).

6 *In the late nineteenth century:* G. Stanley Hall, *Adolescence: Its Psychology and Its Relations to Physiology, Anthropology, Sociology, Sex, Crime, and Religion* (Kowloon, Hong Kong: Hesperides Press, 2013 [Kindle version]).

7 *Throughout the twentieth century:* Sigmund Freud, *The Interpretation of Dreams: The Complete and Definitive Text* (New York: Basic Books, 2010); A. Freud, *The Ego and the Mechanism of Defense* (New York: International Universities Press, 1948); Erik H. Erikson, *Identity and the Life Cycle* (New York: W. W. Norton & Company, 1994); John Bowlby, *Maternal Care and Mental Health* (Lanham, MD: Jason Aronson, Inc., 1995); Jean Piaget, *The Child's Conception of the World* (Scotts Valley, CA: CreateSpace Independent Publishing Platform, 2015); N. Tinbergen, *Social Behavior in Animals with Special Reference to Vertebrates* (London: Psychology Press, 2013).

7 *Marian Diamond's work:* Marian Cleeves Diamond, *Enriching Heredity: The Impact of the Environment on the Anatomy of the Brain* (New York: Free Press, 1988); Robert Sapolsky, *Behave: The Biology of Humans at Our Best or Worst* (City of Westminster, UK: Penguin Books, 2018); Frances E. Jensen and Amy Ellis Nutt, *The Teenage Brain: A Neuroscientist's Survival Guide to Raising Adolescents and Young Adults* (New York: Harper Paperbacks, 2016); Sarah-Jayne Blakemore, *Inventing Ourselves: The Secret Life of the Teenage Brain* (New York: PublicAffairs, 2018); Hanna Damasio and Antonio R. Damasio, *Lesion Analysis in Neuropsychology* (Oxford, UK: Oxford University Press, 1989); Linda Spear, *The Behavioral Neuroscience of Adolescence* (New York: W. W. Norton & Company, 2009); Judy Stamps, "Behavioural processes affecting development: Tinbergen's fourth question comes of age," *Animal Behaviour* 66 (2003): doi: 10.1006/anbe.2003.2180; Laurence Steinberg, *Age of Opportunity: Lessons from the New Science of Adolescence* (Boston: Mariner Books, 2015); Jeffrey Jensen Arnett, *Adolescence and Emerging Adulthood: A Cultural Approach* (London: Pearson, 2012).

8 *Great white sharks go through puberty:* Lisa J. Natanson and Gregory B. Skomal, "Age and growth of the white shark, Carcharodon, carcharias, in the western Northern Atlantic Ocean," *Marine and Freshwater Research* 66 (2015): 387–98; Christopher P. Kofron, "The reproductive cycle of the Nile crocodile (Crocodylus nilotkus)," *Journal of Zoology* (1990): 477–88; John L. Gittleman, "Are the Pandas Successful Specialists or Evolutionary Failures?" *BioScience* 44 (1994): 456–64; Erica Taube et al., "Reproductive biology and postnatal development in sloths, Bradypus and Choloepus: Review with original data from the field (French Guiana) and from captivity," *Mammal Review* 31 (2001): 173–88; A. J. Hall-Martin and J. D. Skinner, "Observations on puberty and pregnancy in female giraffe," *South African Journal of Wildlife Research* 8 (1978): 91–94; Sam P. S. Cheong et al., "Evolution of Ecdysis and Metamorphosis in Arthropods: The Rise of Regulation of Juvenile Hormone," *Integrative and Comparative Biology* 55 (2015): 878–90; Smithsonian National Museum of Natural History, "*Australopithecus afarensis*," http://humanorigins.si.edu/evidence/human-fossils/species/australopithecus-afarensis; Antonio Rosas et al., "The growth pattern of Neandertals, reconstructed from a juvenile skeleton from El Sidrón (Spain)," *Science* 357 (2017): 1282–287; Christine Tardieu, "Short adolescence in early hominids: Infantile and adolescent growth of the human femur," *American Journal of Physical Anthropology* 107 (1998): 163–78; Meghan Bartels, "Teenage Dinosaur Fossil Discovery Reveals What Puberty Was Like for a Tyrannosaur," *Newsweek*, October 20, 2017, https://www.newsweek.com/teenage-dinosaur-fossil-discovery-reveals-puberty-tyrannosaur-689448; Society of Vertebrate Paleontology, "Press Release—Adolescent T. Rex Unraveling Controversy About

Growth Changes in Tyrannosaurus," October 21, 2015, http://vertpaleo.org/Society-News/SVP-Paleo-News/Society-News,-Press-Releases/Press-Release-Adolescent-T-rex-unraveling-controve.aspx; Laura Geggel, "Meet Jane, the Most Complete Adolescent T. Rex Ever Found," LiveScience, October 19, 2015, https://www.livescience.com/52510-adolescent-t-rex-jane.html.

8 *The same hormones kick it:* Erica Eisner, "The relationship of hormones to the reproductive behaviour of birds, referring especially to parental behaviour: A review," *Animal Behaviour* 8 (1960): 155–79; Satoshi Kusuda et al., "Relationship between gonadal steroid hormones and vulvar bleeding in southern tamandua, *Tamandua tetradactyla*," *Zoo Biology* 30 (2011): 212–17; O. J. Ginther et al., "Miniature ponies: 2. Endocrinology of the oestrous cycle," *Reproduction, Fertility and Development* 20 (2008): 386–90.

8 *snails and slugs:* N. Treen et al., "Mollusc gonadotropin-releasing hormone directly regulates gonadal functions: A primitive endocrine system controlling reproduction," *General and Comparative Endocrinology* 176 (2012): 167–72; Ganji Purna Chandra Nagaraju, "Reproductive regulators in decapod crustaceans: an overview," *The Journal of Experimental Biology* 214 (2011): 3–16.

8 *protozoal puberty:* Arthur M. Talman et al., "Gametocytogenesis: The puberty of Plasmodium falciparum," *Malar J.* 3 (2004), doi: 10/1186/1475-2875-3-24.

8 *Hearts grow:* Kathleen F. Janz, Jeffrey D. Dawson, and Larry T. Mahoney, "Predicting Heart Growth During Puberty: The Muscatine Study," *Pediatrics* 105 (2000): e63.

9 *their deadliest bites:* T. L. Ferrara et al., "Mechanics of biting in great white and sandtiger sharks," *Journal of Biomechanics* 44 (2011): 430–35; eScience News, "Teenage Great White Sharks Are Awkward Biters," Biology & Nature News, December 2, 2010, http://esciencenews.com/articles/2010/12/02/teenage.great.white.sharks.are.awkward.biters.

9 *a fish called* Tiktaalik*:* Correspondence with Neil Shubin, March 5, 2019.

10 *Transitioning "teenage" brains:* L. P. Spear, "The adolescent brain and age-related behavioral manifestations," *Neuroscience and Biobehavioral Reviews* 24 (2000): 417–63; Linda Patia Spear, "Neurobehavioral Changes in Adolescence," *Current Directions in Psychological Science* 9 (2000): 111–14; Linda Patia Spear, "Adolescent Neurodevelopment," *Journal of Adolescent Health* 52 (2013): S7–13; Robert Sapolsky, *Behave: The Biology of Humans at Our Best or Worst* (City of Westminster: Penguin Books, 2018).

10 *the "reminiscence bump":* Khadeeja Munawar, Sara K. Kuhn, and Shamsul Haque, "Understanding the reminiscence bump: A systematic review," *PLoS ONE* 13 (2018): e0208595.

11 *birds have a brain region:* Tadashi Nomura and Ei-Ichi Izawa, "Avian birds: Insights from development, behavior and evolution," *Develop Growth Differ*

59 (2017): 244–57; O. Gunturkun, "The avian 'prefrontal cortex' and cognition," *Current Opinion in Neurobiology* 15 (2005): 686–93.

11 *The brains of adolescent orcas and dolphins:* Sam H. Ridgway, Kevin P. Carlin, and Kaitlin R. Van Alstyne, "Dephinid brain development from neonate to adulthood with comparisons to other cetaceans and artiodactyls," *Marine Mammal Science* 34 (2018): 420–39.

11 *other primates and smaller mammals:* Spear, "The adolescent brain and age-related behavioral manifestations."

11 *Even adolescent reptiles show unique neurological shifts:* Daniel Jirak and Jiri Janacek, "Volume of the crocodilian brain and endocast during ontogeny," *PLoS ONE* 12 (2017): e0178491; Matthew L. Brien et al., "The Good, the Bad, and the Ugly: Agonistic Behaviour in Juvenile Crocodilian," *PLoS ONE* 8 (2013): e80872.

14 *The term "adolescentia":* Steven Mintz, *Huck's Raft: A History of American Childhood* (Cambridge, MA: Harvard University Press, 2006), 196.

14 *In North America, the New England Puritans:* Ross W. Beales, "In Search of the Historical Child: Miniature Adulthood and Youth in Colonial New England," in eds. N. Ray Hiner and Joseph M. Hawes, *Growing Up in America: Children in Historical Perspective* (Champaign: University of Illinois Press, 1985), 20.

14 *The word "teenager" first appeared:* Ben Cosgrove, "The Invention of Teenagers: LIFE and the Triumph of Youth Culture," *Time*, September 28, 2013, http://time.com/3639041/the-invention-of-teenagers-life-and-the-triumph-of-youth-culture/.

PART I: SAFETY

Ursula's story is based on research by and interviews with Klemens Pütz of the Antarctic Research Trust and Phil N. Trathan of the British Antarctic Survey. Additional insights on penguin behavior and descriptions of Antarctica were drawn from interviews with Bill Fraser of Palmer Long-Term Ecological Research as well as Fen Montaigne's book Fraser's Penguins: A Journey to the Future in Antarctica *(New York: Henry Holt and Co., 2010).*

CHAPTER 1: DANGEROUS DAYS

21 *As a fluffy nestling warm under her parents':* The Cornell Lab of Ornithology: Neotropical Birds, "King Penguin Aptenodytes patagonicus," https://neotropical.birds.cornell.edu/Species-Account/nb/species/kinpen1/behavior.

22 zugunruhe*, which is German for "migration-anxiety":* ScienceDirect, "Zugunruhe," https://www.sciencedirect.com/topics/agricultural-and-biological-sciences/zugunruhe; J. M. Cornelius et al., "Contributions of endo-

crinology to the migration life history of birds," *General and Comparative Endocrinology* 190 (2013): 47–60.

23 *Lurking offshore:* Lisa M. Hiruki et al., "Hunting and social behaviour of leopard seals (*Hydruga Leptonyx*) at Seal Island, South Shetland Islands, Antarctica," *Journal of the Zoological Society of London* 249 (1999): 97–109; Australian Antarctic Division: Leading Australia's Antarctic Program, "Leopard Seals," http://www.antarctica.gov.au/about-antarctica/wildlife/animals/seals-and-sea-lions/leopard-seals.

23 *Led by Klemens Pütz:* Klemens Pütz et al., "Post-Fledging Dispersal of King Penguins (*Aptenodytes patagonicus*) from Two Breeding Sites in South Atlantic," *PLoS ONE* 9 (2014): e97164.

24 *Of the thousands:* Pütz et al., "Post-Fledging Dispersal of King Penguins"; interview with Klemens Pütz, August 14, 2017; interview with Dr. Phil Trathan, head of conservation biology, British Antarctic Survey, August 7, 2017.

24 *In the wild they crash:* Bo Ebenman and Johnny Karlsson, "Urban Blackbirds (Turdus merula): From egg to independence," *Annales Zoologici Fennici* (1984): 21:249–51; F. L. Bunnell and D. E. N. Tait, "Mortality rates of North American bears," *Arctic* 38, no. 4 (December 1985): 316–23; David G. Ainley and Douglas P. DeMaster, "Survival and mortality in a population of Adélie penguins," *Ecology* 6, no. 3 (1980): 522–30; Wayne F. Kasworm and Timothy J. Their, "Adult black bear reproduction, survival, and mortality sources in northwest Montana," *International Conference on Bear Research and Management* 9, no. 1 (1994): 223–30; Charles J. Jonkel and Ian McT. Cowan, "The black bear in the Spruce-Fir forest," *Wildlife Monographs* 27, no. 27 (December 1971): 3–57; José Alejandro Scolaro, "A model life table for Magellanic penguins (*Spheniscus magellanicus*) at Punta Tombo, Argentina," *Journal of Field Ornithology* 58 (1987): 432–41; Norman Owen-Smith and Darryl R. Mason, "Comparative changes in adult vs. juvenile survival affecting population trends of African ungulates," *Journal of Animal Ecology* 74 (2005): 762–73; Krzysztof Schmidt and Dries P. J. Kuijper, "A 'death trap' in the landscape of fear," *Mammal Research* 60 (2015): 275–84.

24 *Fortunately, when human adolescents:* World Health Organization, "Adolescents: Health Risks and Solutions," May 2017 Fact Sheet.

25 *A nearly 200 percent increase in mortality:* "Environmental Influences on Biobehavioral Processes," presentation at the Science of Adolescent Risk-Taking: Workshop at the National Academies/National Institutes of Health, http://nationalacademies.org/hmd/~/media/Files/Activity%20Files/Children/AdolescenceWS/Workshop%202/1%20Dahl.pdf; Agnieszka Tymula et al., "Adolescents' risk-taking behaviour is driven by tolerance to ambiguity," *PNAS* 109 (2012): 17135–140.

25 *Adolescents drive faster than adults:* CDC Motor Vehicle Safety (Teen Drivers):

https://www.cdc.gov/motorvehiclesafety/teen_drivers/index.html; Laurence Steinberg, "Risk-taking in adolescence: What changes and why?" *Annals of the New York Academy of Sciences* (2004): 51–58; Bruce J. Ellis et al., "The Evolutionary Basis of Risky Adolescent Behavior: Implications for Science, Policy, and Practice," *Developmental Psychology* 48 (2012): 598–623; Kenneth A. Dodge and Dustin Albert, "Evolving science in adolescence: Comment on Ellis et al (2012)," *Developmental Psychology* 48 (2012): 624–27; Adriana Galván, "Insights about adolescent behavior, plasticity, and policy from neuroscience research," *Neuron* 83 (2014): 262–65; David Bainbridge, *Teenagers: A Natural History* (London: Portobello, 2010).

25 *Robert Sapolsky, the Stanford neuroscientist:* Robert Sapolsky, *Behave: The Biology of Humans at Our Best or Worst* (City of Westminster: Penguin Books, 2018), 155.

26 *inexperienced, unsuspecting young animals:* Andrew Sih et al., "Predator–prey naïveté, antipredator behavior, and the ecology of predator invasions," *OIKOS* 119 (2010): 610–21.

27 *Adolescent risk-taking is seen throughout the animal world:* L. P. Spear, "The adolescent brain and age-related behavioral manifestations," *Neuroscience and Biobehavioral Reviews* 24 (2000): 417–63; Linda Patia Spear, "Neurobehavioral Changes in Adolescence," *Current Directions in Psychological Science* 9 (2000): 111–14; Debra A. Lynn and Gillian R. Brown, "The Ontology of Exploratory Behavior in Male and Female Adolescent Rats (Rattus norvegicus)," *Developmental Psychobiology* 51 (2009): 513–20; Giovanni Laviola et al., "Risk-Taking behavior in adolescent mice: psychobiological determinants and early epigenetic influence," *Neuroscience and Behavioral Reviews* 27 (2003): 19–31; Kristian Overskaug and Jan P. Bolstad, "Fledging Behavior and Survival in Northern Tawny Owls," *The Condor* 101 (1999): 169–74; Melanie Dammhahn and Laura Almeling, "Is risk taking during foraging a personality trait? A field test for cross-context consistency in boldness," *Animal Behavior* 84 (2012): 1131–39; Theodore Garland, Jr., and Stevan J. Arnold, "Effects of a Full Stomach on Locomotory Performance of Juvenile Garter Snakes," *Copeia* 1983 (1983): 1092–96; Svein Lokkeborg, "Feeding behaviour of cod, Gadus morhua: Activity rhythm and chemically mediated food search," *Animal Behaviour* 56 (1998): 371–78; Gerald Carter et al., "Distress Calls of a Fast-Flying Bat (Molossus molossus) Provoke Inspection Flights but Not Cooperative Mobbing," *PLoS ONE* 10 (2015): e0136146.

CHAPTER 2: THE NATURE OF FEAR

29 *The video shows a roly-poly mother panda:* "Sneezing Baby Panda, Original Video," https://www.youtube.com/watch?v=93hq0YU3Gqk.

30 *fast escapes:* J. A. Walker et al., "Do faster starts increase the probability of evading predators?" *Functional Ecology* 19 (2005): 808–15.

30 *Clever octopuses have devised:* Robert Sanders, "Octopus shows unique hunting, social and sexual behavior," *Berkeley Research News*, August 12, 2015, https://news.berkeley.edu/2015/08/12/octopus-shows-unique-hunting-social-and-sexual-behavior/.

30 *Charles Darwin noted:* Charles Darwin, *The Expression of the Emotions in Man and Animals* (London: Harper Perennial, 2009), 45, 304.

30 *This is called "fear-conditioning":* Tanja Jovanovic, Karin Maria Nylocks, and Kaitlyn L. Gamwell, "Translational neuroscience measures of fear conditioning across development: applications to high-risk children and adolescents," *Biology of Mood & Anxiety Disorders* 3 (2013): doi: 10.1186/2045-5380-3-17; J. J. Kim and M. W. Jung, "Neural circuits and mechanisms involved in Pavlovian fear conditioning: a critical review," *Neuroscience & Biobehavioral Reviews* 30 (2006): 188–202.

31 *"As penguins get older":* Interview (phone) with Dr. Phil Trathan, head of conservation biology, British Antarctic Survey, August 7, 2017.

31 *Along with a light-brown vest:* Porcupine fish skin helmet from Oceania/Republic of Kiribati, Catalog 00-8-70/55612, Peabody Museum, Harvard University; Imperial War Museum, "Equipment: Body Armour (Sappenpanzer): German," https://www.iwm.org.uk/collections/item/object/30110403; Seth Stern, "Body Armor Could Be a Technological Hero of War in Iraq," *Christian Science Monitor*, April 2, 2003, https://www.csmonitor.com/2003/0402/p04s01-usmi.html.

31 *Armor design:* Imperial War Museum, "Equipment: Body Armour (Sappenpanzer): German," https://www.iwm.org.uk/collections/item/object/30110403; Seth Stern, "Body Armor Could Be a Technological Hero of War in Iraq," *Christian Science Monitor*, April 2, 2003, https://www.csmonitor.com/2003/0402/p04s01-usmi.html.

32 *First conceptualized by turn-of-the-twentieth-century psychoanalysts:* A. Freud, *The Ego and the Mechanisms of Defense* (New York: International Universities Press, 1948).

33 *Wild fish, reptiles, amphibians, birds:* Karen M. Warkentin, "The development of behavioral defenses: A mechanistic analysis of vulnerability in red-eyed tree frog hatchlings," *Behavioral Ecology* 10 (1999): 251–62; Lois Jane Oulton, Vivian Haviland, and Culum Brown, "Predator Recognition in Rainbowfish, Melanotaenia duboulayi, Embryos," *PLoS ONE* (2013), doi: 10.1371.journal.pones.0076061.

34 *Island tameness:* Maren N. Vitousek et al., "Island tameness: An altered cardiovascular stress response in Galápagos marine iguanas," *Physiology & Behavior* 99, no. 4 (2010): 544–48; D. T. Blumstein, "Moving to suburbia:

Ontogenetic and evolutionary consequences of life on predator-free islands," *Journal of Biogeography* 29 (2002): 685–92; D. T. Blumstein, "The multipredator hypothesis and the evolutionary persistence of anti-predator behaviour," *Ethology* 112 (2006): 209–17; D. T. Blumstein and J. C. Danielm, "The loss of anti-predator behaviour following isolation on islands," *Proceedings of the Royal Society B* 272 (2005): 1663–68.

34 *When Darwin explored:* Charles Darwin, *Journal of Researches into the Natural History and Geology of the Countries Visited During the Voyage of the H.M.S. Beagle Round the World, under the Command of Capt. Fitz Roy, R.N.* (New York: D. Appleton and Company, 1878), http://darwin-online.org.uk/converted/pdf/1878_Researches_F33.pdf.

35 *Yellowstone elk are a classic example:* J. W. Laundré et al., "Wolves, elk, and bison: Re-establishing the 'landscape of fear' in Yellowstone National Park, U.S.A.," *Canadian Journal of Zoology* 79 (2001): 1401–9.

35 *relaxation in safe sex practices seen:* Seth C. Kalichman et al., "Beliefs about treatments for HIV/AIDS and sexual risk behaviors among men who have sex with men, 1997 to 2006," *Journal of Behavioral Medicine* 30 (2007): 497–503.

36 *David Strachan wondered:* D. P. Strachan, "Hay fever, hygiene, and household size," *BMJ* 299 (1989): 1259–60.

36 *Lars Svendsen, a Norwegian philosopher:* Lars Svendsen, *A Philosophy of Fear*, 2nd ed. (London: Reaktion Books, 2008).

36 *Franklin Roosevelt:* History Matters, "FDR's First Inaugural Address," http://historymatters.gmu.edu/d/5057/.

CHAPTER 3: KNOWING YOUR PREDATORS

40 *Every time they set out to make a kill:* Caro, *Antipredator Defenses in Birds and Mammals*; also see "Costs & Benefits" and "Opportunity Costs" in William E. Cooper, Jr., and Daniel T. Blumstein, *Escaping from Predators: An Integrative View of Escape Decisions* (Cambridge, UK: Cambridge University Press, 2015).

40 *Off Kodiak, Alaska:* Eva Saulitis et al., "Biggs killer whale (Orcinus orca) predation on subadult humpback whales (Megaptera novaeangliae) in lower cook inlet and Kodiak, Alaska," *Aquatic Mammals* 41 (2015): 341–44.

41 *cheetahs and an antelope called kudu:* Douglas F. Makin and Graham I. H. Kerley, "Selective predation and prey class behaviour as possible mechanisms explaining cheetah impacts on kudu demographics," *African Zoology* 51 (2016): 217–20.

41 *the remnants of tuco-tuco:* Aldo I. Vassallo, Marcelo J. Kittlein, and Cristina Busch, "Owl Predation on Two Sympatric Species of Tuco-Tucos (Rodentia: Octodontidae)," *Journal of Mammology* 75 (1994): 725–32.

41 *Even lowly sardine adolescents:* Richard B. Sherley et al., "The initial journey of an Endangered penguin: Implications for seabird conservation," *Endangered Species Research* 21 (2013): 89–95.

41 *People who hunt deer know:* Lindsay Thomas, Jr., "QDMA's Guide to Successful Deer Hunting," Quality Deer Management Association, 2016 (eBook).

42 *The evolutionary biologist Richard Wrangham:* Interview with Dr. Richard Wrangham, August 30, 2017.

42 *But deception can be flipped:* Caro, *Antipredator Defenses in Birds and Mammals*; Tim Clutton-Brock, *Mammal Societies* (Hoboken, NJ: Wiley-Blackwell, 2016).

43 *being shy helps young fish:* Robert J. Lennox, "What makes fish vulnerable to capture by hook? A conceptual framework and a review of key determinants," *Fish and Fisheries* 18 (2017): 986–1010.

43 *Sharks, for example, learn to approach:* C. Huveneers et al., "White Sharks Exploit the Sun during Predatory Approaches," *American Naturalist* 185 (2015): 562–70.

43 *Pharaoh cuttlefish can change their colors:* Koehi Okamoto et al., "Unique arm-flapping behavior of the pharaoh cuttlefish, Sepia pharaonis: Putative mimicry of a hermit crab," *Journal of Ethology* 35 (2017): 307–11.

43 *A ten-year analysis by the National:* National Center for Missing and Exploited Children, "A 10-Year Analysis of Attempted Abductions and Related Incidents," June 2016, http://www.missingkids.com/content/dam/pdfs/ncmec-analysis/attemptedabductions10yearanalysisjune2016.pdf.

44 *As a former sex trafficker:* K5 News, "A Pimp's Playbook: Galen Harper's Story," November 9, 2017, https://www.king5.com/video/news/investigations/selling-girls/a-pimps-playbook-galen-harpers-story/281-2796032.

46 *youth "would rather always do noble deeds":* Aristotle, *The Essential Aristotle* (New York: Simon & Schuster, 2013).

46 *A magnetic sonic device:* Compound Security Systems, "CSS Mosquito M4K," https://www.compoundsecurity.co.uk/security-equipment-mosquito-mk4-anti-loitering-device.

46 *Credit card companies:* The Balance, "Why Credit Card Companies Target College Students," September 10, 2018, https://www.thebalance.com/credit-card-companies-love-college-students-960090.

46 *College sports programs:* Kareem Abdul-Jabbar, "It's Time to Pay the Tab for America's College Athletes," *Guardian*, January 9, 2018, https://www.theguardian.com/sport/2018/jan/09/its-time-to-pay-the-tab-for-americas-college-athletes; Doug Bandow, "End College Sports Indentured Servitude: Pay 'Student Athletes,'" *Forbes*, February 21, 2012, https://www.forbes.com/sites/dougbandow/2012/02/21/end-college-sports-indentured-servitude-pay-student-athletes/#8676bd23db6c.

46 *On police forces, rookie cops:* Andrew Fan, "The Most Dangerous Neighborhood, the Most Inexperienced Cops," Marshall Project, September 20, 2016, https://www.themarshallproject.org/2016/09/20/the-most-dangerous-neighborhood-the-most-inexperienced-cops.

46 *Agence France-Presse reported in 2015:* "China's 'Young and Inexperienced' Firefighters in Spotlight After Blasts," *Straits Times*, August 20, 2015, https://www.straitstimes.com/asia/east-asia/chinas-young-and-inexperienced-firefighters-in-spotlight-after-blasts.

47 *ships' boys for the British Royal Navy:* Roland Pietsch, "Ships' Boys and Youth Culture in Eighteenth-Century Britain: The Navy Recruits of the London Marine Society," *The Northern Mariner/Le marin du nord* 14 (2004): 11–24.

47 *real-life story of Peter Williamson:* Mintz, *Huck's Raft*.

47 *In 1988, the R.J. Reynolds tobacco company:* Stanford Research into the Impact of Tobacco Advertising, "Cigarettes Advertising Themes: Targeting Teens," http://tobacco.stanford.edu/tobacco_main/images.php?token2=fm_st138.php&token1=fm_img4072.php&theme_file=fm_mt015.php&theme_name=Targeting.

47 *e-cigarette use:* Centers for Disease Control and Prevention, "Quick Facts on the Risk of E-cigarettes for Kids, Teens, and Young Adults," https://www.cdc.gov/tobacco/basic_information/e-cigarettes/Quick-Facts-on-the-Risks-of-E-cigarettes-for-Kids-Teens-and-Young-Adults.html.

48 *Alessandro Minelli, believes it is impeding:* Alessandro Minelli, "Grand challenges in evolutionary development biology," *Frontiers in Ecology and Evolution* 2 (2015): doi: 10.3389/fevo.2014.00085.

48 *"the wild west" of cancer referral:* Interviews with Dr. Joshua Schiffman, professor in the Department of Pediatrics and adjunct professor in the Department of Oncological Sciences in the School of Medicine at the University of Utah, September 21, 2018, and December 25, 2018.

49 *Female reproductive cycles:* Katherine A. Liu and Natalie A. Dipietro Mager, "Women's involvement in clinical trials: historical perspective and future implications," *Pharmacy Practice (Granada)* 14 (2016): 708; M. E. Burke, K. Albritton, and N. Marina, "Challenges in the recruitment of adolescents and young adults to cancer clinical trials," *Cancer* 110 (2007): 2385–93; M. Shnorhavorian et al., "Knowledge of clinical trial availability and reasons for nonparticipation among adolescent and young adult cancer patients: A population-based study," *American Journal of Clinical Oncology* 41 (2018): 581–87; S. J. Rotz et al., "Challenges in the treatment of sarcomas of adolescents and young adults," *Journal of Adolescent and Young Adult Oncology* 6 (2017): 406–13; A. L. Potosky et al., "Use of appropriate initial treatment among adolescents and young adults with cancer," *Journal of the National Cancer Institute* 106 (2014), doi: 10/1093/jnci/dju300.

49 *Adolescents have been denied transplants:* P. Rianthavorn and R. B. Ettenger, "Medication non-adherence in the adolescent renal transplant recipient: a clinician's viewpoint," *Pediatric Transplant* 9 (2005): 398–407; Cyd K. Eaton et al., "Multimethod assessment of medication nonadherence and barriers in adolescents and young adults with solid organ transplants," *Journal of Pediatric Psychology* 43 (2018): 789–99.

49 *avian adolescence varies:* Andrew U. Luescher, *Manual of Parrot Behavior* (Hoboken, NJ: Blackwell, 2008); Lafeber Company, "Indian Ring-Necked Parakeet," https://lafeber.com/pet-birds/species/indian-ring-necked-parakeet/#5.

50 *More than half of all dogs:* M. D. Salman et al., "Human and animal factors related to the relinquishment of dogs and cats in 12 selected animal shelters in the United States," *Journal of Applied Animal Welfare Science J* (1998): 207–26.

50 *Wild, older songbirds:* Kari Koivula, Seppo Rytkonen, and Marukku Orell, "Hunger-dependency of hiding behaviour after a predator attack in dominant and subordinate willow tits," *Ardea* 83 (1995): 397–404.

51 *"puppy license":* Interview with James Ha, February 25, 2019.

51 *The importance of puppy license for humans:* Alexa C. Curtis, "Defining adolescence," *Journal of Adolescent and Family Health* 7 (2–15): issue 2, article 2, https://scholar.utc.edu/jafh/vol7/iss2/2.

53 *"Their mothers are kicking them out":* Interview with Joe Hamilton, September 1, 2017.

54 *Experienced ground squirrels:* Tim Caro, *Antipredator Defenses in Birds and Mammals* (Chicago: University of Chicago Press, 2005), 15.

54 *Moose can differentiate:* Ibid.

54 *Garter snakes can tell:* Ibid.

54 *the "Predator's Sequence":* Ibid.; D. T. Blumstein, "Fourteen Security Lessons from Antipredator Behavior," in *Natural Security: A Darwinian Approach to a Dangerous World* (2008); Clutton-Brock, *Mammal Societies*; Gerald Carter et al., "Distress calls of a fast-flying bat (Molossus molossus) provoke inspection flights but not cooperative mobbing," *PLoS ONE* 10 (2015): e0136146; Andrew W. Bateman et al., "When to defend: Antipredator defenses and the predation sequence," *American Naturalist* 183 (2014): 847–55.

57 *In fact, it's one of the fastest:* Carter et al., "Distress calls of a fast-flying bat (Molossus molossus) provoke inspection flights but not cooperative mobbing."

58 *The confusion effect has:* Maria Thaker et al., "Group Dynamics of Zebra and Wildebeest in a Woodland Savanna: Effects of Predation Risk and Habitat Density," *PLoS ONE* 5 (2010): e12758.

58 *To study the oddity effect:* Hans Kruuk, *The Spotted Hyena: A Study of Pre-*

dation and Social Behavior (Brattleboro, VT: Echo Point Books and Media, 2014); Rebecca Dannock, "Understanding the behavioral trade-off made by blue wildebeest (*Connochaetes taurinus*): The importance of resources, predation, and the landscape," thesis, University of Queensland, School of Biological Sciences (2016).

59 *dyed some minnows blue:* Christopher W. Theodorakis, "Size segregation and the effects of oddity on predation risk in minnow schools," *Animal Behaviour* 38 (1989): 496–502; Laurie Landeau and John Terborgh, "Oddity and the 'confusion effect' in predation," *Animal Behavior* 34 (1986): 1372–80.

59 *a study of the social ostracism of albino catfish:* Ondrej Slavik, Pavel Horky, and Matus Maciak, "Ostracism of an albino individual by a group of pigmented catfish," *PLoS ONE* 10 (2015): e0128279.

59 *Group decision-making*: David J. Sumpter, *Collective Animal Behavior* (Princeton, NJ: Princeton University Press, 2010).

60 *Common during adolescence:* Michaela M. Bucchianeri et al., "Youth experiences with multiple types of prejudice-based harassment," *Journal of Adolescence* 51 (2016): 68–75.

60 *"overestimate risk, limit their exposure":* Blumstein, "Fourteen Security Lessons from Antipredator Behavior."

61 *when they detect rattlesnakes:* Caro, *Antipredator Defense in Birds and Mammals*, 248–49; Charles Martin Drabek, "Ethoecology of the Round-Tailed Ground Squirrel, Spermophilus Tereticaudus," University of Arizona Dissertation, PhD in Zoology, 1970.

62 *Diana monkeys in Cote d'Ivoire:* Klaus Zuberbuhler, Ronald Noe, and Robert M. Seyfarth, "Diana monkey long-distance calls: Messages for conspecifics and predators," *Animal Behaviour* 53 (1997): 589–604.

62 You and Your Adolescent*:* Laurence Steinberg, *You and Your Adolescent, New and Revised Edition: The Essential Guide for Ages 10–25* (New York: Simon & Schuster, 2011).

62 *kangaroo rats start drumming:* Jan A. Randall, "Evolution and Function of Drumming as Communication in Mammals," *American Zoologist* 41 (2001): 1143–56; Jan A. Randall and Marjorie D. Matocq, "Why do kangaroo rats (Dipodomys spectabilis) footdrum at snakes?" *Behavioral Ecology* 8 (1997): 404–13.

62 *Another unprofitability signal, called "stotting":* C. D. FitzGibbon and J. H. Fanshawe, "Stotting in Thomson's gazelles: An honest signal of condition," *Behavioral Ecology Sociobiology* 23 (1988): 69; "Stotting," *Encyclopedia of Ecology and Environmental Management* (New York: Blackwell, 1998); José R. Castelló, *Bovids of the World: Antelopes, Gazelles, Cattle, Goats, Sheep, and Relatives* (Princeton, NJ: Princeton University Press, 2016).

63 *Their unique escape song lasts thirteen:* Tim Caro and William L. Allen,

"Interspecific visual signalling in animals and plants: A functional classification," *Philosophical Transactions of the Royal Society B* 372 (2017), doi: 10.1098/rstb.2016.0344; Caro, *Antipredator Defenses in Birds and Mammals.*

65 *Her disc-shaped face:* "How Does an Owl's Hearing Work: Super Powered Owls," BBC Earth, March 23, 2016, https://www.youtube.com/watch?v=8SI73-Ka51E.

65 *a sudden, dramatic slowing of your heart:* The cardiac effects of fear were the subject of Chapter Two, "The Feint of Heart," in our first book, *Zoobiquity.*

65 *Human beings retain:* Barbara Natterson-Horowitz and Kathryn Bowers, "The Feint of Heart," in *Zoobiquity* (New York: Vintage, 2013), 25–39.

65 *kill rates:* James Fair, "Hunting Success Rates: How Predators Compare," *Discover Wildlife*, December 17, 2015, http://www.discoverwildlife.com/animals/hunting-success-rates-how-predators-compare.

CHAPTER 4: THE SELF-CONFIDENT FISH

68 *"A naive young animal":* Bennett G. Galef, Jr., and Kevin N. Laland, "Social learning in animals: Empirical studies and theoretical models," *BioScience* 55 (2005): 489–500.

68 *Ring-tailed lemurs:* Mel Norris, "Oh Yeah? Smell This! Or, Conflict Resolution, Lemur Style," Duke Lemur Center, March 16, 2012, https://lemur.duke.edu/oh-yeah-smell-this-or-conflict-resolution-lemur-style/.

68 *When a wolf attacks a mother bison:* Caro, *Antipredator Defenses in Birds and Mammals*, 27.

69 *Nestling Japanese* Parus major *birds:* Indrikis Krams, Tatjana Krama, and Kristine Igaune, "Alarm calls of wintering great tits Parus major: Warning of mate, reciprocal altruism or a message to the predator?" *Journal of Avian Biology* 37 (2006): 131–36.

69 *Mobbing:* Caro, *Antipredator Defenses in Birds and Mammals.*

71 *When Atlantic salmon hit adolescence:* Torbjorn Jarvi and Ingebrigt Uglem, "Predator Training Improves Anti-Predator Behaviour of Hatchery Reared Atlantic Salmon (Salmo salar) Smolt," *Nordic Journal of Freshwater Research* 68 (1993): 63–71. Predator training has been studied in a range of animals. See, for example, B. Smith and D. Blumstein, "Structural consistency of behavioural syndromes: Does predator training lead to multi-contextual behavioural change?" *Behaviour* 149 (2012): 187–213; D. M. Shier and D. H. Owings, "Effects of predator training on behavior and post-release survival of captive prairie dogs (*Cynomys ludovicianus*)," *Biological Conservation* 132 (2006): 126–35; Rafael Paulino et al., "The role of individual behavioral distinctiveness in exploratory and anti-predatory behaviors of red-browed Amazon parrot (*Amazona rhodocorytha*) during pre-release training," *Applied Animal*

Behaviour Science 205 (2018): 107–14; R. Lallensack, "Flocking Starlings Evade Predators with 'Confusion Effect,'" *Science*, January 17, 2017, https://www.sciencemag.org/news/2017/01/flocking-starlings-evade-predators-confusion-effect?r3f_986=https://www.google.com/; Rebecca West et al., "Predator exposure improves anti-predator responses in a threatened mammal," *Journal of Applied Ecology* 55 (2018): 147–56; Andrea S. Griffin, Daniel T. Blumstein, and Christopher S. Evans, "Training captive-bred or translocated animals to avoid predators," *Conservation Biology* 14 (2000): 1317–26; Janelle R. Sloychuk et al., "Juvenile lake sturgeon go to school: Life-skills training for hatchery fish," *Transactions of the American Fisheries Society* 145 (2016): 287–94; Ian G. McLean et al., "Teaching an endangered mammal to recognise predators," *Biological Conservation* 75 (1996): 51–62; Desmond J. Maynard et al., "Predator avoidance training can increase post-release survival of chinook salmon," in R. Z. Smith, ed., *Proceedings of the 48th Annual Pacific Northwest Fish Culture Conference* (Gleneden Beach, OR: 1997), 59–62; Alice R. S. Lopes et al., "The influence of anti-predator training, personality, and sex in the behavior, dispersion, and survival rates of translocated captive-raised parrots," *Global Ecology and Conservation* 11 (2017): 146–57.

74 *Fish do the hourglass:* D. Noakes et al., eds., "Predators and Prey in Fishes: Proceedings of the 3rd biennial conference on behavioral ecology of fishes held at Normal, Illinois, U.S.A.," Dr W. Junk Publishers, May 19–22, 1981; R. V. Palumbo et al., "Interpersonal Autonomic Physiology: A Systematic Review of the Literature," *Personality and Social Psychology Review* 22 (2017): 99–141; Viktor Muller and Ulman Linderberger, "Cardiac and Respiratory Patterns Synchronized Between Persons During Choir Singing," *PLoS ONE* 6 (2011): e24893; Maria Elide Vanutelli et al., "Affective Synchrony and Autonomic Coupling During Cooperation: A Hyperscanning Study," *BioMed Research International* 2017, doi: 10.1155/2017/3104564.

74 *Researchers tracking the heartbeats:* Björn Vickhoff et al., "Music structure determines heart rate variability of singers," *Frontiers in Psychology* 4 (2013): 334.

74 *Anthropologists at UCLA led by Dan Fessler:* Daniel M. T. Fessler and Colin Holbrook, "Friends Shrink Foes: The Presence of Comrades Decreases the Envisioned Physical Formidability of an Opponent," *Psychological Science* 24 (2013): 797–802; Daniel M. T. Fessler and Colin Holbrook, "Synchronized behavior increases assessments of the formidability and cohesion of coalitions," *Evolution and Human Behavior* 37 (2016): 502–9; Meg Sullivan, "In sync or in control," UCLA Newsroom, August 26, 2014, http://newsroom.ucla.edu/releases/in-sync-and-in-control.

CHAPTER 5: SCHOOL FOR SURVIVAL

79 *Social learning with peers:* A. S. Griffin, "Social learning about predators: A review and prospectus," *Learning and Behavior* 1 (2004): 131–140; Galef Jr. and Laland, "Social Learning in Animals: Empirical Studies and Theoretical Models."

79 *Adolescent guppies from a river in Trinidad:* Jennifer L. Kelley et al., "Back to school: Can antipredator behaviour in guppies be enhanced through social learning?" *Animal Behaviour* 65 (2003): 655–62.

80 *In April 2017, a group:* Hannah Natanson, "Harvard Rescinds Acceptances for At Least Ten Students for Obscene Memes," *The Harvard Crimson*, June 5, 2017, https://www.thecrimson.com/article/2017/6/5/2021-offers-rescinded-memes/.

80 *One study of starlings showed:* Julia Carter et al., "Subtle cues of predation risk: Starlings respond to a predator's direction of eye-gaze," *Proceedings of the Royal Society B* 275 (2008): 1709–15.

80 *A combination of scent molecules:* Ferris Jabr, "Scary Stuff: Fright Chemicals Identified in Injured Fish," *Scientific American*, February 23, 2012, https://www.scientificamerican.com/article/fish-schreckstoff/.

81 *"predator inspection":* Tim Caro, *Antipredator Defenses in Birds and Mammals* (Chicago: University of Chicago Press, 2005); Jean-Guy J. Godin and Scott A. Davis, "Who dares, benefits: Predator approach behaviour in the guppy (Poecilia reticulata) deters predator pursuit," *Proceedings of the Royal Society B* 259 (1995): 193–200; Carter et al., "Distress Calls of a Fast-Flying Bat (Molossus molossus) Provoke Inspection Flights but Not Cooperative Mobbing"; Maryjka B. Blaszczyk, "Boldness towards novel objects predicts predator inspection in wild vervet monkeys," *Animal Behavior* 123 (2017): 91–100; C. Crockford et al., "Wild chimpanzees inform ignorant group members of danger," *Current Biology* 22 (2012): 142–46; Anne Marijke Schel et al., "Chimpanzee Alarm Call Production Meets Key Criteria for Intentionality," *PLoS ONE* 8 (2013): e76674; Beauchamp Guy, "Vigilance, alarm calling, pursuit deterrence, and predator inspection," in William E. Cooper, Jr., and Daniel T. Blumstein, eds., *Escaping from Predators: An Integrative View of Escape Decisions* (Cambridge: Cambridge University Press, 2015); Michael Fishman, "Predator inspection: Closer approach as a way to improve assessment of potential threats," *Journal of Theoretical Biology* 196 (1999): 225–35.

82 *One study of juvenile minnows:* T. J. Pitcher, D. A. Green, and A. E. Magurran, "Dicing with death: Predator inspection behaviour in minnow shoals," *Journal of Fish Biology* 28 (1986): 439–48.

82 *Slender Thomson's gazelles skip:* Clare D. FitzGibbon, "The costs and benefits

of predator inspection behaviour in Thomson's gazelles," *Behavioral Ecology and Sociobiology* 34 (1994): 139–48.

84 *painfully stinging* Hormiga veinticuatro *ants:* Vilma Pinchi et al., "Dental Ritual Mutilations and Forensic Odontologist Practice: A Review of the Literature," *Acta Stomatologica Croatica* 49 (2015): 3–13; Rachel Nuwer, "When Becoming a Man Means Sticking Your Hand into a Glove of Ants," Smithsonian.com, October 27, 2014, https://www.smithsonianmag.com/smart-news/brazilian-tribe-becoming-man-requires-sticking-your-hand-glove-full-angry-ants-180953156/.

85 *"The juvenile male was lean, in poor condition":* M. N. Bester et al., "Vagrant leopard seal at Tristan da Cunha Island, South Atlantic," *Polar Biology* 40 (2017): 1903–5.

86 *As the scientists noted, dispersal is:* Pütz et al., "Post-fledging dispersal of king penguins (*Aptenodytes patagonicus*) from two breeding sites in South Atlantic."

86 *Bill Fraser, an ecologist and penguin expert:* Interview with Dr. William R. Fraser, president and lead investigator, Polar Oceans Research Group, November 30, 2017, and December 7, 2017.

PART II: STATUS

Shrink's story comes from Oliver Höner of the Leibniz Institute for Zoo and Wildlife Research in Berlin and the Spotted Hyena Project, Ngorongoro Crater, Tanzania.

CHAPTER 6: THE AGE OF ASSESSMENT

93 *Females rule most hyena clans:* Laurence G. Frank, "Social organization of the spotted hyaena Crocuta crocuta. II. Dominance and reproduction," *Animal Behaviour* 34 (1986): 1510–27.

93 *Of all the carnivores:* Hyena Project Ngorongoro Crater, https://hyena-project.com/.

94 *Shrink's story came to us:* Interviews with Oliver Höner (Berlin), May 3, 2018, and October 4, 2018.

94 *Their Spotted Hyena Project:* Hyena Project Ngorongoro Crater, https://hyena-project.com/.

96 *What the boy, at age ten:* Jack El-Hai, "The Chicken-Hearted Origins of the 'Pecking Order—The Crux," *Discover*, July 5, 2016, http://blogs.discovermagazine.com/crux/2016/07/05/chicken-hearted-origins-pecking-order/#.XIShIShKg2w; Thorleif Schjelderup-Ebbe, "Weitere Beiträge zur Sozial und psychologie des Haushuhns," *Zeitschrift für Psychologie* 88 (1922): 225–52.

96 *the same one that stratifies:* Elizabeth A. Archie et al., "Dominance rank rela-

tionships among wild female African elephants, *Loxodonta africana*," *Animal Behaviour* 71 (2006): 117–27; Justin A. Pitt, Serge Lariviere, and Francois Messier, "Social organization and group formation of raccoons at the edge of their distribution," *Journal of Mammalogy* 89 (2008): 646–53; Logan Grosenick, Tricia S. Clement, and Russel D. Fernald, "Fish can infer social rank by observation alone," *Nature* 445 (2007): 427–32; Bayard H. Brattstrom, "The evolution of reptilian social behavior," *American Zoologist* 14 (1974): 35–49; Steven J. Portugal et al., "Perch height predicts dominance rank in birds," *IBIS* 159 (2017): 456–62.

96 *never more intensely than during wildhood:* S. J. Blakemore, "Development of the social brain in adolescence," *Journal of the Royal Society of Medicine* 105 (2012): 111–16.

96 *rest of their lives:* Ying Shi and James Moody, "Most likely to succeed: Long-run returns to adolescent popularity," *Social Currents* 4 (2017): 13–33.

97 *social rank—an individual's place:* Michael Sauder, Freda Lynn, and Joel Podolny, "Status: Insights from organizational sociology," *Annual Review of Sociology* 38 (2012): 267–83.

97 *The highest-ranking rooster in the coop:* Tsuyoshi Shimmura, Shosei Ohashi, and Takashi Yoshimura, "The highest-ranking rooster has priority to announce the break of dawn," *Nature Scientific Reports* 5 (2015): 11683.

97 *Dominant female hamsters:* U. W. Huck et al., "Progesterone levels and socially induced implantation failure and fetal resorption in golden hamsters (*Mesocricetus auratus*)," *Physiology and Behavior* 44 (1988): 321–26.

97 *High-ranking crayfish claim thermally:* Glenn J. Tattersall et al., "Thermal games in crayfish depend on establishment of social hierarchies," *The Journal of Experimental Biology* 215 (2012): 1892–1904.

97 *The top-ranking homing pigeon:* Portugal et al., "Perch height predicts dominance rank in birds."

98 *And top-ranking fish swim near:* P. Domenici, J. F. Steffensen, and S. Marras, "The effect of hypoxia on fish schooling," *Philosophical Transactions of the Royal Society of London B: Biological Sciences* 372 (2017), doi: 10/1098/rstb.2016.0236.

98 *Lower-ranking fish are pushed toward the "domain of danger":* Stefano Marras and Paolo Domenici, "Schooling fish under attack are not all equal: Some lead, others follow," *PLoS ONE* 6 (2013): e65784; Lauren Nadler, "Fish schools: Not all seats in the class are equal," *Naked Scientists*, October 22, 2014, https://www.thenakedscientists.com/articles/science-features/fish-schools-not-all-seats-class-are-equal; Domenici, Steffensen, and Marras, "The effect of hypoxia on fish schooling."

98 *Group living offers benefits:* Tzo Zen Ang and Andrea Manica, "Aggression, segregation and stability in a dominance hierarchy," *Proceedings of the Royal Society B: Biological Sciences*, 277 (2010): 1337–43.

98 *brain systems alert:* Noriya Watanabe and Miyuki Yamamoto, "Neural mechanisms of social dominance," *Frontiers in Neuroscience* 9 (2015): doi: 10.3389/fnins.2015.00154.

99 *theologians were ranking the angels in heaven:* Nicolas Verdier, "Hierarchy: A short history of a word in Western thought," HAL archives-ouvertes.fr, https://halshs.archives-ouvertes.fr/halshs-00005806/document; R. H. Charles, *The Book of Enoch* (Eugene, OR: Wipf & Stock Publishers, 2002), 390.

99 *He saw that when new chickens:* Thorleif Schjelderup-Ebbe, "Social Behavior of Birds," in C. Murchison, ed., *Handbook of Social Psychology* (Worcester, MA: Clark University Press, 1935), 947–72.

99 *As Marc Bekoff:* Marc Bekoff, ed., *Encyclopedia of Animal Behavior*, vol. 1: *A–C* (Westport, CT: Greenwood Press, 2004); Marc Bekoff, ed., *Encyclopedia of Animal Behavior*, vol. 2: *D–P* (Westport, CT: Greenwood Press, 2004); Marc Bekoff, ed., *Encyclopedia of Animal Behavior*, vol. 3: *R–Z* (Westport, CT: Greenwood Press, 2004).

100 *Status, in contrast, isn't an objective measure:* C. Norman Alexander Jr., "Status perceptions," *American Sociological Review* 37 (1972): 767–73.

100 *Animal behaviorists now routinely measure:* Isaac Planas-Sitjà and Jean-Louis Deneubour, "The role of personality variation, plasticity and social facilitation in cockroach aggregation," *Biology Open* 7 (2018): doi: 10.1242/bio.036582; Takao Tasaki et al., "Personality and the collective: Bold homing pigeons occupy higher leadership ranks in flocks," *Philosophical Transactions of the Royal Society B* 373 (2018): 20170038.

101 *Animals can deduce status relationships:* Stephan Keckers et al., "Hippocampal Activation During Transitive Inference in Humans," *Hippocampus* 14 (2004): 153–62; Logan Grosenick, Tricia S. Clement, and Russell D. Fernald, "Fish can infer social rank by observation alone," *Nature* 445 (2007): 427–32; Shannon L. White and Charles Gowan, "Brook trout use individual recognition and transitive inference to determine social rank," *Behavioral Ecology* 24 (2013): 63–69; Guillermo Paz-y-Mino et al., "Pinyon jays use transitive inference to predict social dominance," *Nature* 430 (2004), doi: 10.1038/nature02723.

101 *Mammals, birds, and fish all can use TRI:* Heckers et al., "Hippocampal activation during transitive inference in humans"; Grosenick, Clement, and Fernald, "Fish can infer social rank by observation alone"; Paz-y-Mino et al., "Pinyon jays use transitive inference to predict social dominance."

102 *According to public health sources, the twenty-first century:* Centers for Disease Control and Prevention, "Mental Health Conditions: Depression and Anxiety," https://www.cdc.gov/tobacco/campaign/tips/diseases/depression-anxiety.html; Centers for Disease Control and Prevention, "Key Findings: U.S. Children with Diagnosed Anxiety and Depression," https://

www.cdc.gov/childrensmentalhealth/features/anxiety-and-depression .html; Centers for Disease Control and Prevention, "Suicide Rising Across the US," https://www.cdc.gov/vitalsigns/suicide/index.html.

102 *Spotted hyenas in the Ngorongoro Crater:* Interviews with Oliver Höner (Berlin), May 3, 2018, and October 4, 2018.

103 *nurse for less time:* Interviews with Oliver Höner (Berlin), May 3, 2018, and October 4, 2018.

103 *Within litters of Eurasian lynx:* A. L. Antonevich and S. V. Naidenko, "Early intralitter aggression and its hormonal correlates," *Zhurnal Obshchei Biologii* 68 (2007): 307–17.

103 *High-ranking canary mothers anoint their eggs:* Aurelie Tanvez et al., "Does maternal social hierarchy affect yolk testosterone deposition in domesticated canaries?" *Animal Behaviour* 75 (2008): 929–34.

103 *In the yolks of fish eggs:* Tim Burton et al., "Egg hormones in a highly fecund vertebrate: Do they influence offspring social structure in competitive conditions?" *Oecologia* 160 (2009): 657–65.

CHAPTER 7: THE RULES OF GROUPS

105 *In the communal den:* Interviews with Oliver Höner (Berlin), May 3, 2018, and October 4, 2018.

106 *After being left by their mothers:* Interview with Dr. Kay Holekamp, professor, Department of Integrative Biology, Program in Ecology, Evolution, Biology & Behavior, Michigan State University, May 1, 2018.

106 *From birds to fish, crustaceans to mammals, and even in some spiders, being larger:* Alain Jacob et al., "Male dominance linked to size and age, but not to 'good genes' in brown trout (Salmo trutta)," *BMC Evolutionary Biology* 7 (2007): 207; Advances in Genetics, "Dominance Hierarchy," 2011, ScienceDirect Topics, https://www.sciencedirect.com/topics/agricultural-and-biological-sciences/dominance-hierarchy; Jae C. Choe and Bernard J. Crespi, *The Evolution of Social Behaviour in Insects and Arachnids* (Cambridge, UK: Cambridge University Press, 1997), 469.

107 *for a female hyena, family ties and social networks:* Interviews with Oliver Höner (Berlin), May 3, 2018, and October 4, 2018.

107 *Getting older is a status booster in many animals:* Clutton-Brock, *Mammal Societies*, 473–74; Roberto Bonanni et al., "Age-graded dominance hierarchies and social tolerance in packs of free-ranging dogs," *Behavioral Ecology* 28 (2017): 1004–20; Simona Cafazzo et al., "Dominance in relation to age, sex, and competitive contexts in a group of free-ranging domestic dogs," *Behavioral Ecology* 21 (2010): 443–55; Jacob et al., "Male dominance linked to size and age"; Rebecca L. Holberton, Ralph Hanano, and Kenneth P. Able,

"Age-related dominance in male dark-eyed juncos: Effects of plumage and prior residence," *Animal Behaviour* 40 (1990): 573–79; Stephanie J. Tyler, "The behaviour and social organization of the new forest ponies," *Animal Behaviour Monographs* 5 (1972): 87–196; Karen McComb, "Leadership in elephants: The adaptive value of age," *Proceedings of the Royal Society B: Biological Sciences* 278 (2011): 3270–76; Steeve D. Côté, "Dominance hierarchies in female mountain goats: Stability, aggressiveness and determinants of rank," *Behaviour* 137 (2000): 1541–66; T. H. Clutton-Brock et al., "Intrasexual competition and sexual selection in cooperative mammals," *Nature* 444 (2006): 1065–68; Steffen Foerster, "Chimpanzee females queue but males compete for social status," *Scientific Reports* 6 (2016): 35404; Amy Samuels and Tara Gifford, "A quantitative assessment of dominance relations among bottlenose dolphins," *Marine Mammal Science* 13 (1997): 70–99.

107 *Seniority—hanging around long enough:* Janis L. Dickinson, "A test of the importance of direct and indirect fitness benefits for helping decisions in western bluebirds," *Behavioral Ecology* 15 (2004): 233–38; Bernard Stonehouse and Christopher Perrins, *Evolutionary Ecology* (London: Palgrave, 1979), 146–47.

107 *Males have to wait years to work:* Interviews with Oliver Höner (Berlin), May 3, 2018, and October 4, 2018.

107 *Attractiveness—even physical beauty—can raise status:* Tonya K. Frevert and Lisa Slattery Walker, "Physical Attractiveness and Social Status," *Social Psychology and Family* 8 (2014): 313–23; Richard O. Prum, *The Evolution of Beauty: How Darwin's Forgotten Theory of Mate Choice Shapes the Animal World—and Us* (New York: Doubleday, 2017), https://books.google.com/books?id=AinWDAAAQBAJ&q=a+taste+for+the+beautiful#v=snippet&q=a%20taste%20for%20the%20beautiful&f=false.

107 *For flamingos:* Marina Koren, "For Some Species, You Really Are What You Eat," Smithsonian.com, April 24, 2013, https://www.smithsonianmag.com/science-nature/for-some-species-you-really-are-what-you-eat-40747423; J. A. Amat et al., "Greater flamingos *Phoenicopterus roseus* use uropygial secretions as make-up," *Behavioral Ecology and Sociobiology* 65 (2011): 665–73.

108 *A black swan's elaborately curled wing feathers:* Ken Kraaijeveld et al., "Mutual ornamentation, sexual selection, and social dominance in the black swan," *Behaviour Ecology* 15 (2004): 380–89.

108 *Schjelderup-Ebbe recognized the grooming gap:* John S. Price and Leon Sloman, "Depression as yielding behavior: An animal model based on Schjelderup-Ebbe's pecking order," *Ethology and Sociobiology* 8 (1987), 92S.

108 *top hyenas are less blemished:* Interviews with Oliver Höner (Berlin), May 3, 2018, and October 4, 2018.

108 *Finally, an animal's sex (what has been called gender in humans):* Char-

lotte K. Hemelrijk, Jan Wantia, and Karin Isler, "Female dominance over males in primates: Self-organisation and sexual dimorphism," *PLoS ONE* 3 (2008): e2678; Laura Casas et al., "Sex change in clownfish: Molecular insights from transcriptome analysis," *Scientific Reports* 6 (2016): 35461; J. F. Husak, A. K. Lappin, R. A. Van Den Bussche, "The fitness advantage of a high-performance weapon," *Biological Journal of the Linnean Society* 96 (2009): 840–45; Clutton-Brock, *Mammal Societies*; Julie Collet et al., "Sexual selection and the differential effect of polyandry," *Proceedings of the National Academy of Sciences* 109 (2012): 8641–45.

109 *Schools of colorful tropical clownfish:* Casas et al., "Sex change in clownfish: Molecular insights from transcriptome analysis."

109 *prefer high-ranking social companions:* Cheney and Seyfarth, *How Monkeys See the World*, 37–38, 545; Barbara Tiddi, Filippo Aureli and Gabriele Schino, "Grooming up the hierarchy: The exchange of grooming and rank-related benefits in a new world primate," *PLoS ONE* 7 (2012): e36641; T. H. Friend and C. E. Polan, "Social rank, feeding behavior, and free stall utilization by dairy cattle," *Journal of Dairy Science* 57 (1974): 1214–20; Kelsey C. King et al., "High society: Behavioral patterns as a feedback loop to social structure in Plains bison (Bison bison bison)," *Mammal Research* (2019): 1–12, doi: 10.1007/s13364-019-00416-7; Norman R. Harris et al., "Social associations and dominance of individuals in small herds of cattle," *Rangeland Ecology & Management* 60 (2007): 339–49.

110 *These animal social signifiers, called "status badges" by biologists:* Cody J. Dey, "Manipulating the appearance of a badge of status causes changes in true badge expression," *Proceedings of the Royal Society B: Biological Sciences* 281 (2014): 20132680.

110 *A fiddler crab that loses:* Simon P. Lailvaux, Leeann T. Reaney, and Patricia R. Y. Backwell, "Dishonesty signalling of fighting ability and multiple performance traits in the fiddler crab *Uca mjoebergi*," *Functional Ecology* 23 (2009): 359–66.

110 *In a display case in the Mesoamerican hall at Harvard's Peabody Museum:* "Incised carving of human figure upon bone," Catalog 92-49-20/C921, Peabody Museum, Harvard University.

111 *Mayans, who lived around one to four thousand years ago:* Stephen Houston, *The Gift Passage: Young Men in Classic Maya Art and Text* (New Haven, CT: Yale University Press, 2018).

111 Pitz *players were usually elites:* Mary Miller and Stephen Houston, "The Classic Maya ballgame and its architectural setting: A study of relations between text and image," *Anthropology and Aesthetics* 14 (1987): 46–65; Mary Ellen Miller, "The Ballgame," *Record of the Art Museum, Princeton University* 48 (1989): 22–31; "Maya: Ballgame," William P. Palmer III Collection, Uni-

versity of Maine Library, https://library.umaine.edu/hudson/palmer/Maya/ballgame.asp.

111 *Anthropologist and archaeologist Stephen Houston:* Stephen Houston, *The Gift Passage: Young Men in Classic Maya Art and Text* (New Haven, CT: Yale University Press, 2018), 67.

112 *hyena queens must be first to offer their lives:* Interviews with Oliver Höner (Berlin), May 3, 2018, and October 4, 2018.

112 *A low-status wolf betrays his rank:* David L. Mech and Luigi Boitani, *Wolves: Behavior, Ecology, and Conservation* (Chicago: University of Chicago Press, 2007), 93.

112 *Higher-ranking individuals like Meregesh:* Interviews with Oliver Höner (Berlin), May 3, 2018, and October 4, 2018.

113 *Frans de Waal, the author and primatologist:* Frans de Waal, *Our Inner Ape: A Leading Primatologist Explains Why We Are Who We Are* (New York: Riverhead Books, 2006), 59.

113 *Spotted hyenas are famous for a particular vocalization:* Federic Theunissen, Steve Glickman, and Suzanne Page, "The spotted hyena whoops, giggles and groans. What do the groans mean?" Acoustics.org, July 3, 2008, http://acoustics.org/pressroom/httpdocs/155th/theunissen.htm.

114 *these systems are called the Social Brain Network (SBN):* K. P. Maruska et al., "Social descent with territory loss causes rapid behavioral, endocrine and transcriptional changes in the brain," *Journal of Experimental Biology* 216 (2013): 3656–66.

114 *In mammals, the SBN is housed:* Joan Y. Chiao, "Neural basis of social status hierarchy across species," *Current Opinion* 20 (2010), doi: 10.1016/j.comb.2010.08.006; K. P. Maruska et al., "Social descent with territory loss causes rapid behavioral, endocrine and transcriptional changes in the brain."

114 *abnormalities in SBN development:* Vivek Misra, "The social Brain network and autism," *Annals Neuroscience* 21 (2014): 69–73.

115 *A dog's social brain:* Attila Andics et al., "Voice-sensitive regions in the dog and human brain and revealed by comparative fMRI," *Current Biology* 24 (2014): 574–78.

115 *The social brains of babies sit primed:* Karen Wynn, "Framing the Issues," "Infant Cartographers," and "Social Acumen: Its Role in Constructing Group Identity and Attitude" in Jeannete McAfee and Tony Attwood, eds., *Navigating the Social World* (Arlington, TX: Future Horizons, 2013), 8, 24–25, 323.

115 *As toddler cartographers, they construct mental status maps:* Ibid.

115 *monkeys will pass up the sweet juice:* R. O. Deaner, A. V. Khera, and M. L. Platt, "Monkeys pay per view: Adaptive valuation of social images by rhesus macaques," *Current Biology* 15 (2005): 543–48.

116 *Compared with adults and children, she writes, "adolescents are more sociable":* Blakemore, "Development of the social brain in adolescence."

116 *"Peer relations are never more salient":* Dustin Albert, Jason Chein, and Laurence Steinberg, "Peer influences on adolescent decision making," *Current Directions in Psychological Science* 22 (2013): 114–20.

116 *The preoptic area pulls in visuals:* Joan Y. Chiao, "Neural basis of social status hierarchy across species," *Current Opinion* 20 (2010), doi: 10.1016/j.comb.2010.08.006; Maruska et al., "Social descent with territory loss causes rapid behavioral, endocrine and transcriptional changes in the brain."

117 *By late adolescence, SBN development is nearly fully complete:* Blakemore, "Development of the social brain in adolescence."

117 *A person is said to have prestige:* Jon K. Maner, "Dominance and prestige: A tale of two hierarchies," *Current Directions in Psychological Science* (2017): doi: 10.1177/0963721417714323; Joey T. Cheng et al., "Two ways to the top: Evidence that dominance and prestige are distinct yet viable avenues to social rank and influence," *Journal of Personality and Social Psychology* 104 (2013): 103–25.

118 *A process called "niche picking" emerges:* Lisa J. Crockett, "Developmental Paths in Adolescence: Commentary," in Lisa Crockett and Ann C. Crouter, eds., *Pathways Throughout Adolescence: Individual Development in Relation to Social Contexts*, Penn State Series on Child and Adolescent Development (London: Psychology Press, 1995), 82.

CHAPTER 8: PRIVILEGED CREATURES

119 *In communal dens, hyena cub hierarchy reshuffles after:* Interview with Dr. Kay Holekamp, professor, Department of Int grative Biology, Program in Ecology, Evolution, Biology & Behavior, Michigan State University, May 1, 2018.

120 *maternal rank inheritance guarantees that the sons:* Kay E. Holekamp and Laura Smale, "Dominance Acquisition During Mammalian Social Development: The 'Inheritance' of Maternal Rank," *Integrative and Comparative Biology* 31 (1991): 306–17.

120 *From red deer to snow monkeys:* T. H. Clutton-Brock, S. D. Albon, and F. E. Guinness, "Maternal dominance, breeding success and birth sex ratios in red deer," *Nature* 308 (1984): 358–60; Nobuyuki Kutsukake, "Matrilineal rank inheritance varies with absolute rank in Japanese macaques," *Primates* 41 (2000): 321–35.

120 *For these privileged creatures—sperm whales, domestic swine, wild spider monkeys:* Hal Whitehead, "The behaviour of mature male sperm whales on the Galapagos Islands breeding grounds," *Canadian Journal of Zoology* 71 (1993): 689–99; Clutton-Brock, Albon, and Guinness, "Maternal domi-

nance, breeding success and birth sex ratios in red deer"; G. B. Meese and R. Ewbank, "The establishment and nature of the dominance hierarchy in the domesticated pig," *Animal Behaviour* 21 (1973): 326–34; Douglas B. Meikle et al., "Maternal dominance rank and secondary sex ratio in domestic swine," *Animal Behaviour* 46 (1993): 79–85; M. McFarland Symington, "Sex ratio and maternal rank in wild spider monkeys: When daughters disperse," *Behavioral Ecology and Sociobiology* 20 (1987): 421–25.

120 *The offspring of the high-ranking and well-connected:* Kenneth J. Arrow and Simon A. Levin, "Intergenerational resource transfers with random offspring numbers," *PNAS* 106 (2009): 13702–6; Shifra Z. Goldenberg, Ian Douglas-Hamilton, and George Wittemyer, "Vertical Transmission of Social Roles Drives Resilience to Poaching in Elephant Networks," *Current Biology* 26 (2016): 75–79; Amiyaal Ilany and Erol Akcay, "Social inheritance can explain the structure of animal social networks," *Nature Communications* 7 (2016), https://www.nature.com/articles/ncomms12084.

120 *Among species, notably birds:* Robert Moss, Peter Rothery, and Ian B. Trenholm, "The inheritance of social dominane rank in red grouse (*Lagopus Lagopush scoticus*)," *Aggressive Behavior* 11 (1985): 253–59.

121 *With chimpanzees living in Gombe National Park in Tanzania:* A. Catherine Markham et al., "Maternal rank influences the outcome of aggressive interactions between immature chimpanzees," *Animal Behaviour* 100 (2015): 192–8.

121 *These mothers turn out to be less effective:* Interview with Dr. Kay Holekamp, May 1, 2018.

121 *A hyena expert at Michigan State University:* Ibid.

122 *the "loser effect"—a process in which social losing:* Lee Alan Dugatkin and Ryan L. Earley, "Individual recognition, dominance hierarchies and winner and loser effects," *Proceedings of the Royal Society B: Biological Sciences* 271 (2004): 1537–40; Lee Alan Dugatkin, "Winner and loser effects and the structure of dominance hierarchies," *Behavioral Ecology* 8 (1997): 583–87.

122 *A vivid primate example of this kind of training behavior:* Tim Clutton-Brock, *Mammal Societies* (Hoboken, NJ: Wiley-Blackwell, 2016), 263.

124 *Like human dynasts, advantaged Eurasian beavers:* Katrin Hohwieler, Frank Rossell, and Martin Mayer, "Scent-marking behavior by subordinate Eurasian beavers," *Ethology* 124 (2018): 591–99; Ruairidh D. Campbell et al., "Territory and group size in Eurasian beavers (Castor fiber): Echoes of settlement and reproduction?" *Behavioral Ecology* 58 (2005): 597–607.

124 *Juvenile pikas, red foxes, and scrub jays also inherit their parents' territories:* Charles Brandt, "Mate choice and reproductive success of pikas," *Animal Behaviour* 37 (1989): 118–32; Clutton-Brock, *Mammal Societies*; Philip J. Baker, "Potential fitness benefits of group living in the red fox, *Vulpes vulpes*," *Animal*

Behaviour 56 (1998): 1411–24; Glen E. Woolfenden and John W. Fitzpatrick, "The inheritance of territory in group-breeding birds," *BioScience* 28 (1978): 104–8.

124 *And self-sacrificing red squirrel mothers:* Karen Price and Stan Boutin, "Territorial bequeathal by red squirrel mothers," *Behavioral Ecology* 4 (1992): 144–50.

124 *Within a malignant tumor:* L. Stanley, A. Aktipis, and C. Maley, "Cancer initiation and progression within the cancer microenvironment," *Clininal & Experimental Metastasis* 35 (2018): 361–67; Athena Aktipis, "Principles of cooperation across systems: From human sharing to multicellularity and cancer," *Evolutionary Applications* 9 (2016): 17–36.

125 *the Ngorongoro teems with food—one study counted a whopping:* Oliver P. Höner et al., "The effect of prey abundance and foraging tactics on the population dynamics of a social, territorial carnivore, the spotted hyena," *OIKOS* 108 (2005): 544–54; interviews with Oliver Höner (Berlin), May 3, 2018, and October 4, 2018; Bettina Wachter, et al., "Low aggression levels and unbiased sex ratios in a prey-rich environment: No evidence of siblicide in Ngorongoro spotted hyenas (Crocuta crocuta)," *Behavioral Ecology and Sociobiology* 52 (2002): 348–56.

126 *determined by winning fights:* Clutton-Brock, *Mammal Societies*, 470.

127 *Studies of meerkats:* Norbert Sachser, Michael B. Hennessy, and Sylvia Kaiser, "Adaptive modulation of behavioural profiles by social stress during early phases of life and adolescence," *Neuroscience & Biobehavioral Reviews* 35 (2011): 1518–33; A. Thornton and J. Samson, "Innovative problem solving in wild meerkats," *Animal Behaviour* 83 (2012): 1459–68.

CHAPTER 9: THE PAIN OF SOCIAL DESCENT

129 *In the United States in the 1950s, five people became severely depressed:* Edward D. Freis, "Mental Depression in Hypertensive Patients Treated for Long Periods with Large Doses of Reserpine," *New England Journal of Medicine* 251 (1954): 1006–8.

129 *depression is caused by:* D. A. Slattery, A. L. Hudson, D. J. Nutt, "The evolution of antidepressant mechanisms," *Fundamental and Clinical Pharmacology* 18 (2004): 1–21.

129 *Serotonin, the neurotransmitter:* James M. Ferguson, "SSRI antidepressant medications: Adverse effects and tolerability," *Primary Care Companion to the Journal of Clinical Psychiatry* 3 (2001): 22–27.

130 *When a lobster is born:* Nathalie Paille and Luc Bourassa, "American Lobster," St. Lawrence Global Observatory, https://catalogue.ogsl.ca/dataset/46a463f8-8d55-4e38-be34-46f12d5c2b33/resource/c281bcd4-2bde

-4f3e-adbe-dd3ee01fb372/download/american-lobster-slgo.pdf; J. Emmett Duffy and Martin Thiel, *Evolutionary Ecology and Social and Sexual Systems: Crustaceans as Model Organisms* (Oxford, UK: Oxford University Press, 2007), 106–7; Francesca Gherardi, "Visual recognition of conspecifics in the American lobster, *Homarus americanus*," *Animal Behaviour* 80 (2010): 713–19; D. H. Edwards and E. A. Kravitz, "Serotonin, social status and aggression," *Current Opinion in Neurobiology* 7 (1997): 812–19; Robert Huber et al., "Serotonin and aggressive motivation in crustaceans: Altering the decision to retreat," *Proceedings of the National Academy of Sciences* 94 (1997): 5939–42.

130 *Scientists studying rank relationships among these crustaceans:* J. Duffy and Thiel, *Evolutionary Ecology and Social and Sexual Systems*, 106–7.

130 *A similar study in crayfish showed the same thing:* S. R. Yeh, R. A. Fricke, and D. H. Edwards, "The effect of social experience on serotonergic modulation of the escape circuit of crayfish," *Science* 271 (1996): 366–69.

131 *In the early twentieth century:* Thorleif Schjelderup-Ebbe, "Social behavior of birds," in C. Murchison, ed., *Handbook of Social Psychology* (Worcester, MA: Clark University Press, 1935), 955, 966.

131 *V. C. Wynne-Edwards, an English zoologist:* John S. Price and Leon Sloman, "Depression as yielding behavior: An animal model based on Schjelderup-Ebbe's pecking order," *Ethology and Sociobiology* 8 (1987): 85–98.

132 *Albert Demaret noticed that birds with territory:* John S. Price et al., "Territory, Rank and Mental Health: The History of an Idea," *Evolutionary Psychology* 5 (2007): 531–54.

132 *elixirs of neurochemicals:* Christopher Bergland, "The neurochemicals of happiness," *Psychology Today*, November 29, 2012, https://www.psychologytoday.com/us/blog/the-athletes-way/201211/the-neurochemicals-happiness.

133 *Newer research on status and serotonin in anole lizards:* Cliff H. Summers and Svante Winberg, "Interactions between the neural regulation of stress and aggression," *Journal of Experimental Biology* 209 (2006): 4581–89; Olivier Lepage et al., "Serotonin, but not melatonin, plays a role in shaping dominant-subordinate relationships and aggression in rainbow trout," *Hormones and Behavior* 48 (2005): 233–42; Earl T. Larson and Cliff H. Summers, "Serotonin reverses dominant social status," *Behavioural Brain Research* 121 (2001): 95–102; Huber et al., "Serotonin and aggressive motivation in crustaceans"; Yeh, Fricke, and Edwards, "The effect of social experience on serotonergic modulation of the escape circuit of crayfish"; Varenka Lorenzi et al., "Serotonin, social status and sex change in bluebanded goby Lythrypnus dalli," *Physiology and Behavior* 97 (2009): 476–83.

133 *the combination of the adolescents' increased sensitivity to social status:* Leah H. Somerville, "The teenage brain: Sensitivity to social evaluation," *Current Directions in Psychological Science* 22 (2013): 121–27.

134 *In one study, her team imaged the brains of adolescents:* Naomi I. Eisenberger et al., "Does Rejection Hurt? An fMRI Study of Social Exclusion," *Science* 302 (2003): 290–92; Naomi I. Eisenberger, "The neural bases of social pain: Evidence for shared representations with physical pain," *Psychosomatic Medicine* 74 (2012): 126–35.

134 *Eisenberger has also linked social pain to opiate addiction and overdose:* Naomi I. Eisenberger and Matthew D. Lieberman, "Why It Hurts to be Left Out: The Neurocognitive Overlap Between Physical and Social Pain" (2004), http://www.scn.ucla.edu/pdf/Sydney(2004).pdf; Naomi I. Eisenberger, "Why Rejection Hurts: What Social Neuroscience Has Revealed About the Brain's Response to Social Rejection," in Greg J. Norman, John T. Cacioppo, and Gary G. Berntson, eds., *The Oxford Handbook of Social Neuroscience* (Oxford, UK: Oxford University Press, 2001), https://sanlab.psych.ucla.edu/wp-content/uploads/sites/31/2015/05/39-Decety-39.pdf.

134 *It's worth noting that:* Centers for Disease Control and Prevention, "Teen Substance Use and Risks," https://www.cdc.gov/features/teen-substance-use/index.html.

134 *In a related study, Eisenberger has shown that acetaminophen:* Eisenberger, "The neural bases of social pain"; C. N. Dewall et al., "Acetaminophen reduces social pain: Behavioral and neural evidence," *Psychological Science* 21 (2010): 931–37.

135 *High-ranking mothers continue to intercede:* Interviews with Oliver Höner (Berlin), May 3, 2018, and October 4, 2018.

135 *what's called a target animal:* Clutton-Brock, *Mammal Societies*, 104, 269, 272.

135 *Studies of social defeat in mice:* V. Klove et al., "The winner and loser effect, serotonin transporter genotype, and the display of offensive aggression," *Physiology & Behavior* 103 (2001): 565–74. Stephan R. Lehner, Claudia Rutte, and Michael Taborsky, "Rats benefit from winner and loser effects," *Ethology* 117 (2011): 949–60.

136 *Studies of lobster dominance:* Rachel L. Rutishauser et al., "Long-term consequences of agonistic interactions between socially naïve juvenile American lobsters (*Homarus americanus*)," *Biological Bulletin* 207 (December 2004): 183–87.

136 *Seriously depressed adolescents:* Stephanie Dowd, "What Are the Signs of Depression?" Child Mind Institute, https://childmind.org/ask-an-expert-qa/im-16-and-im-feeling-like-there-is-something-wrong-with-me-i-may-be-depressed-but-im-not-sure-please-help/.

136 *Worthlessness is among the symptoms:* American Psychiatric Association, "What Is Depression?" https://www.psychiatry.org/patients-families/depression/what-is-depression; Julio C. Tolentino and Sergio L. Schmidt,

"DSM-5 criteria and depression severity: Implications for clinical practice," *Front Psychiatry* 9 (2018): 450.

136 *Schjelderup-Ebbe in 1935 described:* Thorleif Schjelderup-Ebbe, "Social behavior of birds," in Murchison, ed., *Handbook of Social Psychology*, 955.

137 *Experiments on hierarchy stability show that removing individual fish:* Rui F. Oliveira and Vitor C. Almada, "On the (in)stability of dominance hierarchies in the cichlid fish Oreochromis mossambicus," *Aggressive Behavior* 22 (1996): 37–45; E. J. Anderson, R. B. Weladji, and P. Paré, "Changes in the dominance hierarchy of captive female Japanese macaques as a consequence of merging two previously established groups," *Zoo Biology* 35 (2016): 505–12.

137 *The veterinarian who cares for these Saskatchewan bison:* Interview with Todd Shury, Parks Canada, Office of the Chief Ecosystem Scientist, wildlife health specialist, adjunct professor, Department of Veterinary Pathology, University of Saskatchewan, August 20, 2014.

138 *studies of depressed adolescents:* Vanja Putarek and Gordana Kerestes, "Self-perceived popularity in early adolescence," *Journal of Social and Personal Relationships* 33 (2016): 257–74.

138 *Multiple studies have shown a close association:* Riittakerttu Kaltiala-Heino and Sari Jrodj, "Correlation between bullying and clinical depression in adolescent patients," *Adolescent Health, Medicine and Therapeutics* 2 (2011): 37–44.

138 *A 2005 study that compared bullying:* P. Due et al., "Bullying and symptoms among school-aged children: International comparative cross sectional study in 28 countries," *European Journal of Public Health* 15 (2005): 128–32.

138 *Some 20 percent of school students, grades nine through twelve, in the United States:* NIH Eunice Kennedy Shriver National Institute of Child Health and Human Development, "Bullying," https://www.nichd.nih.gov/health/topics/bullying.

139 *its function and form in animals:* Hogan Sherrow, "The Origins of Bullying," *Scientific American Guest Blog*, December 15, 2011, https://blogs.scientificamerican.com/guest-blog/the-origins-of-bullying/.

141 *According to a 2018 report released by the nonprofit YouthTruth:* YouthTruth Student Survey, "Bullying Today," https://youthtruthsurvey.org/bullying-today/.

142 *This is a process sociologists:* Alan Bullock and Stephen Trombley, "Othering," in *The New Fontana Dictionary of Modern Thought*, Third Edition (New York: HarperCollins, 2000), 620.

142 *Nazi Germany's portrayal:* United States Holocaust Memorial Museum, "Defining the Enemy," https://encyclopedia.ushmm.org/content/en/article/defining-the-enemy; "Rwanda Jails Man Who Preached Genocide of Tutsi

'Cockroaches,'" BBC News, April 15, 2016, https://www.bbc.com/news/world-africa-36057575.

143 *"We don't punish fear"*: Lecture by Robin Foster, August 4, 2012.

143 *lack of socialization:* Interview with James Ha, February 26, 2019.

144 *A study of mice demonstrated how status descent impairs learning:* C. J. Barnard and N. Luo, "Acquisition of dominance status affects maze learning in mice," *Behavioural Processes* 60 (2002): 53–59.

144 *A study of rhesus macaques showed a different way:* Christine M. Drea and Kim Wallen, "Low-status monkeys 'play dumb' when learning in mixed social groups," *Proceedings of the National Academy of Sciences* 96 (1999): 12965–69.

CHAPTER 10: THE POWER OF AN ALLY

147 *Jaana Juvonen, an adolescent bullying expert:* Jaana Juvonen, "Bullying in the Pig Pen and on the Playground," Zoobiquity Conference, September 29, 2012, https://www.youtube.com/watch?v=tD8ajvbwKSQ.

147 *"social coalition walking"*: Interviews with Oliver Höner (Berlin), May 3, 2018, and October 4, 2018.

148 *Karl Groos, a German philosopher:* Karl Groos, *The Play of Animals* (New York: D. Appleton and Company, 1898), 75, https://archive.org/details/playofanimalsoogroouoft/page/ii.

148 *parents who bring "toys"*: Clutton-Brock, *Mammal Societies*, 202.

148 *A wild African cat called a serval engages in "angling play"*: Gordon M. Burghardt, *The Genesis of Animal Play: Testing the Limits* (Cambridge, MA: A Bradford Book/The MIT Press, 2006), 101.

149 *Adolescent killer whales play at beaching themselves:* Christophe Guinet, "Intentional stranding apprenticeship and social play in killer whales (Orcinus orca)," *Canadian Journal of Zoology* 69 (1991): 2712–16.

149 *For bald eagles, whose pre-mating rituals involve a harrowing:* Patricia Edmonds, "For Amorous Bald Eagles, a 'Death Spiral' Is a Hot Time," *National Geographic*, July 2016, https://www.nationalgeographic.com/magazine/2016/07/basic-instincts-bald-eagle-mating-dance/.

149 *Australian wombats and tiger quolls . . . sparring, and kicking:* Burghardt, *The Genesis of Animal Play*, 220; Duncan W. Watson and David B. Croft, "Playfighting in Captive Red-Necked Wallabies, Macropus rufogriseus banksianus," *Behaviour* 126 (1993): 219–45.

150 *As University of Massachusetts Amherst biologist Judith Goodenough:* Judith Goodenough and Betty McGuire, *Perspectives on Animal Behavior* (Hoboken, NJ: Wiley, 2009).

150 *Young male white-tailed deer spend summers:* Interview with Joe Hamilton and Matt Ross, September 1, 2017.

150 *Play-fighting trains young animals:* Gordon M. Burghardt, *The Genesis of Animal Play: Testing the Limits* (Cambridge, MA: A Bradford Book/The MIT Press, 2006).

151 *New research confirms that video gaming isn't necessarily socially isolating:* Helena Cole and Mark D. Griffiths, "Social Interactions in Massively Multiplayer Online Role-Playing Gamers," *CyberPsychology and Behavior* 10 (2007), doi: 10/1089/cpb.2007.9988; Eshrat Zamani, "Comparing the social skills of students addicted to computer games with normal students," *Addiction and Health* 2 (2010): 59–65.

153 *Mismatch disorders arise:* Elisabeth Lloyd, David Sloan Wilson, and Elliott Sober, "Evolutionary mismatch and what to do about it: A basic tutorial," *Evolutionary Applications* (2011): 2–4.

154 *As Kathy Krupa, a certified equine assisted therapist, told the* New York Times*:* Bill Finley, "Horse Therapy for the Troubled," *New York Times*, March 9, 2008, https://www.nytimes.com/2008/03/09/nyregion/nyregionspecial2/09horse nj.html.

155 *In Philadelphia, a nonprofit organization called Hand2Paw:* Interview with Rachel Cohen, Hand2Paw founder, May 5, 2017.

156 *2018 memoir,* Educated*:* Tara Westover, *Educated: A Memoir* (New York: Random House, 2018).

157 *"working in her father's junkyard"*: Tara Westover, "Bio," https://tarawestover .com/bio.

157 *"[t]hings that I now recognize as"* and *"You can love someone and still":* Louise Carpenter, "Tara Westover: The Mormon Who Didn't Go to School (but Now Has a Cambridge PhD)," *Times of London*, February 10, 2018, https://www.thetimes.co.uk/article/tara-westover-the-mormon-who-didnt-go -to-school-but-now-has-a-cambridge-phd-pxwgtz7pv.

157 *For Thorleif Schjelderup-Ebbe:* El-Hai, "The Chicken-Hearted Origins of the 'Pecking Order'—The Crux."

PART III: SEX

Salt's story is from the Center for Coastal Studies, Provincetown, Massachusetts, whose scientists have been tracking her since the mid-1970s.

CHAPTER 11: ANIMAL ROMANCE

162 *Above a public park:* New York State Department of Environmental Conservation, "Watchable Wildlife: Bald Eagle," https://www.dec.ny.gov/animals/63144 .html; Patricia Edmonds, "For Amorous Bald Eagles, a 'Death Spiral' Is a Hot Time," *National Geographic*, July 2016, https://www.nationalgeographic.com /magazine/2016/07/basic-instincts-bald-eagle-mating-dance/.

163 *In tropical Australia two flying foxes:* Nicola Markus, "Behaviour of the Black Flying Fox Pteropus alecto: 2. Territoriality and Courtship," *Acta Chiropterologica* 4 (2002): 153–66.

163 *Along a creek in Grayson County, Virginia:* Leslie A. Dyal, "Novel Courtship Behaviors in Three Small Easter Plethodon Species," *Journal of Herpetology* 40 (2006): 55–65.

163 *And right there, in the banana bowl:* Interviews with Dr. Michael Crickmore and Dr. Dragana Rogulja, December 6, 2018.

163 *One journal described it as "complex, ritualistic":* Danielle Simmons, "Behavioral Genomics," *Nature Education* 1 (2008): 54.

164 *With multiple partners:* Natterson-Horowitz and Bowers, "The Koala and the Clap," in *Zoobiquity*, 249–72.

165 *a word the Yaghan people of Tierra del Fuego:* Anna Bitong, "Mamihlapinatapai: A Lost Language's Untranslatable Legacy," BBC Travel, April 3, 2018, http://www.bbc.com/travel/story/20180402-mamihlapinatapai-a-lost-languages-untranslatable-legacy; Thomas Bridge, "Yaghan Dictionary: Language of the Yamana People of Tierra del Fuego," 1865, https://patlibros.org/yam/ey.php, 182a.

166 *In 1978, the study of humpback chorusing:* H. E. Winn and L. K. Winn, "The song of the humpback whale *Megaptera novaeangliae* in the West Indies," *Marine Biology* 47 (1978): 97–114.

166 *But a University of Hawaii humpback expert thought:* Louis M. Herman et al., "Humpback whale song: Who sings?" *Behavioral Ecology and Sociobiology* 67 (2013): 1653–63; L. M. Herman, "The multiple functions of male song within the humpback whale (Megaptera novaeangliae) mating system: Review, evaluation and synthesis," *Biological Reviews of the Cambridge Philosophical Society* 92 (2017): 1795–1818.

167 *Among singing bat species:* Mirjam Knornschild et al., "Complex vocal imitation during ontogeny in a bat," *Biology Letter* 6 (2010): 156–59; Yosef Prat, Mor Taub, and Yossi Yovel, "Vocal learning in a social mammal: Demonstrated by isolation and playback experiments in bats," *Science Advances* 1 (2015): e1500019.

167 *For songbirds, receiving musical instruction:* Todd M. Freeberg, "Social transmission of courtship behavior and mating preferences in brown-headed cowbirds, Molothrus ater," *Animal Learning and Behavior* 32 (2004): 122–30; Haruka Wada, "The development of birdsong," *Nature Education Knowledge* 3 (2010): 86.

167 *Male and female leopard seals vocalize over vast ocean distances to find mates:* T. L. Rogers, "Age-related differences in the acoustic characteristics of male leopard seals, Hydrurga leptonyx," *Journal of the Acoustical Society of America* 122 (2007): 596–605; Voice of the Sea, "The Leop-

ard Seal," http://cetus.ucsd.edu/voicesinthesea_org/species/pinnipeds/leopardSeal.html.

168 *In many birds and a few mammals, erotically stimulating songs:* Jennifer Minnick, "Bioacoustics: Listening to the Animals," Zoological Society of San Diego, Institutional Interviews, June 2009, http://archive.sciencewatch.com/inter/ins/pdf/09junZooSanDgo.pdf.

168 *specific "sexy syllables":* Jay Withgott, "The Secret to Seducing a Canary," *Science Magazine*, November 7, 2001, https://www.sciencemag.org/news/2001/11/secret-seducing-canary.

168 *"[S]inging may function in part to synchronize ovulation":* C. Scott Baker and Louis M. Herman, "Aggressive behavior between humpback whales (Megaptera novaeangliae) wintering in Hawaiian waters," *Canadian Journal of Zoology* 62 (1984): 1922–37.

168 *Humpback songs also attract lone males:* L. M. Herman, "The multiple functions of male song within the humpback whale (*Megaptera novaeangliae*) mating system: Review, evaluation and synthesis," *Biological Reviews of the Cambridge Philosophical Society* 92 (2017): 1795–818; Adam A. Pack et al., "Penis extrusions by humpback whales (*Megaptera novaeangliae*)," *Aquatic Mammals* 28.2 (2002): 131–46; James D. Darling and Martine Berube, "Interactions of singing humpback whales with other males," *Marine Mammal Science* 17 (2001): 570–84; Phillip J. Clapham and Charles A. Mayo, "Reproduction and recruitment of individually identified humpback whales, *Megaptera novaeangliae*, observed in Massachusetts Bay, 1979–1985," *Canadian Journal of Zoology* 65 (1987): 2853–963.

CHAPTER 12: DESIRE & RESTRAINT

171 The Lion King *comes a scene of teenage love:* DisneyMusicVEVO, "The Lion King—Can You Feel the Love Tonight," YouTube, https://www.youtube.com/watch?v=25QyCxVkXwQ.

172 *nature documentary, a genre invented and largely produced by mature men:* Brett Mills, "The animals went in two by two: Heteronormativity in television wildlife documentaries," *European Journal of Cultural Studies* 16 (2013): 100–114.

173 *Daylight and breeding season:* Colorado State University, Equine Reproduction Laboratory, "SEE THE LIGHT—Advancing the Breeding Season for Early Foals—Press Release," http://csu-cvmbs.colostate.edu/Documents/case-advancing-breeding-season.pdf.

173 *fur seals might not start breeding until they're seven:* M. N. Bester, "Reproduction in the male sub-Antarctic fur seal *Arctocephalus tropicalis*," *Journal of Zoology* (1990): 177–85.

173 *Dominant female mammals:* Clutton-Brock, *Mammal Societies*, 268.

173 *In wolf packs:* P. Hradecky, "Possible pheromonal regulation of reproduction in wild carnivores," *Journal of Chemical Ecology* 11 (1985): 241–50.

174 *In wild Bornean orangutans:* P. R. Marty et al., "Endocrinological correlates of male bimaturism in wild bornean orangutans," *American Journal of Primatology* 77 (November 2015) (11): 1170–78, doi: 10.1002/ajp.22453. Epub 2015 Jul 31.

174 *After a ten-year puberty, male sperm whales:* National Oceanic and Atmospheric Administration/NOAA Fisheries, "Sperm Whale," https://www.fisheries.noaa.gov/species/sperm-whale.

174 *male elephants too develop physically:* H. B. Rasmussen et al., "Age- and tactic-related paternity success in male African elephants," *Behavioral Ecology* 19 (2008): 9–15; J. C. Beehner and A. Lu, "Reproductive suppression in female primates: a review," *Evolutionary Anthropology* 22 (2013): 226–38.

174 *Fish who incubate eggs in their mouths:* Barbara Taborsky, "The influence of juvenile and adult environments on life-history trajectories," *Proceedings of the Royal Society B: Biological Sciences* 273 (2006): 741–50.

174 *First-time sheep mothers:* Jimmy D. Neill, "Volume 2," *Knobil and Neill's Physiology of Reproduction*, 3rd Edition (Cambridge, MA: Academic Press, 2005), 1957; A. Zedrosser et al., "The effects of primiparity on reproductive performance in the brown bear," *Oecologia* 160 (2009): 847–54; Andrew M. Robbins et al., "Age-related patterns of reproductive success among female mountain gorillas," *American Journal of Physical Anthropology* 131 (2006): 511–21; G. Schino and A. Troisi, "Neonatal abandonment in Japanese macaques," *American Journal of Physical Anthropology* 126 (2005): 447–52.

174 *Young mandrill mothers have infants:* Stanton et al., "Maternal Behavior by Bird Order in Wild Chimpanzees (Pan troglodytes): Increased Investment by First-Time Mothers."

174 *young mother marmosets and rhesus monkeys:* Ibid.

175 *"[A]cross human societies, adverse pregnancy outcomes":* Margaret A. Stanton et al., "Maternal behavior by bird order in wild chimpanzees (Pan troglodytes): Increased investment by first-time mothers," *Current Anthropology* 55 (2014): 483–89; K. L. Kramer and J. B. Lancaster, "Teen motherhood in cross-cultural perspective," *Annals of Human Biology* 37 (2010): 613–28.

175 *The World Health Organization reports that children:* World Health Organization, "Adolescent Pregnancy Fact Sheet," https://www.who.int/news-room/fact-sheets/detail/adolescent-pregnancy/.

175 *Low-ranking adolescent and young adult bird parents:* Steven J. Portugal et al., "Perch height predicts dominance rank in birds," *IBIS* 159 (2017): 456–62.

175 *Horse breeders know that stallions and mares:* Katherine A. Houpt, *Domestic*

Animal Behavior for Veterinarians and Animal Scientists, 5th edition (Hoboken, NJ: Wiley-Blackwell, 2010), 114–15.

176 *coercive first sexual experiences are alarmingly common:* R. L. T. Lee, "A systematic review on identifying risk factors associated with early sexual debut and coerced sex among adolescents and young people in communities," *Journal of Clinical Nursing* 27 (2018): 478–501.

176 *Over the past twenty years:* Gilda Sedgh et al., "Adolescent pregnancy, birth, and abortion rates across countries: Levels and recent trends," *Journal of Adolescent Health* 56 (2015): 223–30; Centers for Disease Control and Prevention, "Reproductive Health: Teen Pregnancy," https://www.cdc.gov/teenpregnancy/about/index.htm.

176 *According to a Harvard study that:* Richard Weissbourd et al., "The Talk: How Adults Can Promote Young People's Healthy Relationships and Prevent Misogyny and Sexual Harassment," Making Caring Common Project, Harvard Graduate School of Education, 2017, https://mcc.gse.harvard.edu/reports/the-talk.

177 *milled a herd of Milu deer:* Visit to the Wilds and interview with Dr. Barbara Wolfe, June 26, 2014.

178 *Bowerbirds, native to New Guinea and northern Australia:* Gerard L. Hawkins, Geoffrey E. Hill, and Austin Mercadante, "Delayed plumage maturation and delayed reproductive investment in birds," *Biological Reviews* 87 (2012): 257–74; "The Crazy Courtship of Bowerbirds," BBC Earth, November 20, 2014, http://www.bbc.com/earth/story/20141119-the-barmy-courtship-of-bowerbirds.

179 *Delayed plumage maturation:* Hawkins, Hill, and Mercadante, "Delayed plumage maturation and delayed reproductive investment in birds."

179 *Jennifer Hirsch, a Columbia public health professor:* Jennifer S. Hirsch and Holly Wardlow, *Modern Loves: The Anthropology of Romantic Courtship and Companionate Marriage* (Ann Arbor: University of Michigan Press, 2006).

180 *An artwork created by a Plains Warrior:* Oglala Sioux blanket strip, Catalog 985-27-10/59507, Peabody Museum, Harvard University.

180 *"when a young man":* Peabody Museum of Archaeology & Ethology at Harvard University, "Love Blooms Among the Lakota," *Inside the Peabody Museum*, February 2012, https://www.peabody.harvard.edu/node/762.

181 *Teen life coach Cyndy Etler:* Cyndy Etler, "Young People Can Tell You the Kind of Sex Ed They Really Need," CNN Opinion, October 31, 2018, https://www.cnn.com/2018/10/31/opinions/sex-assault-controversies-prove-we-need-better-sex-ed-etler/index.html.

181 *The psychologist Richard Weissbourd offers:* Interview with Dr. Richard Weissbourd, Harvard psychologist, February 14, 2018; Weissbourd et al., "The Talk."

182 *list of "Most Challenged Books":* American Library Association, "Infographics," Banned and Challenged Books, http://www.ala.org/advocacy/bbooks/frequentlychallengedbooks/statistics.

182 *The ALA, a professional organization:* American Library Association, "About ALA," http://www.ala.org/aboutala/.

182 *sexual content is by far the most common complaint:* American Library Association, "Infographics."

182 *social learning is extremely powerful:* Andrew Whiten and Erica van de Waal, "The pervasive role of social learning in primate lifetime development," *Behavioral Ecology and Sociobiology* 72 (2018): 80.

182 *echo throughout adult life, for humans:* C. V. Smith and M. J. Shaffer, "Gone but not forgotten: Virginity loss and current sexual satisfaction," *Journal of Sex & Marital Therapy* 39 (2013): 96–111.

183 *As described to us by Mia-Lana Lührs:* Interview with Dr. Mia-Lana Lührs, October 16, 2017.

185 *For fossas, wildhood:* Clare E. Hawkins et al., "Transient masculinization in the fossa, Cryptoprocta ferox (Carnivora, Viverridae)," *Biology of Reproduction* 66, no. 3 (March 1, 2002): 610–15.

185 *Andrew Solomon explores:* Andrew Solomon, *Far from the Tree: Parents, Children, and the Search for Identity* (New York: Scribner, 2012).

CHAPTER 13: THE FIRST TIME

187 *It's part of a worldwide conservation effort:* Jen Fields, "The Wilds Celebrates Births of Three At-Risk Species," Columbus Zoo and Aquarium Press Release, March 27, 2018, https://www.columbuszoo.org/home/about/press-releases/press-release-articles/2018/03/27/the-wilds-celebrates-births-of-three-at-risk-species; Association of Zoos and Aquariums, "Species Survival Plan Programs," https://www.aza.org/species-survival-plan-programs.

188 *Severely inclement weather can inhibit sexual interest:* Houpt, *Domestic Animal Behavior for Veterinarians and Animal Scientists.*

189 *Two whale-watch boat captains:* Conscious Breath Adventures, "About Humpback Whales: Rowdy Groups," https://consciousbreathadventures.com/rowdy-groups/.

190 *Tony Wu, an underwater photographer:* Tony Wu, "Humpback Whales in Tonga 2014, Part 3," http://www.tonywublog.com/journal/humpback-whales-in-tonga-2014-part-3.

190 *Another cameraman, Roger Munns, documented:* Matt Walker, "Epic Humpback Whale Battle Filmed," BBC Earth News, October 23, 2009, http://news.bbc.co.uk/earth/hi/earth_news/newsid_8318000/8318182.stm.

190 *In 2010, a photographer reported having:* "Photographer First to Capture Humpbacks' Magic Moment," *NZ Herald*, June 22, 2012, https://www.nzherald.co.nz/nz/news/article.cfm?c_id=1&objectid=10814498; Malcolm Holland, "The Tender Mating Ritual of the Humpback Whale Captured on Camera for the First Time," *Daily Telegraph*, June 20, 2012, https://www.dailytelegraph.com.au/news/nsw/the-tender-mating-ritual-of-the-humpback-whale-captured-ion-camera-for-the-first-time/news-story/175cc74142e7b85fbac49150fcf2035f?sv=f9df3726babb600fd5d3a784a82d6160.

191 *Shannon Farrell was studying courtship behavior:* Shannon L. Farrell and David A. Andow, "Highly variable male courtship behavioral sequence in a crambid moth," *Journal of Ethology* 35 (2017): 221–36; Panagiotis G. Milonas, Shannon L. Farrell, and David A. Andow, "Experienced males have higher mating success than virgin males despite fitness costs to females," *Behavioral Ecology Sociobiology* 65 (2011): 1249–56.

192 *essentially the same patterned behavior:* Barbara Natterson-Horowitz and Kathryn Bowers, "Roar-gasm," in *Zoobiquity: The Astonishing Connection Between Human and Animal Health* (New York: Vintage, 2012), 70–110.

192 *Gaining sexual experience:* Interview with Dr. Richard Weissbourd, Harvard psychologist, February 14, 2018.

193 *Some animals remain together:* Judith Goodenough and Betty McGuire, *Perspectives on Animal Behavior* (Hoboken, NJ: Wiley, 2009), 371; Brandon J. Aragona et al., "Nucleus accumbens dopamine differentially mediates the formation and maintenance of monogamous pair bonds," *Nature Neuroscience* 9 (2006): 133–39.

193 *Titi monkeys:* Benjamin J. Ragen et al., "Differences in titi monkey (*Callicebus cupreus*) social bonds affect arousal, affiliation, and response to reward," *American Journal of Primatology* 74 (2012): 758–69.

193 *Some birds, like swans and some hawks:* Nathan J. Emergy et al., "Cognitive adaptations of social bonding in birds," *Philosophical Transactions of the Royal Society of London Biological Sciences* 362 (2007): 489–505; William J. Mader, "Ecology and breeding habits of the Savanna hawk in the Llanos of Venezuela," *Condor: Ornithological Applications* 84 (1982): 261–71.

193 *One animal, a seahorse relative:* Judith Goodenough and Betty McGuire, *Perspectives on Animal Behavior* (Hoboken, NJ: Wiley, 2009), 371–72.

CHAPTER 14: COERCION & CONSENT

195 *Michael Crickmore, who, with his colleague Dragana Rogulja:* Interviews with Dr. Michael Crickmore and Dr. Dragana Rogulja, December 6, 2018; Stephen X. Zhang, Dragana Rogulja, and Michael A. Crickmore, "Dopaminergic Circuitry

Underlying Mating Drive," *Neuron* 91 (2016): 168–81; ScienceDaily, "Neurobiology of Fruit Fly Courtship May Shed Light on Human Motivation," *Science News*, July 13, 2018, https://www.sciencedaily.com/releases/2018/07/180713220147.htm.

197 *Oliver Sacks's book* Awakenings*:* Oliver Sacks, *Awakenings* (1973; rev. ed. New York: Vintage, 1999) (Kindle version, location 1727–1825).

197 *Dopamine, Crickmore explained:* Interviews with Dr. Michael Crickmore and Dr. Dragana Rogulja, December 6, 2018.

198 *George Murray Levick, a scientist with the 1910–13 Scott:* William J. L. Sladen and David G. Ainley, "Dr. George Murray Levick (1876–1956): Unpublished notes on the sexual habits of the Adelie penguin," *Polar Record* (2012), doi: 10.1017/S0032247412000216.

199 *Male sheep, turkeys, fur seals:* Denis Reale, Patrick Bousses, and Jean-Louis Chapuis, "Female-biased mortality induced by male sexual harassment in a feral sheep population," *Canadian Journal of Zoology* 74 (1996): 1812–18; David A. Wells et al., "Male brush-turkeys attempt sexual coercion in unusual circumstances," *Behavioural Processes* 106 (2014): 180–86; P. J. Nico de Bruyn, Cheryl A. Tosh, and Marthan N. Bester, "Sexual harassment of a king penguin by an Antarctic fur seal," *Journal of Ethology* 26 (2008): 295–97; Silu Wang, Molly Cummings, and Mark Kirkpatrick, "Coevolution of male courtship and sexual conflict characters in mosquitofish," *Behavioral Ecology* 26 (2015): 1013–20; Silvia Cattelan et al., "The effect of sperm production and mate availability on patterns of alternative mating tactics in the guppy," *Animal Behaviour* 112 (2016): 105–10; Heather S. Harris et al., "Lesions and behavior associated with forced copulation of juvenile Pacific harbor seals (*Phoca vitulina richardsi*) by southern sea otters (*Enhydra lutris nereis*)," *Aquatic Mammals* 36 (2010): 331–41.

199 *Male Amazonian red-necked turtles signal:* Camila Rudge Ferrara et al., "The role of receptivity in the courtship behavior of Podocnemis erythrocephala in captivity," *Acta Ethologica* 12 (2009): 121–25.

200 *courtship seems to be completely:* Yasuhisa Henmi, Tsunenori Koga, and Minoru Murai, "Mating behavior of the San Bubbler Crab Scopimera globosak," *Journal of Crustacean Biology* 13 (1993): 736–44; Paul Verrell, "The Sexual Behaviour of the Red-Spotted Newt, Notophthalmus Viridescens (Amphibia : Urodela : Salamandridae)," *Animal Behaviour* 30 (1982): 1224–36.

200 *University of Cambridge professor Tim Clutton-Brock described:* T. H. Clutton-Brock and G. A. Parker, "Sexual coercion in animal societies," *Animal Behavior* 49 (1995): 1345–65.

200 *a male Antarctic fur seal:* Barcoft TV, "Scientists Capture Unique Footage of Seals Attempting to Mate with Penguins," YouTube, November 18, 2014,

https://www.youtube.com/watch?v=ABM8RTVYaVw&t=3s; Harris et al., "Lesions and behavior associated with forced copulation of juvenile Pacific harbor seals (*Phoca vitulina richardsi*) by southern sea otters (*Enhydra lutris nereis*)."

201 *Female white-cheeked pintails:* Martin L. Lalumière, et al., "Forced Copulation in the Animal Kingdom," in *The Causes of Rape: Understanding Individual Differences in Male Propensity for Sexual Aggression* (Washington, DC: American Psychological Association, 2005), 32.

201 *According to a Canadian study from 2005:* Ibid., 294.

201 *Some males may persistently badger:* Mariana Freitas Nery and Sheila Marina Simao, "Sexual coercion and aggression towards a newborn calf of marine tucuxi dolphins (*Sotalia guianensis*)," *Marine Mammal Science* 25 (2009): 450–54; Reale, Bousses, and Chapuis, "Female-biased mortality induced by male sexual harassment in a feral sheep population"; Kamini N. Persaud and Bennett G. Galef, Jr., "Female Japanese quail (*Coturnix Japonica*) mated with males that harassed them are unlikely to lay fertilized eggs," *Journal of Comparative Psychology* 119 (2005): 440–46; Jason V. Watters, "Can the alternative male tactics 'fighter' and 'sneaker' be considered 'coercer' and 'cooperator' in coho salmon?" *Animal Behaviour* 70 (2005): 1055–62.

201 *Sexually harassed elephant seals:* T. H. Clutton-Brock and G. A. Parker, "Sexual coercion in animal societies," *Animal Behavior* 49 (1995): 1345–65.

201 *Primatologist Richard Wrangham:* Martin N. Muller et al., "Sexual coercion by male chimpanzees show that female choice may be more apparent than real," *Behavioral Ecology and Sociobiology* 65 (2011): 921–33; Martin N. Muller and Richard W. Wrangham, eds., *Sexual Coercion in Primates and Humans: Evolutionary Perspective on Male Aggression against Females* (Cambridge, MA: Harvard University Press, 2009); Martin N. Muller, Sonya M. Kahlenberg, Melissa Emery Thompson, and Richard W. Wrangham, "Male coercion and the cost of promiscuous mating for female chimpanzees," *Proceedings of the Royal Society B: Biological Sciences* 274 (2007): 1009–14.

202 *captive female gorillas who reduced the likelihood:* Clutton-Brock and Parker, "Sexual coercion in animal societies."

202 *The 2017 #MeToo movement:* Jessica Bennett, "The #MeToo Moment: When the Blinders Come Off," *New York Times*, November 30, 2017, https://www.nytimes.com/2017/11/30/us/the-metoo-moment.html; Stephanie Zacharek, Eliana Dockterman, and Haley Sweetland Edwards, "TIME Person of the Year 2017: The Silence Breakers," *Time*, http://time.com/time-person-of-the-year-2017-silence-breakers/.

203 *For example, young and adolescent guinea pigs:* Norbert Sachser, Michael B. Hennessy, and Sylvia Kaiser, "Adaptive modulation of behavioural profiles

by social stress during early phases of life and adolescence," *Neuroscience & Biobehavioral Reviews* 35 (2011): 1518–33.

203 *Rats raised without play:* G. J. Hole, D. F. Einon, and H. C. Plotkin, "The role of social experience in the development of sexual competence in *Rattus Norvegicus*," *Behavioral Processes* 12 (1986): 187–202.

203 *young American mink—both males and females:* Stephanie Craig, "Research relationships focus on mink mating," *Ontario Agricultural College, University of Guelph*, February 14, 2017, https://www.uoguelph.ca/oac/news/research-relationships-focus-mink-mating.

203 *"Having absolutely no social interactions":* Houpt, *Domestic Animal Behavior for Veterinarians and Animal Scientists*, 5th ed.

204 *One of those complications in the twenty-first century is hookup culture:* Justin R. Garcia et al., "Sexual hookup culture: A review," *Review of General Psychology* 16 (2012): 161–76, https://www.ncbi.nlm.nih.gov/pmc/articles/PMC3613286/pdf/nihms443788.pdf.

204 *"We're dealing with a culture":* Binghamton University, State University of New York, "College Students' Sexual Hookups More Complex than Originally Thought," *Science News*, October 17, 2012, https://www.sciencedaily.com/releases/2012/10/121017122802.htm.

205 *"Hookups, although increasingly socially acceptable":* Garcia et al., "Sexual hookup culture: A review," 20.

205 *Sociologist Lisa Wade says:* Lisa Wade, *American Hookup: The New Culture of Sex on Campus* (New York: W. W. Norton and Co., 2017).

205 *Richard Weissbourd agrees:* Interview with Dr. Richard Weissbourd, February 14, 2018.

206 *Alcohol and drug use:* Garcia et al., "Sexual hookup culture: A review," 14.

PART IV: SELF-RELIANCE

Slavc's story was recounted for us in interviews with Hubert Potočnik, documented in "The Wolf Who Traversed the Alps," in James Cheshire and Oliver Uberti, *Where the Animals Go: Tracking Wildlife with Technology in 50 Maps and Graphics* (New York: W. W. Norton & Company, 2017), 62–65, and supplemented with reporting by Henry Nicholls in the *Guardian*; PJ's story was gathered from news reports.

CHAPTER 15: LEARNING TO LAUNCH

211 *It's both a moment and a behavior called "dispersal":* Clutton-Brock, *Mammal Societies*, 94–122, 401–26; Bruce N. McLellan and Frederick W. Hovey, "Natal dispersal of grizzly bears," *Canadian Journal of Zoology* 79 (2001): 838–44; Martin Mayer, Andreas Zedrosser, and Frank Rosell, "When to leave: The

timing of natal dispersal in a large, monogamous rodent, the Eurasian beaver," *Animal Behaviour* 123 (2017): 375–82; Jonathan C. Shaw et al., "Effect of population demographics and social pressures on white-tailed deer dispersal ecology," *Journal of Wildlife Management* 70 (2010): 1293–301; Eric S. Long et al., "Forest cover influences dispersal distance of white-tailed deer," *Journal of Mammology* 86 (2005): 623–29; Yun Tao, Luca Börger, and Alan Hastings, "Dynamic range size analysis of territorial animals: An optimality approach," *American Naturalist* 188 (2016): 460–74.

212 *Modern human adolescents and young adults around the world disperse for the first time in a number of ways:* For lively and historical accounts of home-leaving around the world, see Steven Mintz's *The Prime of Life: A History of Modern Adulthood* (Cambridge, MA: Harvard University Press, 2015), Prologue, 1–18, Chapter 1: The Tangled Transition to Adulthood, 19–70, as well as his book *Huck's Raft: A History of American Childhood.* In addition, Jeffrey Jensen Arnett's work on emerging adulthood, including *Adolescence and Emerging Adulthood: A Cultural Approach* (London: Pearson, 2012), contains extensive discussions of the phase of life in which adolescents leave home—or don't. Also see Arnett's *Emerging Adulthood: The Winding Road from the Late Teens Through the Twenties,* 2nd ed. (New York: Oxford, 2015).

212 *At one extreme are Australian possums:* Interview with Dr. Hannah Bannister, February 6, 2018.

212 *At the other are* Parus varius *birds:* Hiroyoshi Higuchi and Hiroshi Momose, "Deferred independence and prolonged infantile behaviour in young varied tits, Parus varius, of an island population," *Animal Behaviour* 28 (1981): 523–24.

212 *males tend to:* Russell C. Van Horn, Teresa L. McElinny, and Kay E. Holekamp, "Age estimation and dispersal in the spotted hyena (*Crocuta crocuta*)," *Journal of Mammology* 84 (2003): 1019–30; Axelle E. J. Bono et al., "Payoff- and sex-biased social learning interact in a wild primate population," *Current Biology* 28 (2018): P2800–2805; Gerald L. Kooyman and Paul J. Ponganis, "The initial journey of juvenile emperor penguins," *Aquatic Conservation: Marine and Freshwater Ecosystems* 17 (2008): S37–S43; Robin W. Baird and Hal Whitehead, "Social organization of mammal-eating killer whales: Group stability and dispersal patterns," *Canadian Journal of Zoology* 78 (2000): 2096–105; P. A. Stephens et al., "Dispersal, eviction, and conflict in meerkats (*Suricata suricatta*): An evolutionarily stable strategy model," *American Naturalist* 165 (2005): 120–35.

213 *females who seek their fortunes:* Namibia Wild Horse Foundation, "Social Structure," http://www.wild-horses-namibia.com/social-structure/; Frans B. M. De Waal, "Bonobo Sex and Society," *Scientific American,* June 1, 2006,

https://www.scientificamerican.com/article/bonobo-sex-and-society-2006-06/.

213 *Biologists believe that the drive to disperse:* Martha J. Nelson-Flower et al., "Inbreeding avoidance mechanisms: Dispersal dynamics in cooperatively breeding southern pied babblers," *Journal of Animal Ecology* 81 (2012): 876–83; Nils Chr. Stenseth and William Z. Lidicker, Jr., *Animal Dispersal: Small Mammals as a Model* (Dordrecht, Netherlands: Springer Science+Business Media, 1992).

213 *Slavc, the wolf:* James Cheshire, "The Wolf Who Traversed the Alps," 62–65, in *Where the Animals Go*; interview with Hubert Potočnik, February 20, 2019.

214 *Many mammals, birds, and even fish species:* J. Michael Reed et al., "Informed Dispersal," in *Current Ornithology* 15, ed V. Nolan, Jr., and Charles F. Thompson (New York: Springer, 1999): 189–259; J. Clobert et al., "Informed dispersal, heterogeneity in animal disperal syndromes and the dynamics of spatially structured populations," *Ecology Letters* 12 (2009): 197–209.

214 *Possums are a good example:* Interview with Dr. Hannah Bannister, February 6, 2018.

215 *Wolves, with their exquisitely complex societies:* L. David Mech and Luigi Boitani, *Wolves: Behavior, Ecology, and Conservation* (Chicago: University of Chicago Press, 2007), 12.

215 *hunting school—or what wolf expert:* Ibid., 52.

215 *pumas are solitary as adults:* Kenneth A. Logan and Linda L. Sweanor, *Desert Puma: Evolutionary Ecology and Conservation of an Enduring Carnivore* (Washington, DC: Island Press, 2001), 143, 278; T. M. Caro and M. D. Hauser, "Is there teaching in nonhuman animals?" *Quarterly Review of Biology* 67 (1992): 151–74; L. Mark Elbroch and Howard Quigley, "Observations of wild cougar (*Puma concolor*) kittens with live prey: Implications for learning and survival," *Canadian Field-Naturalist* 126 (2012): 333–35.

216 *African elephants are orphaned:* Shifra Z. Goldenberg and George Wittemyer, "Orphaned female elephant social bonds reflect lack of access to mature adults," *Scientific Reports* 7 (2017): 14408; Shifra Z. Goldenberg and George Wittemyer, "Orphaning and natal group dispersal are associated with social costs in female elephants," *Animal Behaviour* 143 (2018): doi: 10.1016/j.anbehav.2018.07.002.

216 *Nestling barn owls have distinctive soft white plumage:* Alexandre Roulin, "Delayed maturation of plumage coloration and plumage spottedness in the Barn Owl (Tyto alba)," *Journal fur Ornithologie* 140 (1999): 193–97.

216 *Humans who revert:* Hermioni N. Lokko and Theodore A. Stern, "Regression: Diagnosis, evaluation, and management," *Primary Care Companion for CNS Disorders* 17 (2015): doi: 10.408/PCC.14f01761.

217 *so-called nest helpers:* Walter D. Koenig et al., "The Evolution of Delayed

Dispersal in Cooperative Breeders," *The Quarterly Review of Biology* 67 (1992): 111–50; Lyanne Brouwe, David S. Richardson, and Jan Komdeur, "Helpers at the nest improve late-life offspring performance: Evidence from a long-term study and a cross-foster experiment," *PLoS ONE* 7 (2012): e33167; J. L. Brown, *Helping Communal Breeding in Birds* (Princeton, NJ: Princeton University Press, 2014), 91–101.

217 *Wolves have been seen:* L. David Mech and H. Dean Cluff, "Prolonged intensive dominance behavior between gray wolves, *Canis lupus*," *Canadian Field-Naturalist* 124 (2010): 215–18.

217 *Young adult rodents sometimes put up a fight:* Clutton-Brock, *Mammal Societies*, 186; Robert L. Trivers, "Parent-offspring conflict," *American Zoologist* 14 (1974): 249–64; Bram Kujiper and Rufus A. Johnstone, "How dispersal influences parent-offspring conflict over investment," *Behavioral Ecology* 23 (2012): 898–906.

217 *Puma mothers may also:* Logan and Sweanor, *Desert Puma*, 143.

218 *Since the 1970s, biologists:* Robert L. Trivers, "Parental Investment and Sexual Selection," 52–95, in Bernard Campbell, ed., *Sexual Selection and the Descent of Man, 1871–1971* (Chicago: Aldine, 1972), http://roberttrivers.com/Robert_Trivers/Publications_files/Trivers%201972.pdf.

218 *how and when animals leave:* Trivers, "Parent-offspring conflict"; Kujiper and Johnstone, "How dispersal influences parent-offspring conflict over investment." For more on parent-offspring conflict, see also Phil Reed, "A transactional analysis of changes in parent and chick behavior prior to separation of herring gulls (*Larus argentatus*): A three-term contingency model," *Behavioural Processes* 118 (2015): 21–27; T. H. Clutton-Brock and G. A. Parker, "Punishment in animal societies," *Nature* 373 (1995): 209–16.

219 *The Spanish imperial eagles of Spain's:* Juan Carlos Alonso et al., "Parental care and the transition to independence of Spanish Imperial Eagles *Aquila heliaca* in Doñana National Park, southwest Spain," *IBIS* 129 (1987): 212–24.

220 Born a Crime, *his memoir:* Trevor Noah, *Born a Crime: Stories from a South African Childhood* (New York: Spiegel and Grau, 2016), 255.

221 *A study of free-ranging dogs:* Manabi Paul et al., "Clever mothers balance time and effort in parental care—a study on free-ranging dogs," *Royal Society Open Science* 4 (2017): 160583.

222 *"In the dry mountains of the Granite Range":* Tim Clutton-Brock, *Mammal Societies* (Hoboken, NJ: Wiley-Blackwell, 2016), 94.

223 *in vervet monkey societies females:* Dorothy L. Cheney and Robert M. Seyfarth, "Nonrandom dispersal in free-ranging vervet monkeys: Social and genetic consequences," *American Naturalist* 122 (1983): 392–412.

223 *in the words of one vervet expert, "depressed":* Interview with Lynn Fairbanks, May 3, 2011.

223 *Biochemically, mammals like us experience:* J. Kolevská, V. Brunclík, and M. Svoboda, "Circadian rhythm of cortisol secretion in dogs of different daily activity," *Acta Veterinaria Brunensis* 72 (2002), doi: 10/2754/abc200372040599; Mark S. Rea et al., "Relationship of morning cortisol to circadian phase and rising time in young adults with delayed sleep times," *International Journal of Endocrinology* (2012), doi://10.115/2012/74940; R. Thun et al., "Twenty-four-hour secretory pattern of cortisol in the bull: Evidence of episodic secretion and circadian rhythm," *Endocrinology* 109 (1981): 2208–12.

224 *Motor vehicles are the leading cause:* World Health Organization, "Adolescent Health Epidemiology," https://www.who.int/maternal_child_adolescent/epidemiology/adolescence/en/.

224 *motor vehicles ravage animal populations:* John Boulanger and Gordon B. Stenhouse, "The impact of roads on the demography of grizzly bears in Alberta," *PLoS ONE* 9 (2014): e115535; Amy Haigh, Ruth M. O'Riordan, and Fidelma Butler, "Hedgehog *Erinaceus europaeus* mortality on Irish roads," *Wildlife Biology* 20 (2014): 155–60; Ronald L. Mumme et al., "Life and death in the fast lane: Demographic consequences of road mortality in the Florida scrub-jay," *Conservation Biology* 14 (2000): 501–12; Brenda D. Smith-Patten and Michael A. Patten, "Diversity, seasonality, and context of mammalian roadkills in the Southern Great Plains," *Environmental Management* 41 (2008): 844–52; Brendan D. Taylor and Ross L. Goldingay, "Roads and wildlife: Impacts, mitigation and implications for wildlife management in Australia," *Wildlife Research* 37 (2010): 320–31; Amy Haigh et al., "Non-invasive methods of separating hedgehog (*Erinaceus europaeus*) age classes and an investigation into the age structure of road kill," *Acta Theriologica* 59 (2014): 165–71; Richard M. F. S. Sadleir and Wayne L. Linklater, "Annual and seasonal patterns in wildlife road-kill and their relationship with traffic density," *New Zealand Journal of Zoology* 43 (2016): 275–91; Evan R. Boite and Alfred J. Mead, "Application of GIS to a baseline survey of vertebrate roadkills in Baldwin County, Georgia," *Southeastern Naturalist* 13 (2014): 176–90; Changwan Seo et al., "Disentangling roadkill: The influence of landscape and season on cumulative vertebrate mortality in South Korea," *Landscape and Ecological Engineering* 11 (2015): 87–99; interview with Andy Alden, senior research associate, Virginia Tech, August 23, 2017; interview with Bridget Donaldson, a senior scientist with Virginia Transportation Research Council and expert on wildlife crossings, August 14, 2017; interview with Fraser Shilling, co-director of the UC Davis Road Ecology Center, August 9, 2017.

224 *On American highways:* Malia Wollan, "Mapping Traffic's Toll in Wildlife," *New York Times*, September 12, 2010, https://www.nytimes.com/2010/09/13/technology/13roadkill.html.

224 *Vehicles take out native pukeko birds:* Richard M. F. S. Sadleir and Wayne

L. Linklater, "Annual and seasonal patterns in wildlife road-kill and their relationship with traffic density," *New Zealand Journal of Zoology* 43 (2016): 275–91.

224 *possums in Australia:* R. A. Giffney, T. Russell, and J. L. Kohen, "Age of road-killed common brushtail possums (Trichosurus vulpecula) and common ringtail possums (Pseudocheirus peregrinus) in an urban environment," *Australian Mammalogy* 31 (2009): 137–42.

224 *elephant seals along Highway 1:* Kerry Klein, "Largest US Roadkill Database Highlights Hotspots on Bay Area Highways," *Mercury News* (San Jose), May 5, 2015, https://www.mercurynews.com/2015/05/05/largest-u-s-roadkill-database-highlights-hotspots-on-bay-area-highways/.

224 *meerkats in the Kalahari:* Nicolas Perony and Simon W. Townsend, "Why did the meerkat cross the road? Flexible adaptation of phylogenetically-old behavioural strategies to modern-day threats," *PLoS ONE* (2013), doi: 10.1371/journal.pone.0052834.

224 *Even adolescent whales growing up:* Whale and Dolphin Conservation, "Boat Traffic Effects on Whales and Dolphins," https://us.whales.org/issues/boat-traffic; A. Szesciorka et al., "Humpback whale behavioral response to ships in and around major shipping lanes off San Francisco, CA," Abstract (Proceedings) 21st Biennial Conference on the Biology of Marina Mammals, San Francisco, California, December 14–18, 2015; Karen Romano Young, *Whale Quest: Working Together to Save Endangered Species* (Brookfield, CT: Millbrook Press, 2017).

224 *Some urban coyotes have even learned:* Christine Dell'Amore, "Downtown Coyotes: Inside the Secret Lives of Chicago's Predator," *National Geographic*, November 21, 2014, https://news.nationalgeographic.com/news/2014/11/141121-coyotes-animals-science-chicago-cities-urban-nation/.

225 *motor vehicle is the single deadliest activity:* Centers for Disease Control and Prevention, "Motor Vehicle Safety (Teen Drivers)," https://www.cdc.gov/motorvehiclesafety/teen_drivers/index.html; Centers for Disease Control and Prevention, "Motor Vehicle Crash Deaths," https://www.cdc.gov/vitalsigns/motor-vehicle-safety/index.html; Children's Hospital of Philadelphia, "Seat Belt Use: Facts and Stats," https://www.teendriversource.org/teen-crash-risks-prevention/rules-of-the-road/seat-belt-use-facts-and-stats.

225 *Texting while driving is a dangerous:* National Highway Traffic Safety Administration, "Overview of the National Highway Traffic Safety Administration's Driver Distraction Program," https://www.nhtsa.gov/sites/nhtsa.dot.gov/files/811299.pdf.

225 *according to a 2012 report:* National Highway Traffic Safety Administration, "U.S. DOT and NHTSA Kick Off 5th Annual U Drive. U Text. U Pay. Cam-

paign," April 5, 2018, https://www.nhtsa.gov/press-releases/us-dot-and-nhtsa-kick-5th-annual-u-drive-u-text-u-pay-campaign.

226 *News reports and eyewitnesses later described:* Angel Jennings, "Mountain Lion Killed in Santa Monica Was Probably Seeking a Home," *Los Angeles Times*, May 24, 2012, http://articles.latimes.com/2012/may/24/local/la-me-0524-mountain-lion-20120524.

228 *Potočnik had worried all night:* Interview with Hubert Potočnik, February 20, 2019.

229 *Conservation scientists who monitor:* James Cheshire and Oliver Uberti, *Where the Animals Go: Tracking Wildlife with Technology in 50 Maps and Graphics* (New York: W. W. Norton & Company, 2017); Doug P. Armstrong et al., "Using radio-tracking data to predict post-release establishment in reintroduction to habitat fragments," *Biological Conservation* 168 (2013): 152–60.

CHAPTER 16: MAKING A LIVING

232 *According to the Urban Institute:* Susan J. Popkin, Molly M. Scott, and Martha Galvez, "Impossible Choices: Teen and Food Insecurity in America," Urban Institute, September 2016, https://www.urban.org/sites/default/files/publication/83971/impossible-choices-teens-and-food-insecurity-in-america_1.pdf; Mkael Symmonds et al., "Metabolic state alters economic decision making under risk in humans," *PLoS ONE* 5 (2010): e11090; No Kid Hungry, "Hunger Facts," https://www.nokidhungry.org/who-we-are/hunger-fact.

232 *hungry individuals are forced to take greater risks:* Stan Boutin, "Hunger makes apex predators do risky things," *Journal of Animal Ecology* 87 (2018): 530–32; Andrew D. Higginson et al., "Generalized optimal risk allocation: Foraging and antipredator behavior in a fluctuating environment," *American Naturalist* 180 (2012): 589–603; Michael Crossley, Kevin Staras, and György Kemenes, "A central control circuit for encoding perceived food value," *Science Advances* 4 (2018), doi: 10.1126/sciadv.aau9180; Kari Koivula, Seppo Rytkonen, and Marukku Orell, "Hunger-dependency of hiding behaviour after a predator attack in dominant and subordinate willow tits," *Ardea* 83 (1995): 397–404; Benjamin Homberger et al., "Food predictability in early life increases survival of captive grey partridges (*Perdix perdix*) after release into the wild," *Biological Conservation* 177 (2014): 134–41; Hannah Froy et al., "Age-related variation in foraging behavior in the wandering albatross in South Georgia: No evidence for senescence," *PLoS ONE* 10 (2015): doi: 10.1371/journal.pone.0116415; Daniel O'Hagan et al., "Early life disadvantage strengthens flight performance trade-off in European starlings, *Sturnus vulgaris*," *Animal Behaviour* 102 (2015): 141–48; Harry H. Marshall,

"Lifetime fitness consequences of early-life ecological hardship in a wild mammal population," *Ecology and Evolution* 7 (2017): 1712–24; Clare Andrews et al., "Early life adversity increases foraging and information gathering in European starlings, *Sturnus vulgaris*," *Animal Behaviour* 109 (2015): 123–32; Gerald Kooyman and Paul J. Ponganisk, "The initial journey of juvenile emperor penguins," *Aquatic Conservation: Marine and Freshwater Ecosystems* 17 (2007): S37–S43; Richard A. Phillips et al., "Causes and consequences of individual variability and specialization in foraging and migration strategies of seabirds," *Marine Ecology Progress Series* 578 (2017): 117–50.

234 *some of their genes are activated:* Tiffany Armenta et al., "Gene expression shifts in yellow-bellied marmots prior to natal dispersal," *Behavioral Ecology* ary175 (2018), doi: 10.1083/beheco/ary175.

234 *Before migration, many mammals begin to store fat:* Armenta et al., "Gene expression shifts in yellow-bellied marmots prior to natal dispersal."

235 *A week into his journey:* Interview with Hubert Potočnik, February 20, 2019.

235 *According to wolf expert David Mech:* Mech and Boitani, *Wolves*, 283.

235 *Being able to hunt a deer:* Barbara Natterson-Horowitz's visit to the Endangered Wolf Center in Eureka, Missouri, on April 20, 2018.

235 *Ben Kilham, a black bear conservationist:* Interview with Dr. Ben Kilham in Lyme, New Hampshire, April 1, 2018.

236 *Inexperienced meerkats often fumble:* Alex Thornton, "Variations in contributions to teaching by meerkats," *Proceedings of the Royal Society B: Biological Sciences* 275 (2008): 1745–51.

236 *On average, lions hunting on the Serengeti:* James Fair, "Hunting success rates: how predators compare," Discover Wildlife, December 17, 2015, http://www.discoverwildlife.com/animals/hunting-success-rates-how-predators-compare.

236 *For Indian tigers and Arctic polar bears:* Ibid.

236 *Capuchin monkeys, for example:* Amanda D. Melin et al., "Trichromacy increases fruit intake rates of wild capuchins (Cebus capucinus imitator)," *Proceedings of the National Academy of Sciences* 114 (2017): 10402–7.

236 *Sockeye salmon, when they're juveniles:* Inigo Novales Flamarique, "The Ontogeny of Ultraviolet Sensitivity, Cone Disappearance and Regeneration in the Sockeye Salmon Oncorhynchus Nerka," *Journal of Experimental Biology* 203 (2000): 1161–72.

237 *two Simon Fraser University scientists:* Howard Richardson and Nicolaas A. M. Verbeek, "Diet selection and optimization by Northwestern Crows feeding on Japanese littleneck clams," *Ecology* 67 (1986): 1219–26; Howard Richardson and Nicolaas A. M. Verbeek, "Diet selection by yearling Northwestern Crows (Corvus caurinus) feeding on littleneck clams (Venerupis japonica)," *Auk* 104 (1987): 263–69.

238 *psychologist Angela Duckworth calls "grit":* Angela Duckworth, *Grit: The Power of Passion and Perseverance* (New York: Scribner, 2016).

238 *From hyenas to tropical blackbirds:* Sarah Benson-Amram and Kay E. Holekamp, "Innovative problem solving by wild spotted hyenas," *Proceedings of the Royal Society B* 279 (2012): 4087–95; L. Cauchard et al., "Problem-solving performance is correlated with reproductive success in wild bird population," *Animal Behaviour* 85 (2013): 19–26; Andrea S. Griffin, Maria Diquelou, and Marjorie Perea, "Innovative problem solving in birds: a key role of motor diversity," *Animal Behaviour* 92 (2014): 221–27; A. Thornton and J. Samson, "Innovative problem solving in wild meerkats," *Animal Behaviour* 83 (2012): 1459–68.

238 *Some hyenas continue struggling:* Benson-Amram and Holekamp, "Innovative problem solving by wild spotted hyenas."

239 *Squirrels who kept at the challenging task:* Lisa A. Leaver, Kimberly Jayne, and Stephen E. G. Lea, "Behavioral flexibility versus rules of thumb: How do grey squirrels deal with conflicting risks?" *Behavioural Ecology* 28 (2017): 186–92.

239 *When white-tailed ptarmigan mother hens:* John Whitfield, "Mother hens dictate diet," *Nature* (2001), doi: 10/1038/news010719-18, https://www.nature.com/news/2001/010718/full/news010719-18.html.

239 *When lambs graze with their ewe mothers:* A. G. Thorhallsdottir, F. D. Provenza, D. F. Balph, "Ability of lambs to learn about novel foods while observing or participating with social models," *Applied Animal Behaviour Science* 25 (1990): 25–33; Udita Sanga, Frederick D. Provenza, and Juan J. Villalba, "Transmission of self-medicative behaviour from mother to offspring in sheep," *Animal Behaviour* 82 (2011): 219–27.

240 *Studies of parental influence on eating behavior:* Jennifer S. Savage, Jennifer Orlet Fisher, and Leann L. Birch, "Parental Influence on Eating Behavior: Conception to Adolescence," *Journal of Law, Medicine & Ethics* 35 (2007): 22–34.

240 *Orcas use a technique called "stranding":* Christophe Guinet, "Intentional stranding apprenticeship and social play in killer whales (Orcinus orca)," *Canadian Journal of Zoology* 69 (1991): 2712–16.

241 *Humpback whales like Salt catch fish:* Ari Friedlaender et al., "Underwater components of humpback whale bubble-net feeding behaviour," *Behaviour* 148 (2011): 575–602; Rebecca Boyle, "Humpback Whales Learn New Tricks Watching Their Friends," *Popular Science*, April 25, 2013, https://www.popsci.com/science/article/2013-04/humpback-whales-learn-new-tricks-watching-their-friends#page-2; Jane J. Lee, "Do Whales Have Culture?" National Geographic News, April 27, 2013, https://news.nationalgeographic.com/news/2013/13/130425-humpback-whale-culture-behavior-science-animals/; University of St. Andrews, "Humpback whales able to learn from others,

study finds," Phys.org, April 25, 2013, https://phys.org/news/2013-04-hump back-whales.html#jCp.

241 *Called "lobtail feeding," it involved:* Jenny Allen et al., "Network-based diffusion analysis reveals cultural transmission of lobtail feeding in humpback whales," *Science* 26 (2013): 485–88.

242 *Cheetah mothers bring their kittens:* William J. E. Hoppitt et al., "Lessons from animal teaching," *Trends in Ecology & Evolution* 23 (2008): 486–93, 486; T. M. Caro and M. D. Hauser, "Is there teaching in nonhuman animals?" *Quarterly Review of Biology* 67 (1992): 151–74; T. M. Caro, "Predatory behaviour in domestic cat mothers," *Behaviour* 74 (1980): 128–47; T. M. Caro, "Effects of the mother, object play and adult experience on predation in cats," *Behavioral and Neural Biology* 29 (1980): 29–51; T. M. Caro, "Short-term costs and correlates of play in cheetahs," *Animal Behaviour* 49 (1995): 333–45.

242 *puma mothers provide fawns:* Mark Elbroch, "Fumbling Cougar Kittens: Learning to Hunt," National Geographic Blog, October 22, 2014, https://blog .nationalgeographic.org/2014/10/22/fumbling-cougar-kittens-learning -to-hunt/.

242 *Carnivorous meerkats send their young:* Liz Langley, "Schooled: Animals That Teach Their Young," *National Geographic News*, May 7, 2016, https://news .nationalgeographic.com/2016/05/160507-animals-teaching-parents-sci ence-meerkats/.

242 *Predator training is part:* Alonso et al., "Parental care and the transition to independence of Spanish Imperial Eagles *Aquila heliaca* in Doñana National Park, southwest Spain."

243 *A study of leopards:* Guy A. Balme et al., "Flexibility in the duration of parental care: Female leopards prioritise cub survival over reproductive output," *Journal of Animal Ecology* 86 (2017): 1224–34.

244 *When banded mongooses in Uganda:* J. S. Gilchrist, "Aggressive monopolization of mobile carers by young of a cooperative breeder," *Proceedings of the Royal Society B* 275 (2008): 2491–98.

244 *if young Norway rats are given a choice:* Yutaka Hishimura, "Food choice in rats (*Rattus norvegicus*): The effect of exposure to a poisoned conspecific," *Japanese Psychological Research* 40 (1998): 172–77; Jerry O. Wolff and Paul W. Sherman, eds., *Rodent Societies: An Ecological and Evolutionary Perspective* (Chicago: University of Chicago Press, 2007), 210–11; Bennett G. Galef, Jr., "Social interaction modifies learned aversions, sodium appetite, and both palatability and handling-time induced dietary preferences in rats (*Rattus norvegicus*)," *Journal of Comparative Psychology* 100 (1986): 432–39.

244 *when rats reached puberty:* Pallav Sengupta, "The laboratory rat: Relating its age with human's," *International Journal of Preventive Medicine* 4 (2013): 624–30.

244 *adopting peers' food preferences extended to poison*: Galef Jr., "Social interaction modifies learned aversions, sodium appetite, and both palatability and handling-time induced dietary preferences in rats (*Rattus norvegicus*)."

245 *Rats carry scent clues:* Jerry O. Wolff and Paul W. Sherman, eds., *Rodent Societies: An Ecological and Evolutionary Perspective* (Chicago: University of Chicago Press, 2007), 211.

245 *When male fossas:* Interview with Luke Dollar, wildlife biologist and conservationist, November 10, 2017.

245 *"By pairing up, male fossas":* Interview with Mia-Lana Lührs, University of Gottingen, October 16, 2017.

CHAPTER 17: THE GREAT ALONE

248 *For Inuit boys:* BBC Two, "Apak: North Baffin Island," February 21, 2005, http://news.bbc.co.uk/2/hi/programmes/this_world/4270079.stm; Nina Strochlic, "How to Build an Igloo," *National Geographic*, November 2016, https://www.nationalgeographic.com.au/people/how-to-build-an-igloo.aspx; Richard G. Condon, "Inuit Youth in a Changing World," *Cultural Survival Quarterly Magazine*, June 1988, https://www.culturalsurvival.org/publications/cultural-survival-quarterly/inuit-youth-changing-world.

248 *Traditional Australian aboriginal walkabouts:* Julie Tetel Andresen and Phillip M. Carter, "The Language Loop: The Australian Walkabout," in *Language in the World: How History, Culture, and Politics Shape Language* (Hoboken, NJ: Wiley-Blackwell, 2016), 22.

248 *one stage of the Lakota vision quest:* David Martinez, "The soul of the Indian: Lakota philosophy and the vision quest," *Wicazo Sa Review* (University of Minnesota Press) 19 (2004): 79–104.

248 *Adolescents and young adults in modern militaries:* GoSERE, "SERE: Survival, Evasion, Resistance and Escape," https://www.gosere.af.mil/; National Outdoor Leadership School, "The leader in wilderness education," https://www.nols.edu/en/.

249 *a naturally greater preference for solitude:* Ester S. Buchholz and Rochelle Catton, "Adolescents' perceptions of aloneness and loneliness," *Adolescence* 34 (1999): 203–13.

249 *But an ongoing sense of isolation:* Bridget Goosby et al., "Adolescent loneliness and health in early adulthood," *Sociological Inquiry* 83 (2013): doi: 10.1111/soin.12018.

249 *a risk factor for suicide:* Cheryl A. King and Christopher R. Merchant, "Social and interpersonal factors relating to adolescent suicidality: A review of the literature," *Archives of Suicide Research* 12 (2008): 181–96.

249 *As of 2016, eighteen- to thirty-four-year-olds:* Jonathan Vespa, "A Third of Young

Adults Live with Their Parents," United States Census Bureau, August 9, 2017, https://www.census.gov/library/stories/2017/08/young-adults.html.

249 *More than 60 percent of Polish . . . And in most countries in the Middle East:* "Europe's Young Adults Living with Parents—a Country by Country Breakdown," *The Guardian*, March 24, 2014, https://www.theguardian.com/news/datablog/2014/mar/24/young-adults-still-living-with-parents-europe-country-breakdown; Morgan Winsor, "Why Adults in Different Parts of the Globe Live with Their Parents," ABC News, May 27, 2018, https://abcnews.go.com/International/adults-parts-globe-live-home-parents/story?id=55457188; "Life in Modern Cairo," Liberal Arts Instructional Technology Services, University of Texas at Austin, http://www.laits.utexas.edu/cairo/modern/life/life.html.

249 *twenty-two- to twenty-nine-year-olds:* CBRE, "Asia Pacific Millennials: Shaping the Future of Real Estate," October 2016, page 8, https://www.austchamthailand.com/resources/Pictures/CBRE%20-%20APAC%20Millennials%20Survey%20Report.pdf.

250 *The ecological term for these:* Extended parental care includes feeding, sheltering, protecting, and guiding post-dispersed young. It is seen across a wide range of animal species, but varies in degree depending on offspring needs and parental resources. In general, when environments are dangerous because of predation or lack of resources, more extended care is seen. Eleanor M. Russell, Yoram Yom-Tov, and Eli Geffen, "Extended parental care and delayed dispersal: Northern, tropical and southern passerines compared," *Behavioral Ecology* 15 (2004): 831–38; Andrew N. Radford and Amanda R. Ridley, "Recruitment calling: A novel form of extended parental care in an altricial species," *Current Biology* 16 (2006): 1700–704; Michael J. Polito and Wayne Z. Trivelpiece, "Transition to independence and evidence of extended parental care in the gentoo penguin (*Pygoscelis papua*)," *Marine Biology* 154 (2008): 231–40; P. D. Boersma, C. D. Cappello, and G. Merlen, "First observation of post-fledging care in Galapogos penguins (*Spheniscus mendiculus*)," *Wilson Journal of Ornithology* 129 (2017): 186–91; Martin U. Gruebler and Beat Naef-Daenzer, "Survival benefits of post-fledging care: Experimental approach to a critical part of avian reproductive strategies," *Journal of Animal Ecology* 79 (2010): 334–41.

250 *Steven Mintz, a historian of the human life cycle:* Steven Mintz, *The Prime of Life* (Cambridge, MA: Harvard University Press, 2015).

250 *"Contrary to what many people assume":* Steven Mintz, *The Prime of Life* (Cambridge, MA: Harvard University Press, 2015).

251 *"young men generally had to delay":* Ibid.

251 *In many species of birds and mammals:* Lyanne Brouwe, David S. Richardson, and Jan Komdeur, "Helpers at the nest improve late-life offspring perfor-

mance: Evidence from a long-term study and a cross-foster experiment," *PLoS ONE* 7 (2012): e33167; Tim Clutton-Brock, "Cooperative Breeding," in *Mammal Societies* (Hoboken, NJ: Wiley-Blackwell, 2016), 556–63.

251 *Western bluebird sons:* Janis L. Dickinsin et al., "Delayed dispersal in western bluebirds: Teasing apart the importance of resources and parents," *Behavioral Ecology* 25 (2014): 843–51.

252 *North American red squirrel mothers:* Karen Price and Stan Boutin, "Territorial bequeathal by red squirrel mothers," *Behavioral Ecology* 4 (1992): 144–50.

252 *Parental excursions, a behavior:* De Casteele and Matthysen, "Natal dispersal and parental escorting predict relatedness between mates in a passerine bird," *Molecular Ecology* 15, no. 9 (August 2006), 2557–65.

252 *Like the social-climbing mothers:* Erik Matthysen et al., "Family movements before independence influence natal dispersal in a territorial songbird," *Oecologia* 162 (2010): 591–97; Karen Marchetti and Trevor Price, "Differences in the foraging of juvenile and adult birds: The importance of developmental constraints," *Biological Review* 64 (1989): 51–70; S. Choudhury and J. M. Black, "Barnacle geese preferentially pair with familiar associates from early life," *Animal Behaviour* 48 (1994): 81–88.

252 *A study of white-winged choughs:* I. Rowley, "Communal activities among white-winged choughs, Corcorax melanorhamphus," *IBIS* 120 (1978): 178–96; R. G. Heinsohn, "Cooperative enhancement of reproductive success in white-winged choughs," *Evolutionary Ecology* 6 (1992): 97–114; R. Heinsohn et al., "Coalitions of relatives and reproductive skew in cooperatively breeding white-winged choughs," *Proceedings of the Royal Society of London Series B* 267 (2000): 243–49.

252 *Young Mexican jays:* Jack F. Cully, Jr., and J. David Ligon, "Comparative Mobbing Behavior of Scrub and Mexican Jays," *Auk* 93 (1976): 116–25.

253 *A report from the Harvard Graduate School of Education:* Leah Shafer, "Resilience for Anxious Students," Harvard Graduate School of Education, November 30, 2017, https://www.gse.harvard.edu/news/uk/17/11/resilience-anxious-students.

253 *Mintz puts it like this:* Mintz, *The Prime of Life*.

253 *According to an analysis:* Allison E. Thompson, Johanna K. P. Greeson, and Ashleigh M. Brunsink, "Natural mentoring among older youth in and aging out of foster care: A systematic review," *Children and Youth Services Review* 61 (2016): 40–50.

255 *Mintz puts it even more bluntly:* Mintz, *The Prime of Life*.

255 *Post-release monitoring:* Doug P. Armstrong et al., "Using radio-tracking data to predict post-release establishment in reintroduction to habitat fragments," *Biological Conservation* 168 (2013): 152–60.

255 *Mark Elbroch, a wildlife biologist:* Mark Elbroch, "Fumbling Cougar Kittens:

Learning to Hunt," *National Geographic Blog,* October 22, 2014, https://blog.nationalgeographic.org/2014/10/22/fumbling-cougar-kittens-learning-to-hunt/.

EPILOGUE

263 *King penguins in the wild:* "King Penguins," Penguins-World, https://www.penguins-world.com/king-penguin/.

263 *Shrink's body was found:* Interview with Oliver Höner, October 4, 2018.

263 *Salt has become one of the most beloved:* Philip Hoarse, "'Barnacled Angels': The Whales of Stellwagen Bank—a Photo Essay," *Guardian,* June 20, 2018, https://www.theguardian.com/environment/2018/jun/20/barnacled-angels-the-whales-of-stellwagen-bank-a-photo-essay.

264 *the wolf has been spotted:* Interview with Hubert Potočnik, February 20, 2019.

266 *Tesla is installing its new "Megapack":* Fred Lambert, "Tesla and PG&E Are Working on a Massive 'Up to 1.1 GWh' Powerpack Battery System," Electrek, June 29, 2018, https://electrek.co/2018/06/29/tesla-pge-giant-1-gwh-powerpack-battery-system/.

Note About the Illustrations

Oliver Uberti is a former senior design editor for *National Geographic* and the coauthor of two critically acclaimed books of maps and graphics: *Where the Animals Go* and *London: The Information Capital*, each of which won the top British Cartographic Society Award for cartographic excellence.

Assisted by technologies from satellites to drones, humans can now as never before witness the daily lives of animals. For the maps in *Wildhood*, Uberti used geo-location provided by the scientists who tracked our four main animal protagonists. The stories of Ursula, Shrink, Salt, and Slavc would not be known without the efforts of scientists from the Antarctic Research Trust (antarctic-research.de), the Hyena Project (hyena-project.com), Center for Coastal Studies (coastalstudies.org), and Project Slowolf (volkovi.si). Uberti's graphic illustrating the stages of wildhood is based on our interpretations of our own research and phylogenies of adolescent behavior.

Because Uberti's images are so beautiful—and make understanding the information so effortless—it would be easy to get the impression that the data are obvious or easy to obtain. In fact, every single data point is the result of dedicated, multidecade research by scientists and their teams braving extremes of temperature, terrain, distance, and resources all over the globe. Although technological advances give a new perspective, observing animal behavior in the field still depends on the passion and commitment of human individuals.

Index

abductions, 43–44, 69
aboriginal people, Australia, walkabouts by, 248
addiction, and social descent, 134
adolescence, use of term, 8
adultocentrism, 48–49, 172
African penguins, hunting of sardines by, 41
age
  as predictor of status rank, 107, 134
  pregnancy outcome and, 174–75
Agence France-Presse, 46
aggression. *See also* bullying
  dogs' expression of, 142–43
  sexual coercion using, 200, 201–2
alarm calling, 33, 82
  bats and, 81
  functions of, 69
  learning from parents about, 68–69
  as signal of unprofitability, 62, 64, 69
alarm duetting, 63–64
albatrosses, adolescent banding in, 186
albino animals, and oddity effect, 59
alcohol use and abuse
  adolescents and, 80, 253
  driving and, 225
  sexual activity and, 202, 204, 206
  social pain and, 134
ambush hunters, 23, 61
American Library Association (ALA), 182
American Psychological Association (APA), 204, 205
angels, hierarchies of, 99
angling play, by cats, 148–49
Antarctic Research Trust, 23–24
antler adorning, 177–78
anxiety. *See also* migration anxiety
  assessment overload and, 153–54
  bullying and, 138
  dog's expression of, 143
  hygiene hypothesis and, 36
  island tameness and, 35
  as mismatch disorder, 153
  social hierarchies related to, 102
appearance-based bullying, 60, 141
Aristotle, 46–47
armor, 31–32
Arnett, Jeffrey, 7
assessment, 93–104
  animals' use of, 96–97
  monitoring social rank changes with, 98–99
  nonstop, in human adolescents, 152–53
  status sanctuaries from, 154
assessment overload, 153–54
assess step, Predator's Sequence, 55, 60–64

association with high-status animals, 13, 109–10
athletes
  Freudian sublimation, 33
  status of, 111, 117, 118
attack step, Predator's Sequence, 55, 64–66
attractiveness, and social rank, 109–10
Austen, Jane, 181
autoimmune diseases, 35–36
automobiles, deaths from, 224–25
avoidance training, 68

baboons
  dispersal pattern of, 213
  grooming by, 123
  social rank in, 109
  submission and infertility in, 173
bait shyness, 43
bald eagles, viii
  courtship in, 162–63
  play courtship behavior by, 149
Bannister, Hannah, 215
bats
  courtship in, 163, 167
  dispersal pattern of, 213
  owls' assessment of, 60–61
  owls' attacks on, 64–65
  owls' killing of, 66
  predator inspection by, 81–82
  risk-taking by adolescents in, 28
  ultrasonic hearing of, 57, 65
bears
  breeding behavior in, 174
  farm-raised fish and, 72, 76
beavers, territory inheritance in, 124
behavioral changes, during adolescence, 49–50
Bekoff, Marc, 99–100
Bigg's orcas, 40–41, 44
Binghamton University, 204
bioinspiration, 264–65
birds. *See also specific birds*
  alarm calling by, 69
  behavioral changes in adolescence in, 49
  breeding behavior in, 175
  courtship singing by, 167
  delayed plumage maturation in, 216, 217
  depression in, 136
  dispersal patterns of, 212, 214, 217
  extended parental care by, 250, 252
  fall from dominant position by, 131–32
  grooming in, 108
  group movement against danger by, 74
  hierarchies in, 96
  individuality of, 100–101
  loser effect in, 136
  maternal intervention in, 123
  mobbing by, 69, 70
  motor-vehicle deaths in, 224
  nest-building by, 178–79
  nest helpers in, 217
  older birds' treatment of adolescents in, 50
  parental excursions by, 252
  parental meanness in delayed dispersal and, 219–20, 242–43
  predator deception used by, 42
  privilege and, 123
  rank inheritance in, 120–21
  sexual fluidity in, 185
  social descent and, 132
  social rank in, 101, 106, 108
  startle response in, 29
  zugunruhe (migration anxiety) in, 22
bison, 5
  association with high-status animals by, 109–10
  drinking hierarchy in, 137–38
  learning from wolf attacks on, 68
Blakemore, Sarah-Jayne, 7, 116
bluebirds, extended parental care in, 251

Blumstein, Daniel, 60
boars, sexual activity in, 203
body language, and social status, 112
bonobos
dispersal pattern of, 213
hierarchies in, 101
bottlenose dolphins, social rank in, 107
bowerbirds, nest-building by, 178–79
Bowlby, John, 7
brain
adolescent, 10–11
mood changes and, 132–33
Social Brain Network in, 114–15, 116–17, 131
social pain and, 134
status mapping and, 116–17
breeding. *See also* mating
cheetah example of, 187–88
Species Survival Plan (SSP) in, 186
Bridget Jones's Diary (Fielding), 181
British Antarctic Survey, 31
brown hares, signal of unprofitability by, 61–62
bullying, 138–44, 157–58
allies and friends as mitigating factor in, 147
appearance-based, 60, 141
conformers and, 139–42
depression and anxiety from, 138, 147
dominators and, 138–39
othering and, 142
quality advertisements against, 62
range of actions in, 138
redirection and, 142–44
status and, 139, 144
types of, in animals, 139
Burghardt, Gordon, 148–49
bustards, breeding of, 187
butterflies, courtship behavior in, 191, 201

canaries
courtship singing by, 168
early nutritional endowments in, 103
cancer, survival and relapse rates in adolescents, 48
capuchin monkeys
color vision in, 236
play-fighting by, 149
caribou, partner preference in, 187
cars, deaths from, 224–25
catfish, and oddity effect, 59
cats, viii
angling play with mice by, 148–49
sexual behavior in, 199, 203
sanctuaries for, 155
cattle, status and friendships among, 109
Centers for Disease Control and Protection, 47
cheetahs, viii
gazelles' inspection of, 82, 83, 84
hunting of gazelles by, 39–40, 63, 85
hunting of kudus by, 40
selective breeding of, 187–88
chemo-sensing, 56–57
chickens
pecking order in, 95–96, 99, 101, 108
social rank of roosters and crowing, 97
chimpanzees
fear response in, 30
knowledge of predators by, 42
sexual activity in, 201–2
social rank in, 107
choughs, extended parental care in, 252
clans
humpback whales, 161
spotted hyenas, 95, 105–6, 112
class, in human societies, 123–24
clownfish, hierarchies in, 109
Clutton-Brock, Tim, 122–23, 200, 222–23
coercion in sexual activity, 198–203
humans and, 202–3
signals from female and, 199–200
types of, 200
coho salmon, sexual behavior in, 201

coming-of-age traditions, 17, 47, 84, 232, 248
competence, and niche picking, 118
competitive groups, in humpback whale courtship, 189–90
conformer bullying, 139–42
confusion effect, 58, 59, 60
consent, in courtship, 198
cortisol, 116, 132, 223
courtship, 162–65
  adolescent social experiences needed for, 203
  adults' influence on offspring in, 184–85
  animals raised in isolation and, 202–3
  birds' nest-building in, 178–79
  changing human behaviors in, 179–82
  communication in, 203–4, 206–7
  competitive groups in, 189–90
  consent in, 198
  courtship control centers in, 195–96
  deer antler adorning in, 177–78
  dopamine in, 196–97
  examples of, 162–63
  fictional portrayals of, 181–82
  first sexual encounters in, 191–92
  fossa example of, 183–84, 245
  humpback chorusing in, 165–67, 168–69, 186, 188–89
  humpback whales and, 162, 164–65, 188–90
  intricate motor outputs in, 163
  lack of, and sexual activity, 200
  Lakota traditions in, 180
  learning of complexities of, 164, 176
  motivation in, 197–98
  play behavior for learning, 149
  sexual fluidity in, 185
  singing as part of, 167–69
  social learning in, 182, 184
courtship control centers, 195–96
cows
  breeding of, 188
  social hierarchy in, 137
  status and friendships among, 109
crabs
  courtship in, 200
  cuttlefish's mimicry of, 43
  sexual activity in, 200
  startle response in, 30
  status badges and, 109
crayfish
  dominance in, 130–31
  serotonin and status shift in, 133
  social rank in, 97
Crickmore, Michael, 193, 195, 196–98
crocodiles, viii
  predator deception used by, 43
  puberty in, 8
crowned hawk eagles, signal of unprofitability against, 62
crustaceans
  loser effect in, 136
  social rank in, 106, 130
  startle response in, 29
curiosity, as predator inspection and social learning, 84
cuttlefish, predator deception used by, 43

Damasio, Antonio, 7
danger, 21–28
  adolescents' learning about predators' habits, 53, 54–55
  environmental changes and emergence of new dangers, 37
  fear responses resulting in increase in, 36–37
  group movement against, 74
  king penguins staying safe on first day away from home, 21–24
  oddity effect increasing, 59
  predator inspection and, 83
  risk-taking by adolescents and, 27–28

self-confidence from encounters with, 76, 77
signals of unprofitability to dampen, 64
social learning and communicating about, 79
vagal response to, 57
vulnerable nature of adolescents and, 24–26
Darwin, Charles, 30, 34, 107
dating culture, 179, 180, 204
deaths of animals. *See also* mortality rates
motor vehicles and, 224–25
pregnancy-related complications and, 175
social learning from, 80
deer
antler adorning in, 177–78
challenges of early days without mothers and, 52–53
human hunting of, 41
play-fighting by, 150
rank inheritance in, 120
road-crossing skills of, 224
defense mechanisms, 32–34
animal use of, 33–34
description of types of human use of, 32–33
innate safety knowledge in, 33–34
Demaret, Albert, 132
depression
bullying and, 138, 147
monoamine hypothesis and, 129
privilege and, 126
self-assessment of rank in, 138
social descent and, 131–32
social hierarchies related to, 102
target animals and, 136
worthlessness feelings in, 136
detect step, Predator's Sequence, 55, 56–60
de Waal, Frans, 3, 113
Diamond, Marian, 7
Diana monkeys
alarm calling by, 62
predator-naive, 26
dispersal, 212–23
animal examples of, 212–13
animal reactions to, 222–23
benefits and downsides of, 213
coaching and preparations for, 214–16, 220, 221–22
death in new environment after, 226–27
definition of, 212
delayed, 216–17
environmental conditions affecting outcomes of, 260
human lessons learned from, 259–60
as inciting incident in animals' lives, 212
informed, 214–15, 259
motor-vehicle deaths during, 224
parental meanness in, 218, 219–20
parent-offspring conflict in, 218–19, 220
parents' assessment of offspring's readiness in, 221
prolonged immaturity and, 216
puma example of, 215–16
range of patterns in, 212–13
reaction to leaving home in, 224
repeated relocations after, 257
starvation risk during, 213–14, 215
stress of, 216
uninformed, 214
dogs, viii
behavioral changes in adolescence in, 50
dispersal pattern in, 221–22
fear response in, 30
kennel dog syndrome in, 143, 155
puppy license and, 51
redirection bullying by, 142–43
sexual activity in, 199, 203
shelters for, 155

dogs (*cont.*)
social brain of, and human interaction, 114–15
dolphins
group movement as response to danger by, 74
sexual behavior in, 201
social rank in, 107
dominance. *See also* status hierarchies
bullying for, 139
crayfish and, 130–31
environmental conditions and, 109
gestures related to, 112
lobster meral spread and, 130
prestige's interaction with, 117–18
reaction to fall from, 131–32
sex and, 109
target animals in, 135–38
dominance bullying, 138–39
dominance displays
bullying as, 139, 142
delayed dispersal in wolves and, 217
high-rank females and, 122
dominance hierarchies, regulation of, 117
dopamine, in courtship, 196–97
doves, courtship singing by, 168
drinking. *See* alcohol use and abuse
drug use and abuse
adolescents and, 47, 80, 84, 233, 253
sexual activity and, 202, 206
social pain and, 134

eagles
courtship in, 162–63
parent-offspring conflict in dispersal and, 219, 220
play courtship behavior by, 149
predator training for, 242
signal of unprofitability against, 62
stooping by, 219–20, 242
Eisenberger, Naomi, 134
Elbroch, Mark, 255–56
elephants, viii
adolescent banding in, 186
courtship behavior in, 174
dispersal of, as orphans, 216
hierarchies in, 96
predator-naive, 27
social rank in, 107
elephant seals
motor-vehicle deaths in, 224
sexual activity in, 201
elk, and island tameness, 35
emerging adult, 7, 14, 204–5
emotions
social descent and, 132–33
status and, 98, 102
environmental conditions
delayed plumage maturation and, 217
dispersal outcomes affected by, 260
dominance in hierarchies and, 109
emergence of new dangers from changes in, 37
human adolescents and, 127
hyena destiny affected by, 125
privilege determined by, 125
status related to, 118
ephebiphobia, 49, 206
description of, 45–46
social practices and, 46–47
epidemics, and island tameness effect, 35
Erikson, Erik, 7, 51
Etler, Cyndy, 181
extended parental care, 252–55, 263
animal parents and, 88, 249–50
disadvantages of, 252–53
mentorship and, 253–54
parental excursions during, 252

failure to kill rates, 65–66, 236
faint response, 65
falcons, skylarks' stotting against, 63

families
puppy license for adolescents in, 51
rank inheritance from, 119–21
status change and move away from, 156, 157
Farrell, Shannon, 191–92
fear, 29–37
armoring against, 31–32
defense mechanisms against, 32–34
dogs' redirection bullying and, 143
heart slowing in response to, 65
hygiene hypothesis and, 36
increases in danger from responses to, 36–37
island tameness and, 34–35
learning from, 31
physiology of, 30
sexual coercion by, 200, 202
social learning from experience of, 80
startle response to, 29–31
feeding behavior, learning, 86
ferrets, breeding of, 187
Fessler, Dan, 74, 75
fiddler crabs, status badges and, 109
Fielding, Helen, 181
fighting, play behavior for learning, 149–50
fight-or-flight response, 56, 65
fights
loser effect in, 122, 135–36
maternal intervention in, 122, 135
social rank and, 111–12
financial institutions, and adolescents' inexperience, 46
finishing school, for wolves, 215
fiscal behavior, and island tameness effect, 35
fish. *See also specific fish*
bait shyness of, 43
breeding behavior in, 174
dispersal pattern of, 214
early nutritional endowments in, 103
fishermen's predator deception with, 43
grooming by, 123
hierarchies in, 96, 98, 103
loser effect in, 136
maternal intervention in, 123
mobbing by, 70
oddity effect and, 59
predator inspection by, 82
privilege and, 123
sexual coercion in, 199
sexual fluidity in, 185
social learning by, 79, 80
social rank in, 101, 106, 108, 109, 137
startle response in, 29
flamingos, social rank in, 107
flying foxes, courtship in, 163
food
maternal intervention and, 135
parental teaching about, 252
privilege and fights over, 125
social rank and access to, 97, 98, 103, 109, 122–23, 127
foraging skills, learning, 86
fossa
courtship behavior in, 183–84, 245
sexual fluidity in, 185
Foster, Robin, 143
foster care, 253–54
foxes
signal of unprofitability against, 62
territory inheritance in, 124
Fraser, Bill, 86
freezing behavior
detection avoidance using, 36
of fish, 73, 76
Freud, Anna, 7, 33
Freud, Sigmund, 7, 33
Freudian psychology, 33

friendships
  adolescence and forming of, 17
  humpback whales and, 162
  hyenas and, 147–48
  learning from bad outcomes in, 79, 88
  power of, 147
  safety with, 88
  social status and, 147
  status affecting choices in, 107, 109, 110, 112, 117, 120
friendship walks, 147–48, 156
frogs, inborn defenses of, 33–34
fruit flies, courtship in, 164, 195–96, 197, 198, 205
fur seals
  orcas and breeding patterns of, 173
  sexual coercion in, 199, 200–201

Galef, Bennett G., 68
gangs, recruitment into, 47
garter snakes, predatory periods in, 54
gazelles
  cheetahs' hunting of, 39–40, 63, 85
  inspection of cheetahs by, 82, 83, 84
  predator-naive, 26, 28
  stotting by, 62–63
gestures, status rank shown by, 112–14
giggling, and submissive status, 113
giraffes, puberty in, 8
Goodenough, Judith, 149, 193
gorillas
  breeding behavior in, 174
  sexual behavior in, 173, 202
Greer, Germaine, 181
greeting ceremony, in pipefish mating, 193
grooming
  as predictor of status rank, 107–8
  rank and, 123
  status badges and, 110
Groos, Karl, 148
groups
  advantages of living in, 98
  being member of multiple groups, for social skills, 151
  conformer bullying in, 139–42
  fish shoaling, 41, 59, 73, 74, 75–76, 241
  formation of, as response to danger, 73–74
  heartbeat synchronization in, 74
  hierarchies in, 96, 98
  impact on humans of moving together in, 74
  individuality of each member of, 100–101
  mobbing by, 69–71
  niche picking in, 118
  othering in, 142
  rank versus status in, 100
  rules of, 105–18
  shunning by, 141
  transitive rank inference in, 101–2
guinea pigs
  early-life adversity and later success of, 127
  play-fighting by, 149
  sexual behavior in, 203
guppies
  sexual coercion in, 199
  social learning by, 79

Ha, James, 142
Hall, G. Stanley, 6–7
Hamilton, Joe, 53
hamsters
  social rank in, 97
  submission and infertility in, 173
Hand2Paw, 155
hares, signal of unprofitability by, 61–62
Harvard Graduate School of Education, 253
Harvard University
  adolescent sexual behavior study of, 176

"Coming of Age on Planet Earth" course at, 6
Peabody Museum of. *See* Peabody Museum
risk-taking and student acceptance by, 80
hawks
bonding in, 191
monkey alarm calls about, 173
predatory periods in, 54
headdress, as status badge, 111
heartbeats, synchronized, 74
heart slowing, as attack defense, 65
herds, association with high-status animals in, 109–10
Herman, Louis M., 166, 167
hermit crabs, cuttlefish's mimicry of, 43
hiding, from predators, 57, 60
hierarchies. *See also* status hierarchies
of angels, 99
animals' inborn process of arranging themselves in, 96, 99–100
brain's monitoring of changes in, 98
pecking order and, 96, 99, 101, 108, 136
prestige and, 117–18
range of animals using, 96
types of, 99–100
highways
crossings over, 223–24, 225–26
learning how to cross, 224–25
motor-vehicle deaths on, 224–25
Hirsch, Jennifer, 179–80
Holekamp, Kay E., 122
homing pigeons, social rank in, 98
Höner, Oliver, 94, 95, 103, 112, 113, 125, 127, 140, 147, 148, 150, 155–56, 263
hookup culture, 204–6
horizontal identity, 6, 185
horses
breeding of, 188, 191
courtship signals in, 199
dispersal patterns of, 213, 222–23
human adolescents' relationships with, 154–55
social brain of, and human interaction, 114, 115
status and friendships among, 109
timing of first sexual experiences in, 175–76
Houpt, Katherine, 203
Houston, Stephen, 111
humpback whales, viii, 17
Bigg's orcas' hunting of, 40–41
chorusing by, 165–67, 168–69, 186, 188–89
competitive groups in, 189–90
courtship of, 162, 164–65
mating in, 194
sexual encounters between, 188–90
hunger. *See* starvation
hunters, human
predator deception used by, 42–43
Predator's Sequence used by, 55
hunting techniques
adolescent banding for learning, 186
king penguins' learning of, 86
play mimicry of, 148
predator deception and, 42
pumas and, 215
startle response used in, 30
stooping in eagles and, 219–20, 242
wolves' training for, 215
hyenas. *See also* spotted hyenas
breeding of, 187
bullying in, 139
environmental influences on, 125
maternal intervention by, 121–22
hygiene hypothesis, 36

immune systems
hygiene hypothesis and, 36
social rank and, 108
informed dispersal, 214–15, 259

insects
  puberty in, 8
  sexual coercion in, 198
  startle response in, 29
  zugunruhe (migration anxiety) in, 22
intervention, maternal, 121–23, 135
intimidation, sexual coercion by, 200, 201, 202
Inuit people, coming-of-age traditions among, 248
island tameness, 34–35
  Darwin's experience of, 34
  description of, 34–35
  disease and, 35–36
  Yellowstone elk as non-island example of, 35
isolation, physiologic effects of, 248–49

jays
  dispersal pattern in, 218
  extended parental care in, 252
Jensen, Frances E., 7
joint head swinging, in salamanders, 163
jungle crows, birds' alarm calls against, 69
Juvonen, Jaana, 147

kangaroo rats, quality advertisements by, 62
kangaroos, play-fighting by, 149
kennel dog syndrome, 143, 155
killer whales. *See* orcas
kills
  giggles of submissive animals at, 113
  predator rates of, 65–66, 236
kill step, Predator's Sequence, 55, 66
king penguins, viii, 17, 21–28. *See also* Ursula
  adolescent banding in, 186
  adolescents' departure from birth territory, 24
  feeding and hunting skills learning by, 86
  life expectancy of, 263
  parental care during first year in, 21–22
  play hunting by leopard seals with, 148
  as predators, 54
  sexual coercion of, 200–201
  social learning by, 79
  startle response in, 30, 31
  starvation risk in, 85
  staying safe on first day away from home by, 22–24
  survival rate for fledglings, 24
  transition from predator naive to predator aware by, 26–28
  transponders for tracking behavior of, 23–24, 89
  zugunruhe (migration anxiety) in, 22
Kinsey Institute, 204
Kiribati people, Gilbert Islands, Pacific Ocean, 31
klipspringers, alarm duetting by, 63–64
Krupa, Kathy, 154–55
kudus, cheetahs' hunting of, 40

Lakota people
  courtship among, 180
  vision quest of, 248
learning
  courtship singing and, 167
  from peers. *See* social learning
  social descent and impairment of, 144–45
Leibniz Institute for Zoo and Wildlife Research, Berlin, 94
lemurs, 4
  predator avoidance training by, 68
  risk-taking by, 28
leopards
  kill rates of, 66
  signal of unprofitability against, 62
leopard seals, viii
  courtship singing by, 167–68
  penguins as food for, 23, 30–31, 37, 39, 52, 67, 68, 79, 186

play hunting by, 148
Predator's Sequence of, 54–55
starvation risk in, 85
Levick, George Murray, 198
Lion King, The (movie), 171
lions, viii, 138
courtship and sexual behavior in, 171–72
hyenas and, 112
kill rates of, 66
lobsters, viii
hierarchies in, 130
loser effect in, 136
serotonin and status shift in, 133
loser effect, 122, 135–36
Lührs, Mia-Lana, 183, 184, 245
lynx, early nutritional endowments in, 103

macaques
social rank in, 109, 122–23
status differences and learning in, 144–45
maintenance of pair bonds, 193
mamihlapinatapai, 165, 207
mammals. *See also specific animals*
dispersal pattern of, 214
herd movement against danger by, 74
loser effect in, 136
privilege and, 123
sexual signals in, 199
social rank in, 101, 106, 108
submission and infertility in, 173
zugunruhe (migration anxiety) in, 22
mandrills, courtship behavior in, 174
mapping, of status, 115–17
marmosets, courtship behavior in, 175
maternal intervention, 121–23, 135
maternal rank inheritance, 119–21, 157
mating, 183–85
animal agreement in, 194
birds' nest-building in, 178–79
chemistry during, 187–89, 207
community members' influence on, 173
complexities in nature of, 172–73
courtship before, 164
delaying dispersal and, 217
fossa example of, 183–84, 245
monogamy in, 193
multiple partners over time in, 194
pair bonding in, 193
parental excursions in, 252
partner preference in, 187, 207
physical attractiveness in, 109–10
play and, 203
puberty completion and onset of, 162
sexual inexperience and, 190–92
status and conditions in, 175
timing of first endeavors in, 173, 174–76, 179
wolves and, 260–61
Mayan people, status badges among, 111
Mead, Margaret, 6, 7, 51
meerkats
dispersal pattern of, 213
early-life adversity and later success of, 127
motor-vehicle deaths in, 224
play hunting by, 148
predator inspection by, 82
social rank in, 107
submission and infertility in, 173
mentorship, 253–54
Mexican jays, extended parental care in, 252
mice
cats' angling play with, 148–49
learning and status differences in, 144
migration anxiety (*zugunruhe*)
deer and, 52
king penguins and, 22
Milu deer, antler adorning in, 177–78
Minelli, Alessandro, 48
mink, courtship behavior in, 203

minnows
  early nutritional endowments in, 103
  oddity effect and, 59
  predator inspection by, 82
Mintz, Steven, 250, 251, 253, 255
mismatch disorders, 153
mobbing, 69–71
moles, sexual fluidity in, 185
mollusks, startle response in, 29
Molossus bats. *See* bats
monitoring
  orphaned mountain lion cubs and, 255–56
  post-release, 229, 255–56
  of whales, 23–24, 89
  of wolves, 227–28, 248, 255, 257, 264
monkeys
  breeding behavior in, 174, 175, 193
  maternal intervention and, 122–23
  play-fighting by, 149–59
  preference for watching high-status individuals by, 115
  rank inheritance in, 120
  social rank in, 109, 122–23, 137
  status and learning in, 144–45
monoamine hypothesis, 129–30
mood
  serotonin and, 129–30, 131, 133
  status linked to, 131, 132, 133, 144, 157
moose, and learning predator avoidance, 54
mortality rates
  of human adolescents after leaving home, 24–25
  of new drivers, 225
  in predator inspections, by gazelles, 83
  of salmon during river journey, 72
mosquitofish, sexual coercion in, 199
mothers
  intervention in offspring's lives by, 121–23, 135
  rank inheritance from, 119–21, 157
moths, first sexual encounters in, 191–92
motivation, in courtship, 197–98
motor vehicles, deaths from, 224–25
mountain goats
  risk-taking by adolescents in, 28
  social rank in, 107
mountain lions (pumas), viii. *See also* PJ
  delayed dispersal and, 217
  departure from family (dispersal) by, 211, 213, 215–16
  hunting by, 215
  starvation of orphaned cubs, 255–56
  street-crossing skills of, 226
Muller, Martin, 201–2
mullet fish, social rank in, 98
Munns, Roger, 190

naive. *See also* predator naive
  definition, 26
National Center for Missing and Exploited Children (NCMEC), 43, 69
National Institutes of Health (NIH), 138
nature documentaries, 172
nest-building, in mating behavior, 178–79
nest helpers, in delayed dispersal, 217
newts, courtship behavior in, 200
niche picking, 118
Noah, Trevor, 220–21

obesity as mismatch disorder, 153, 154
octopuses, startle response as hunting technique of, 30, 36
oddity effect, 58–60, 140, 141, 142
opiate addiction, and social descent, 134
orangutans
  fear response in, 30
  mating behavior in, 174
orcas. *See also* Bigg's orcas
  brain development in, 11
  breaching play by, 149

fur seal breeding patterns and, 173
humpback whales and, 186
penguins as food for, 23
stranding during hunting by, 240
surprise used in attacks by, 61
othering, in bullying, 142
otters. *See* sea otters
Outward Bound, 248
ovulation, and courtship singing, 168
owls
assessing of bats by, 60–61
attacking of bats by, 64–65
bats' inspection of, 81–82
birds' social learning about avoiding, 80
delayed plumage maturation in, 216
hunting of tuco-tucos by, 41
killing of bats by, 66

pair bonding, 193
pandas, 4–5
breeding of, 187, 188
puberty in, 8
startle response in, 29
Panthera, 255
parakeets, courtship singing by, 168
parental care, 88, 249–50, 252–55, 263
parental excursions, 252, 259
parental meanness, in delayed dispersal, 218, 219–20, 242–43
parent-offspring conflict, in dispersal, 218–19, 220
partner preference, 187, 207
Parus major birds
alarm calling by, 69
parental excursions by, 252
Parus varius birds, dispersal in, 212
Peabody Museum, Harvard University, 6, 16–17, 64, 248
Art of War exhibition at, 31
Lakota exhibition at, 180
Mayan status pendant in, 110–11
pecking order, 96, 99, 101, 108, 136
peer pressure
decisions made under, 80–81
salmon risk-taking and, 73, 76
peers
dispersing with, 260
importance of, in adolescence, 116
influence of, 116
learning from. *See* social learning
predator inspection and, 82
self-confidence building with, 77
penguins. *See also* king penguins
adolescent banding in, 186
dispersal pattern of, 213
hunting of sardines by, 41
mating by, 191
sexual coercion by, 198
perception
of rank, 100, 117, 131
of self, 136, 138
of status, 138, 139, 157, 188
peregrine falcons, skylarks' stotting against, 63
performance, status differences in, 144–45
pet therapy, 155
pharaoh cuttlefish, predator deception used by, 43
Piaget, Jean, 7
pikas, territory inheritance in, 124
pilot formation training, in groups, 73–74
pipefish, pair bonding in, 193
pitch of voice, status cues in, 113
PJ (mountain lion)
death of, 227, 228
dispersal by, 211, 213, 214
hunting school for, 215
new environment after dispersal and, 226–27
reaction to leaving home by, 224
street-crossing skills of, 226
play
coming tasks of life in, 148
status in, 148–51
video games and, 84, 151

play-fighting, 149–50, 203
polar bears, kill rates of, 66, 236
political leaders, othering and oddity
effect used by, 142
ponies, social rank in, 107
possums
dispersal pattern of, 212, 214–15
motor-vehicle deaths in, 224
post-release monitoring, 229, 255–56
Potočnik, Hubert, 213, 215, 225,
228–29, 247, 248, 255, 257, 264
predator avoidance training, 68
predator deception, 42–45
examples of, 42
predators' tricking of prey using, 42–45
predator inspection, 81–85
bat example of, 81–82
human adolescents and, 83–84
peer influence in, 82
safety improved by, 83
predator naive
adolescents as, 26–27
examples of, 26, 27–28
farm-raised fish and, 72, 76–77
king penguins as, 26, 37, 52
predators' targeting of, 40–42
risk-taking behavior and, 27–28
social learning by, 79
predators
abduction of adolescents by, 43, 69
adolescent risk of, 85–86
adolescents' learning dangerous
times for, 53, 54–55
adolescents' vulnerability to, 50
cost-benefit analysis of kills by, 40, 41
island tameness and lack of, 34, 35
kill rates of, 65–66, 236
safety and ways of knowing, 39–66
sex traffickers as, 44–45
targeting of predator-naïve
adolescents by, 40–42
Predator's Sequence, 55–56
assess step in, 60–64
attack step in, 64–66
description of, 54–55
detect step in, 56–60
four steps in, 55
kill step in, 66
prey's improvisations against, 55–56
pregnancy, factors in outcome of,
174–75
prestige, and status, 117–18
Pride and Prejudice (Austen), 181
primates
bullying in, 139
maternal intervention in, 123
mobbing by, 70
monkey license for, 51
sexual fluidity in, 185
social rank in, 109
privilege, 119–28
animal roots of, 125–26
destinies of adolescents and, 127–28
evolution and, 125–26
maternal intervention and, 121–23
maternal rank inheritance in, 119–21
territory inheritance and, 124–25
pronking, as signal of unprofitability,
63, 64
puberty
ancient occurrence of, 8–9
onset of breeding and completion
of, 162
timing of reproduction and, 172
use of term, 8
pumas. *See* mountain lions
puppy license, 51, 215, 244, 265, 266
Pütz, Klemens, 23–24, 86, 88–89

quail, sexual behavior in, 201
quality advertisements, 62–63, 66
Quality Deer Management Association,
41, 53

raccoons, hierarchies in, 96
radio-tag monitoring
whales, 23–24, 89
wolves, 227–28, 248, 255, 257, 264

rank. *See* social rank
rank inheritance, maternal, 119–21, 157
rats, sexual behavior in, 203
rattlesnakes, predatory periods in, 54
red-eyed tree frogs, inborn defenses of, 33–34
redirection bullying, 142–44
red-necked turtles, courtship signals in, 199–200
red-spotted newts, sexual activity in, 200
red squirrels, territory inheritance in, 124, 252
regression, 217
repression, 32, 33
reproduction. *See* mating
reptiles
  hierarchies in, 96
  privilege and, 123
  social rank in, 108
  startle response in, 29
rhesus monkeys, breeding in, 175
risk-taking
  adolescents' approach to, 27–28
  college applicants and, 80
  group size and degree of, 58
  peer pressure in salmon and, 73, 76
  peers' influence on, 116
  predator inspection and, 81–85
  predator-naïve adolescents and, 27
  social rank related to, 98
  starvation prompting in, 85–86
  subordinate adolescents and, 50
river crossings, by wolves, 247
R.J. Reynolds, 47–48
rodents, delayed dispersal in, 217
Rogulja, Dragana, 193, 195, 196–97
romance. *See* courtship
Roman Empire, 47

Sacks, Oliver, 197
safety, 12, 19–89
  danger and, 21–28
  fear and, 29–37
  group living and, 98
  knowing your predators and, 39–66
  motorway under- or overpasses and, 225–26
  predator inspection resulting in improvements in, 83
  self-confidence and, 67–77
  social learning from peers and, 79–89
  social rank and, 98
safety knowledge
  innate knowledge of, 33–34
  learning throughout an animal's life, 34
salamanders, courtship in, 164
salmon, viii
  peer pressure and risk-taking by, 73, 76
  predator-naive, 26
  predator responses of, 71–73
  self-confidence from encounters with danger by, 76, 77
  sexual behavior in, 201
  shoaling by, 73, 74, 75–77
Salt (humpback whale), 17, 161–62
  annual trips between Silver and Stellwagen Banks by, 161–62
  banding together with other adolescents by, 186
  clan of, 161
  courtship and, 162, 169
  courtship behavior in, 192, 206
  first love of, 162
  first sexual encounter of, 186, 188–90
  journeys of, 160, 161
  map of journey of, 160
  mating by, 194, 206
  social skills of, 161–62
sand-bubbler crabs, sexual activity in, 200
Sapolsky, Robert, 7, 25–26
sardines, penguins' hunting of, 41
Schjelderup-Ebbe, Thorleif, 95–96, 99, 108, 131, 136, 157

schrekstoff, 80
scrub jays, territory inheritance in, 124
seals. *See also* fur seals; leopard seals
  motor-vehicle deaths in, 224
  sexual activity in, 201
  timing of breeding in, 173
sea otters, viii
  alarm calling by, 69
  California home of, 1–3, 266–67
  predator inspection by, 82
  predator-naive, 26, 28
  sexual coercion in, 199, 200–201
sea turtles, predator-naive, 27
selective breeding, 187–88
self-confidence, 67–77
  alarm calling and, 68–69
  mobbing and, 69–71
  movement as group and, 74–77
  peer pressure to help build, 77
  predator avoidance training and, 68
  trial-and-error learning and, 67–68
self-reliance, 12, 209–271
Selling Girls (documentary), 44
seniority, and status rank, 107
senses, in social learning, 80
serotonin, 129–30, 131, 132, 133
servals, angling play by, 148–49
sex, 12, 159–207. *See also* courtship; mating
  activity between different species in 200–201
  adolescent banding and, 186
  adolescent social experiences needed to help understand, 203
  animals raised in isolation and ignorance of, 202–3
  coercion in, 198–203
  community members' influence on, 173
  complexities in nature of, 172–73
  diversity of animal experiences in, 192–93
  first encounter in, 187–94
  hookup culture and, 204–6
  inexperience in, 190–92
  lessons learned from animals about, 207
  nature documentaries' portrayal of, 172
  non-coercive aspects of, 199–200, 201
  partner preference in, 187, 207
  physical maturity and timing of, 171–72
  as predictor of status rank, 108–9
  sexual fluidity in, 185
  signals from female and, 199–200
  timing of first endeavors in, 173, 174–76, 179
sex education
  animal courtship and, 164
  human adolescents and, 164, 181–83
sex traffickers, and predator-naïve victims, 44–45
sexual coercion, 198–203
  by force, 200, 204
  by harassment, 200, 201, 203, 217
  in humans, 202–3
  by intimidation and fear, 200, 201, 204
sexual harassment
  in animals, 200, 201, 217
  in humans, 203
sharks, viii
  chemo-sensing by, 57
  dispersal pattern of, 213
  predator deception used by, 43
  sea otters' alarm calls against, 69
  sea otters' predator inspection of, 82
sheep
  fear response in, 30
  sexual behavior in, 174, 199, 201
shelter animals, 143, 155
ships' boys, 47
shoaling, by fish, 41, 59, 73, 74, 75–77, 241

Shrink (spotted hyena), 17
being in charge of own future, 157
body language of status and, 112
clan queen ruler's cub and, 95, 105–6
communal den life and, 105–6, 119
death of, 263
fighting for his place in group, 93–94, 156, 158
friendship walks by, 147, 148, 156
map of journey of, 92
maternal intervention and, 121, 123, 135
maternal rank inheritance and, 120
mother's place in clan hierarchy, 95
physical appearance and status of, 140
play-fighting by, 149
privilege and, 125
request to queen hyena for help and nursing for, 155–56
Social Brain Network and, 114
social savvy as strength for, 118
social skills and transfer to new group by, 156, 158
status hierarchy place of, 93, 101, 102–4, 109, 156, 158
as target animal, 136
transitive rank inference and, 101
Shubin, Neil, 9–10
shunning
appearance-based bullying and, 141
conformer bullying and, 141
oddity effect and, 59, 60, 140, 141
Siberian jays, dispersal pattern in, 218
signals of unprofitability, 61–64, 66, 70
silver spoon effect, 120
singing
courtship with, 165–69
humpback chorusing, 165–67, 168–69, 186, 188–89
skylark stotting, 63, 64
size
age and social rank related to, 107
as predictor of status rank, 106–7
skylarks, stotting by, 63, 64
Slavc (wolf), 17, 211
adjustment to new environment by, 256
coming-of-age process for, 259
departure from family (dispersal) by, 211, 213, 214
dispersal training of, 215
finding of final home by, 256–57
human monitoring efforts to protect, after dispersal, 228–30
hunting of farm animals by, 256–57
hunting school for, 215
isolation of, 248
journeys of, 210, 247–48, 256, 257
kill sites and diet of, 247
later life of, after adolescence, 263–64
map of journey of, 210
mating by, 260, 261, 264
motorway traffic and, 223–24, 225–26
new environment after dispersal and, 228
radio collar of, 227–28, 248, 255, 257, 264
reaction to leaving home by, 224
river crossing by, 247
sloths, puberty in, 8
smell, in social learning, 80
Smithsonian Tropical Research Station, Barro Colorado Island, Panama, 82
smoking, and protection of adolescents, 47
snakes
birds' alarm calls against, 69
predatory periods in, 54
quality advertisements against, 62
signal of unprofitability against, 61, 62
snow monkeys
breeding behavior in, 174
rank inheritance in, 120

Social Brain Network (SBN), 114–15, 116–17, 118, 131, 134, 152, 153, 154
social descent, 129–45
  adolescents' reaction to, 133
  bullying and, 138–44
  change in dominance and, 131–32
  emotional reactions of animals to, 131–33
  learning impairment from, 144–45
  social pain and, 133–34
  substance use and abuse and, 134
  survival chances and changes in, 133
  target animals in, 135–38
socialization
  developing friends as adults and, 151
  dog aggressive behavior and, 143
  female cat sexual activity and, 203
  kennel dog syndrome and, 155
social learning, 79–89
  courtship and, 182, 184
  examples of, 79–80
  human adolescents' morbid curiosity as, 84
  learning from bad outcomes of peers in, 79, 80
  learning from parents compared with, 80–81
  new beginning after leaving parents and, 86–89
  predator inspection and, 81–85
  senses involved in, 80
  social learning with, 79, 80–81
  survival skills in, 85–86
social media
  hierarchies in, 108
  nonstop use of, 152, 153
social pain, 133–34
social rank
  age as predictor of, 107, 134
  behavior, innate and learned, as factor in, 109
  brain's monitoring of changes in animal's position in, 98
  common criteria of determining, 106
  defense in warfare and fights related to, 111–12
  food access and, 97, 98, 103, 109, 122–23, 127
  gestures showing, 112–14
  grooming and, 107–8, 123
  human toddlers and, 115
  hyena communal dens and, 105–6
  hyena ears and, 140
  impact on animals' lives of, 97–98
  maternal inheritance of, 119–21, 157
  perception of, 100, 117, 131
  play-fighting for, 149
  prestige and, 117–18
  safety related to, 98
  serotonin and, 130–31
  sex as predictor of, 108–9
  as shaper of personal identity for humans, 97
  size as predictor of, 106–7
  Social Brain Network and, 114–15, 131
  social descent in, 131–32
  status differentiated from, 100
  status mapping of, 115–17
  target animals and, 135–38
  transitive rank inference for deducing, 101–2, 131
  voice pitch showing, 113
social skills
  adolescent practice of, 239
  dispersal and, 216
  humpback whales and, 161–62
  navigating hierarchies using, 156, 158
  play for developing, 151
  spotted hyenas and, 156, 158
Solomon, Andrew, 6, 185–86
songbirds, dispersal pattern of, 212
Spanish imperial eagles
  dispersal in, 219, 220
  predator training for, 242
Spear, Linda, 7
Species Survival Plan (SSP), 186

sperm whales
mating behavior in, 174
rank inheritance in, 120
spider monkeys, rank inheritance in, 120
spiders, social rank in, 106
Spotted Hyena Project, Ngorongoro Crater, Tanzania, 94
spotted hyenas, viii, 17. *See also* Shrink
age and status linked in, 107
body language of status and, 112
clear succession in clans of, 112
communal den life and, 105–6, 119
death of, by other predators, 263
early-life adversity and later success of, 127–28
fighting for their place in group, 93–94
friendship walks by, 147–48, 156
grooming in, 108
maternal intervention and, 121–23, 135
maternal rank inheritance in, 119–21
play-fighting by, 149
privilege and, 125
queen ruler of, 94–95
rank related to ears of, 140
sexual fluidity in, 185
size as predictor of social rank of, 107
Social Brain Network (SBN) in, 118
social status and friendships of, 147
status hierarchy in, 4, 93, 94–95, 102–4
submission signs in, 149
target animals and, 136
vocalizations and whoops showing status of, 113–14
squirrels
dangerous predator times and, 54
risk-taking by adolescents in, 28
road-crossing skills of, 224
signal of unprofitability by, 61
territory inheritance in, 124, 252
Stamp, Judy, 7
Stanton, Margaret, 175
starlings
group movement as response to danger by, 74
individuality of each member of flock of, 100–101
social learning by, 80
startle response, 29–31, 34
function of, 29–30
human imagination and, 30
learning to stifle, in predator inspection, 84
panda mother example of, 29
physiology of, 30
predator-naïve adolescents' reliance on, 34
starvation
adolescent risk for, 85–86
dispersal and danger of, 213–14, 215
orphaned mountain lion cubs and, 255–56
status, 12, 91–158
allies and, 147–58
assessment and, 93–104
bullying and, 139, 144
environment affecting, 118
learning impairment related to, 144–45
mapping of, 115–17
mood linked to, 131, 132, 133, 144, 157
niche picking and, 118
perception of, 138, 139, 157, 188
personality and change in, 156–57
play and, 148–51
prestige and, 117–18
privilege and, 119–28
rank differentiated from, 100
rules of groups and, 105–18
sacrifices for changes in, 156
sanctuaries from, 154–55
serotonin and, 131
signs of, 110–12
social descent and, 129–45
social skills for reversing, 156, 158
transitive rank inference for deducing, 101–2, 131

status badges, 110–12
status hierarchies. *See also* hierarchies
  actions causing shifts in, 136
  association with high-status animals in, 13, 109–10
  breeding behavior related to, 175
  clear succession in, 112
  common criteria of individual's place in, 106
  drinking behavior related to, 137–38
  early nutritional endowments related to, 103–4
  factors affecting dominance in, 109
  gestures showing place in, 112–14
  hyenas and, 4, 93, 94–95, 102–4
  impact of social descent in, 131–32
  maternal intervention and, 121–23
  maternal rank inheritance in, 119–21
  preference for watching high-status individuals in, 115
  prestige and, 117–18
  signs of status in, 110–12
  smaller hierarchies within, for learning roles, 151
  Social Brain Network for processing in, 114–15
  social skills for navigating, 156, 158
  status mapping in, 115–17
status sanctuaries, 154–55
Steinberg, Laurence, 7, 62, 116
Stevenson, Robert Louis, 47
stooping, 219–20, 242
stotting, 62–64, 66
Strachan, David, 36
stress hormones, 116, 144, 173, 223
students
  bullying and, 141–42
  hookup culture and, 205
  nonstop assessment by, 153
  risk-taking by, 80
  status differences and performance of, 145
sturm und drang (storm and stress), 6–7
submission. *See also* status hierarchies
  fertility impacted by, 173–74
  giggling at kills as sign of, 113
  hyena's ear movement to indicate, 140
  play-fighting for learning, 149
substance use and abuse
  adolescents and, 10, 27, 176
  social descent and, 132, 134
succession, in hyena clans, 112
surprise, in attacks, 61, 69, 70
survival rates
  of adolescents and young adults with cancer, 48
  of penguin fledglings after leaving home, 24
Svendsen, Lars, 36
swine, rank inheritance in, 120

tail straddling walk, in salamanders, 163
target animals
  bully's choice of, 139, 141, 142
  social descent and, 135–38
teenagers. *See* adolescents
territory inheritance, 124–25
Thomson's gazelles, viii
  predator inspection by, 82, 83, 84
  pronking by, 62–63
Thoreau, Henry David, 250
tigers, kill rates of, 65, 236
Tiktaalik, 9–10
Tinbergen, Nikolaas, 7, 28
titi monkeys, mating in, 193
tonic immobility, 36
tortoiseshell butterflies, sexual behavior in, 201
tough love, 220–21
tranquilizing darts, for animals, 227
transitive rank inference (TRI), 101–2, 131
transplants, adolescents and decisions on, 49
Trathan, Phil, 31

trial-and-error learning
gaining confidence to face fears with, 67
parents' demonstrating techniques in, 67–68
trigger-stacking, 143
tuco-tucos, owls' hunting of, 41
turtles
predator-naive, 27
sexual signals in, 199–200
Twain, Mark, 45

U.S. Department of Agriculture (USDA), 191
U.S. National Highway Traffic Safety Administration, 225
unprofitability, signals of, 61–64, 66, 70
Ursula (king penguin), 17, 21–28
dispersal pattern of, 213, 214, 216
feeding and hunting skills learned by, 86
gaining confidence to face fears by, 67
later life of, after adolescence, 263
map of journey of, 20
new beginning after leaving parents by, 87, 88–89
parental care and protection during first year of, 21–22
predator avoidance training and, 68
as predators, 54
predators' schedules learned by, 54
Predator's Sequence of, 54–55
startle response in, 30, 31, 34
staying safe from predators on the first day away from home by, 22–24
transition from predator naive to predator aware by, 26–28, 37, 52
transponders for tracking behavior of, 23–24, 89
zugunruhe (migration anxiety) in, 22

vagal response, 57
vaping, and protection of adolescents, 47–48
video games, 84, 151
vigilance, against predators, 57–58, 60
vocalization, and hyena status, 113–14
voice
courtship singing and, 168
status cues in pitch of, 113

Wade, Lisa, 205
Weissbourd, Richard, 181, 205
western bluebirds, extended parental care in, 251
Westover, Tara, 156–57
whales. *See also* Bigg's orcas; humpback whales; orcas
breaching play by, 149
deaths from ships and tanks and, 224
dispersal pattern of, 213
mating behavior in, 174
rank inheritance in, 120
white-cheeked pintails, sexual activity in, 201
white-winged choughs, extended parental care in, 252
wildebeest, and oddity effect, 58
wildhood
common ancestors through time, ix
four challenges in, 12
four competencies of, 11–12
use of term, 11
during various animal life spans, viii
Williamson, Peter, 47
wobbling, by fish, 73, 76
wolves. *See also* Slavc
bison attacks by, 68
courtship by, 260–61
culling of, 261

wolves (*cont.*)
dispersal patterns of, 215, 217
gestures showing social status of, 112
hunting school for, 215
predatory periods in, 54
road-crossing skills of, 225–26
submission and infertility in, 173–74
surprise used in attacks by, 61
World Health Organization, 175
Wrangham, Richard, 42, 201–2
Wu, Tony, 190
Wynne-Edwards, V. C., 131–32

Yaghan people, Tierra del Fuego, 165
Yellowstone elk, and island tameness, 35
youth. *See* adolescents
YouthTruth, 141

zebras, viii, 5
dispersal pattern of, 213
Zoobiquity (Natterson-Horowitz and Bowers), 2, 15, 264
zoos, selective breeding in, 187–88
zugunruhe (migration anxiety)
deer and, 52
king penguins and, 22

About the Authors

Barbara Natterson-Horowitz, MD, is a visiting professor in the Department of Human Evolutionary Biology at Harvard University. A professor of medicine in the UCLA Division of Cardiology, she is president of the International Society for Evolution, Medicine, and Public Health.

Kathryn Bowers is a science journalist who has taught courses on animal behavior and writing at UCLA and Harvard. A Future Tense Fellow at New America in Washington, DC, she was previously an editor at Zócalo Public Square in Los Angeles and a staff editor at the *Atlantic Monthly*.

Their previous book, *Zoobiquity*, was a *New York Times* bestseller, *Discover* magazine's Best Book of 2012, and a finalist for the American Association for the Advancement in Science (AAAS)/Subaru SB&F Prize for Excellence in Science Books Award.

WILDHOOD

Barbara Natterson-Horowitz, MD,
and Kathryn Bowers

Conversations about Coming-of-Age for Families, Classrooms, and Communities

OVERVIEW

In ***Wildhood***, Horowitz and Bowers identify four universal tests that shape the adult destiny of every adolescent animal coming of age on earth. Preparing for these tests, they argue, is the purpose of adolescence. Their groundbreaking research presents a new understanding of this highly consequential phase of life, and invites parents, teachers, and adolescents themselves to rethink what it means to prepare for a successful and happy adulthood.

SAFETY

Surviving in a Dangerous and Changing World

1. Adolescent animals are often targeted by predators. Why is that? How do predators spot inexperienced adolescents? Compared to experienced adults, what are adolescents doing—or not doing—that attracts predatory attention? As you read about Ursula, the adolescent king penguin embarking on her first dive into the oceans off Antarctica, what did you want her to know about the dangers ahead?

2. "Predator naive" is the term Horowitz and Bowers use to describe inexperienced adolescents. In what ways are human adolescents predator naive? Was there ever a time you were predator naive? What happened? What did you learn?

3. The authors describe several strategies that can transform a young animal from predator naive to predator aware. These include "predator inspection," "mobbing," giving off "signals of unprofitability," and getting to know the "Predator's Sequence." What are some human versions of those practices? Have you ever "inspected" something that was dangerous? Have you ever communicated "unprofitability" in a situation that felt threatening? Did those experiences keep you safe later in life?

4. As they move into the world, adolescent animals pay less attention to their parents' choices and behaviors and more attention to those of their peers. In fact, some lessons can only be learned from peers. What are examples from the book of either positive or negative peer influence? How does knowing about peer influence in the animal kingdom change how you think about peer pressure on modern teens?

5. Over time, animal species living without dangerous predators may lose their fear responses, a phenomenon called "island tameness." What are some examples of island tameness from the book? Do you think modern humans have lost any important fear responses? How might teenagers be especially impacted by human versions of island tameness?

STATUS

Socializing and Finding Your Place in the Group

1. Horowitz and Bowers tell the story of Shrink, a low-born hyena cub who finds his way into a more advantaged life in the Ngorongoro Crater in Tanzania. What did you learn about pecking orders and the ways that hyenas and other animals create hierarchies? Do humans sort themselves by social rank? In what ways did learning about Shrink's place in his hyena group change how you think about status and rank in people?

2. What surprising facts did you discover about the biology of comparison, such as the Social Brain Network (SBN)? In what ways do people and animals practice being social? Given what you now know about the SBN, what advice would you give to adolescents about making friends? Do you agree with the authors' suggestion that adolescents should join a wide variety of social groups? Why or why not?

3. Wild animals seek higher status because it increases their "fitness" (their chances of surviving and reproducing). As Horowitz and Bowers explain, rising in status triggers a pleasurable neurochemical "reward," while falling in status leads to an unpleasant neurochemical "reprimand." The authors link this ancient neurobiology of status in animals to the modern experience of mood in humans. Try thinking about the ups and downs you feel throughout the day—can you associate these with changes in your status, for example on social media? Do you agree with the authors' claim that adolescents today are experiencing "assessment overload"? If you could design a "status sanctuary," what would it include—or exclude?

4. The authors describe three kinds of animal bullies:
   a. Dominators
   b. Conformers
   c. Redirectors

   What are examples of each type of bully in human life? Can you think of other kinds of bullies besides these three? Where does cyberbullying fit in?

5. In Chapter 8: Privileged Creatures, Horowitz and Bowers demonstrate how a young animal's future is shaped by its parents' status and resources. Offspring of higher status animals receive better nutrition, have stronger immune systems, choose from a wider range of mates, and even get better sleep. Some inherit territory and social networks and even "royal" rank from their parents.

These advantages build up over a lifetime, making it more likely these privileged creatures will succeed as adults. Were you surprised that "there are no level playing fields in nature"? How might we humans use this knowledge to create fairer communities in which all adolescents have opportunities? What are some examples of parental intervention that you've experienced or observed? As an adolescent, how does it feel to have a parent intervene on your behalf?

SEX

Learning to Communicate Desire and Interpret the Romantic Interest of Others

1. The authors state that "developing the back-and-forth of sexual communication starts in early wildhood." How does this play out in the story of Salt, the North Atlantic humpback whale? Growing up, what did you observe about courtship from the adults around you? How were sex and sexuality discussed in your family?

2. The authors write, "Sex is easy. Romance is hard." Think about all the ways in which animals communicate romantic interest, from the singing of humpback whales to the mating tree of fossas, the "dances" of moths and butterflies to the "death spiral" of bald eagles. If you were an animal behaviorist, how would you describe human courtship displays? Which elements seem to be innate, and which are learned?

3. What surprising facts did you learn about animals' first sexual experiences? What are some of the reasons that sexually mature wild animals might "delay" having sex? How does this information fit into sexual decision-making among young adult humans?

4. Horowitz and Bowers describe "transient masculinization" of the bodies and behaviors of young female fossas, temporary feminiza-

tion in adolescent male birds, and male-to-female transformation in some fish. How can this knowledge inform our understanding of gender fluidity and expression among humans?

5. As part of their original research for *Wildhood*, the authors found "an uncomfortable but important truth: coercive copulation (male on female and male on male) is widespread across the animal kingdom." What are some examples from the book of coercive sex among animals? How might "intimidation coercion," noted in some primates and birds, help us better understand the role of power dynamics in sexual relationships? How do these examples inform your understanding of consent and coercion in human sexual relationships?

SELF-RELIANCE

Leaving the Protections of Childhood and Learning to Make a Living

1. On his journey through the Italian Alps, Slavc the wolf faces many challenges. He must find his own food, safely cross roads and rivers, survive periods of isolation, and find a new community. In other words, he must become self-reliant. What does it mean for a person to be self-reliant? Do wild animals have to make a living? In what ways is it similar to or different from what a young person must do?

2. Before he left home, Slavc was living with his wolf pack, observing and learning the basics of safety, socializing, courtship, and hunting. What preparation and training should people expect from parents, teachers, and other adults in their communities during adolescence? What, if anything, do adolescents owe adults in return?

3. Horowitz and Bowers found that, in some species, parent-offspring conflict increases as those offspring get closer to dispersal. In

what ways is human conflict during adolescence similar to what is described in the book? Have you experienced adult-adolescent conflict? As with the Spanish imperial eagles who sometimes show their offspring a form of tough love, can conflict between parents and teens ever be useful?

4. On page 214, Horowitz and Bowers describe "practice dispersals." What is a "practice dispersal"? What are human examples of practice dispersals? Have you ever taken part in one? What did you learn?

5. The authors write that starvation is a central and ever-present danger for animals who are newly independent. At the same time, hunger can motivate an adolescent's drive and persistence. Adolescent meerkats and hyenas, for example, stick with tough tasks longer than adults because they're hungrier. How do adolescent animals learn to find food on their own? In what ways does feeding yourself make you an adult? How can this knowledge be used to help adolescent humans prepare for the adult world?

GENERAL QUESTIONS

1. Which character—Ursula, Shrink, Salt, Slavc—did you relate to the most? What was it about their life story you connected with? Did it make a difference that they were real-life animal adolescents, tracked by scientists over months and years?

2. Horowitz and Bowers suggest that storytelling is a special human form of predator inspection. Think about your favorite coming-of-age movie, book, or story. Did the main characters face the four core challenges of safety, status, sex, and self-reliance? Did they learn the four life skills? Did the protagonists have anything in common with Ursula, Shrink, Salt, or Slavc?

3. What was your initial reaction to the idea that adolescence is at least 600 million years old and that human teen behavior can

be seen in other animals? Did learning this change how you feel about your own transition from child to adult?

4. Watch the videos about animal courtship at www.wildhood.com and consider that each behavior is part of a two-way, back-and-forth conversation. Think about human versions of these conversational steps. Will the conversation be the same for every individual? Will it be the same across regions and cultures? What happens when the conversations are different between people?

5. The authors introduce the term "ephebiphobia," described as the "fear or even hatred of adolescents." Can you think of an example of ephebiphobia drawn from the news or modern life? Have you yourself ever experienced ephebiphobia? In what ways might you have felt ephebiphobia toward others?

6. Which of the four life skills is the hardest for you to confront in yourself? Which are difficult to talk about with family members? Which would you rather talk about with someone else—for example, a doctor, friend, teacher, coach, therapist, or another non-family member?

7. Did reading *Wildhood* change your opinion about the role that adults should play in the lives of teens? What kinds of mentorship or advice prepared you for life as an adult? How much responsibility do you think adolescents should take for learning these skills themselves, versus how much should come from parents, school and teachers, peers, or other adults in the community? Compare the four skills needed for successful adulthood with the tasks that must be done to get into college. In what ways does preparing to get into college overlap with preparation for the four tests?

8. Anthropomorphism is the projection of human characteristics onto animals in a way that's not scientifically justifiable. How

do the authors address the issue of anthropomorphism? Do you agree with their statement, "the bigger danger might be denying humans' real and demonstrable connections with other animals, in body and behavior"? Given what has been learned over the last decades about how much of the human genome and neurophysiology is shared with other animals, which is the more likely scientific error: anthropomorphism or human exceptionalism?

9. How does tradition or spirituality influence your concept of what it means to be an adolescent? Does the science in *Wildhood* clarify or complicate human cultural customs around adolescence? If you could create a coming-of-age ritual, ceremony, or event for yourself that encompasses all four core competencies and demonstrates your familiarity with them, what would it look like? Where would it take place? Who would attend or witness it?

10. Adolescent risky behavior—such as impulsivity, the underestimation of risk, and increased pleasure from new experiences—is often described as being a result of adolescents' undeveloped prefrontal cortexes, aka the teenage brain. But as Horowitz and Bowers show, across species adolescent brains have unique features that drive behaviors. In what ways do adolescent brains benefit human adolescents?